CHICAGO GARDENS

CHICAGO GARDENS

The Early History

CATHY JEAN MALONEY

THE UNIVERSITY OF CHICAGO PRESS

Chicago and London

CATHY JEAN MALONEY is a senior editor at *Chicagoland Gardening*.

Publisher's note: *Chicago Gardens: The Early History* is the twelfth volume in the Center Books on Chicago and Environs series, created and directed by the Center for American Places at Columbia College (www. americanplaces.org). Other volumes in the series are available from the University of Chicago Press.

The University of Chicago Press, Chicago 60637
The University of Chicago Press, Ltd., London
© 2008 by The University of Chicago
All rights reserved. Published 2008
Printed in the United States of America

17 16 15 14 13 12 11 10 09 08   1 2 3 4 5

ISBN-13: 978-0-226-50234-2   (cloth)
ISBN-10: 0-226-50234-1   (cloth)

Library of Congress Cataloging-in-Publication Data

Maloney, Cathy Jean.
    Chicago gardens : the early history / Cathy Jean Maloney.
        p. cm.
    Includes bibliographical references and index.
    ISBN-13: 978-0-226-50234-2 (cloth : alk. paper)
    ISBN-10: 0-226-50234-1 (cloth : alk. paper) 1. Gardens—Illinois—Chicago—History. I. Title.
    SB466.U65C565 2008
    712.09773'11—dc22

                                                                    2007043028

♾ The paper used in this publication meets the minimum requirements of the American National Standard for Information Sciences—Permanence of Paper for Printed Library Materials, ANSI Z39.48–1992.

For Mom and Dad

CONTENTS

Chicago's rich garden heritage is often overlooked by historians and residents alike. Over the decades, many of the region's earliest gardens, then considered unremarkable, were plowed under, subdivided, or simply left to nature's whims. Historic landscape preservation, a relatively new study in America, is only starting to get traction in Chicago.

There is much to savor and celebrate. Not long after the first pioneer log cabins were built, peacocks strutted across many a manicured lawn in Chicago. Croquet was fancied as a fashionable front-yard sport, and rock and water features became the landscape trend du jour. Indoor parlor gardens bloomed with ferns and ivy, and keeping up with the neighbors meant posing outside with the latest reel lawn mower, a newfangled improvement over the sheep or cows that used to graze the grass. Throughout the decades, gardens evolved in response to fashion trends, changing demographics, and technological advances. Many of the garden styles we see today in Chicago are rooted in the early discoveries and traditions of nineteenth-century gardens, as the cyclical nature of garden fashion continues. Today's water gardens, exotic container flowers, native-plant gardens, perennial borders, and rock gardens can be traced back to early Chicagoland landscapes.

Finding existing examples of Chicago's historic landscapes is a challenge, because most of Chicago's gardens from the 1800s and early 1900s have been lost. Real estate speculation consumed outlying towns and their charming villas and gardens. Early gardens were carved up either through subdivision or the changing tastes of subsequent homeowners. Many pioneer Chicagoans dabbling in real estate were either absentee landlords or transient homeowners on their way to the next frontier; investing in or preserving landscapes was not a priority. Additionally, unlike the East Coast or antebellum South, where generations might have lived in one home, houses were rapidly bought and sold in Chicago's red-hot real estate market throughout the second half of the nineteenth century. Furthermore, the technology of lifting and moving homes, perfected by industrialist George Pullman and frequently used when Chicago raised the level of its streets, effectively removed many homes from their original context and landscapes.

Pioneer Chicagoans, unlike their East Coast brethren, did not yet have a sense of their own history. Washington never slept here and, if Lincoln did, it was most likely in a downtown hotel. Land was plentiful, and there seemed to be no need to salvage the simple gardens of Chicago's first merchants and farmers, or even the showcase homes of the city's early elite. Although some of Chicago's later (ca. 1900) country-estate gardens on the North Shore were preserved, many more were lost, and vernacular gardens were completely obliterated. Except for some rare survivors, only vestiges remain of Chicago's early gardens and landscapes. The rest are available to us only through written descriptions and scarce photographs. Reviewing these, particularly in the context of major land-use and garden trends, helps interpret and better appreciate the few legacy gardens that remain.

Fortunately, a few garden remnants remain. At the end of each chapter, I offer my hand-picked selection of those historic gardens and landscapes that demonstrate the chapter's theme. I particularly focus on sites open to the public, so that today's visitors can appreciate firsthand a sense of the wonderful horticultural history of our region. I've not attempted to cover all the historic gardens and landscapes—many other publications have discussed the significance of Chicago's parks and suburban estates, as well as high-profile landscape architects such as Frederick Law Olmsted and Jens Jensen. In this book I hoped to highlight my favorite unsung gardens created in Chicago's first century, from 1833 to 1933. To celebrate Chicago's long tradition of garden history, I submit this sampler, a florilegium if you will, of landscapes representing different time periods, purposes, and aesthetic influences. These are my favorites, and I look forward to hearing from you with any suggestions for more. Please contact me at cjmgardens@ameritech.net if you would like to identify any historic landscapes, gardens, or horticulture-related people from the Chicago area.

ACKNOWLEDGMENTS

This book began, almost twenty years ago, as a personal quest to find out more about my own garden and nearby landscape. This interest blossomed into full-scale research into the rich horticultural history of Chicago-area gardens and landscapes. Along the way, I have received help and encouragement from so many people—the names below are only the tip of the iceberg. Please accept my thanks even if, in error, I've omitted your name below.

I love librarians! And researchers, and archivists, and all who love the pursuit of obscure facts and sources. Many a time I have enlisted the aid of a librarian or archivist, and he or she has become as determined as I to find the answer. Such was the case in uncovering a rare garden plan for the Murray residence. Scrolling through reels of microfilm, I saw the plan referenced, but not on film. Staff from the Chicago History Museum retrieved the original documents from which the microfilm was made and found the garden plan for me. Countless other researchers have shared their expertise in this way, including Frank Lipo, Historical Society of Oak Park and River Forest; Nancy Wilson, Elmhurst Historical Museum; Arthur Miller, Lake Forest College; Michael Stieber, Morton Arboretum; Jane Walsh

Brown, Westchester Township History Museum, Chesterton, Indiana; Daniel Meyer, University of Chicago Library; and staff from the Riverside Library, Riverside Historical Commission, Chicago Public Library, Western Springs Historical Society, Rogers Park Historical Society, Glenview Park District, Tinker Swiss Cottage Museum, and many others.

Many selfless souls read through drafts of sections, and I wish to thank them and absolve them from any remaining errors, which are mine. Ed Hedborn of the Morton Arboretum graciously read through a draft of my chapter on plants. Although I've left many plant names in their original archaic spellings, that is no reflection on Ed—he knows plant nomenclature inside and out. Staff from the Cheney Mansion, the Morton Arboretum, Ragdale, Fabyan Villa, the Frank Lloyd Wright Home and Studio, Glenview Park District, and others have also lent their expertise. Thanks to George Thompson of the Center for American Places for first recognizing this book's potential. My deep gratitude goes to my editor, Christie Henry of the University of Chicago Press, for her expert counsel, encouragement, and professionalism in making this book a reality.

Thanks always to Mike and Tom and my family. Mike is behind every good thing I do, and Tom is my joy, and has unflinchingly accepted too many peanut butter sandwiches instead of "real" dinners.

ONE

INTRODUCTION

Surrounded by mudflats, swamps, and squalid cabins, Chicago's founding fathers defiantly dubbed the young outpost a city in the garden, "Urbs in Horto." Call it boosterism, call it bravado, these scrappy leaders nonetheless accurately presaged Chicago's significant contributions to American gardening. In the one hundred years between Chicago's incorporation as a town in 1833 and the city's 1933 Century of Progress World's Fair, Chicago became a national horticultural leader. Chicago's exponential growth serendipitously coincided with the period often called the golden age of American horticulture. During this era, improvements in railroad and automobile travel forever changed the dynamic of market gardening, and as the nation's transportation hub, Chicago was in the driver's seat. Horticultural societies flourished as a means to exchange information, and Chicago, with its central location, was a favored convention city. By necessity, early Chicago horticulturists became innovators when faced with a challenging climate, topography, and soil, and their newfound knowledge was shared with the world. These contributing factors—Chicago at the center of transportation, communication, and innovation—earned Chicago a well-deserved reputation as a horticultural mecca.

Yet Chicago's early-nineteenth-century gardens were overshadowed by the venerable nurseries and estates of New England and the prosperous plantations of the Old South. Contemporary tastemakers waxed eloquent over the grandeur of Hudson River Valley landscapes, the wonders of New York's Central Park, and the restful beauty of Boston's Mount Auburn; generations of southern gardeners were inspired by Virginia's Mount Vernon and Monticello. Compared with these enshrined garden showcases, upstart Chicago was not taken seriously. Early Chicago settlers were written off as uncouth frontiersmen, incapable of the culture and sophistication needed for fine gardens. Travelogues describing the landscape were often negative, and could hardly tempt a prospective settler. Typical was this 1800s missive from traveler William Keating: "The appearance of the country near Chicago offers but few features upon which the eye of the traveler can dwell with pleasure. There is too much uniformity in the scenery … unenlivened by the spreading canvass, and the fatiguing monotony of which is increased … by the equally undiversified prospect of the land scenery, which affords no relief to the sight, as it consists merely of a plain, in which but few patches of thin and scrubby woods are observed scattered here and there."[1]

Even more discouraging were first impressions from America's horticultural leaders. In 1868, Frederick Law Olmsted, the Father of American Landscape Architecture, described the Chicago environs as "low, flat, miry and forlorn, with a bleak surface, arid soil and exposure to harsh and frigid gusts of wind."[2] Botanist Edward L. Peckham, upon visiting the city in 1857, said, "Chicago, the world renowned Chicago, is as mean a spot as I ever was in, yet."[3] Even Chicago's own boosters were initially downbeat. In a scouting mission to Chicago, the city's first mayor, New Yorker William B. Ogden, decried the region as a poor land investment. John A. Kennicott, Chicago's leading horticulturist in the 1840s and 1850s, admitted in 1852, "Except a very manageable soil, and the glorious expanse of lake, Chicago has few of the natural elements of the garden."[4]

Transplanted East Coast gardeners despaired of taming the windy climate and heavy, clay soils. Transplant shock inevitably occurred when Chicagoans doubted their own ingenuity and tried to slavishly recreate East Coast gardens on western soil. Fondly remembered plants of New England became sickly sticks in the prairie clay. By necessity, Chicagoans soon relearned the art of gardening from the ground up. From truculent soils to treacherous skies, Chicago gardeners redefined the language of the prairie soil. It was this very adversity that challenged local horticulturists to experiment and develop new plants and designs suitable for this strange frontier. With these innovations, and as a popular convention center, Chicago became a hotbed of horticulture, and cross-pollinated ideas from both east and west coasts.

MORE THAN A TEN-CORNERED POTATO PATCH

First, Chicago gardeners needed to understand and come to terms with the region's unique climate and growing conditions. Contemporary writers who had only heard about the "barbarous West" painted Chicago's horticulture with a broad brush that included climates ranging from Michigan to New Mexico. In 1849, John A. Wight, cofounder with John S. Wright of Chicago-based *Prairie Farmer* newspaper, tried to correct these misconceptions for readers of Andrew J. Downing's nationally distributed *Horticulturist*. He began his article,

> In estimating the advantages and disadvantages of our city, Chicago, and its immediate vicinity, for horticultural purposes, it is necessary to put forth in the beginning a word of caution; that is, that this spot must not be compared to any other part of the western country, or supposed to furnish any index to the character of any other part; for, so far as I know, it is a complete exception to all the rest … And yet, eastern people are in the habit of talking about "the west," as though it was some ten cornered potato patch, not a whit bigger than Rhode Island; as though a description of one part must answer for all.[5]

Chicago had its own microclimate, and newcomers soon learned to address its particular challenges in soil, wind, topography, and plantings. With these challenges, Chicago settlers had to adopt an entirely new horticultural vocabulary. Learning to translate the floral code didn't happen overnight, but rather through extensive trial and error. Timely combinations of progressive people, plants, places, and prairie pastimes accelerated Chicago's progress in the horticultural and agricultural world.

Ground-Level Differences

As pioneers moved west, they had to leave behind favored acid-loving plants such as azalea and rhododendron. Heavy clay soils and native prairie plants, with their extensive root systems, resisted plowman and plow alike. As early as the 1840s, horticultural leaders sent away soil samples for analysis, to identify the best means of soil augmentation. New systems were developed for underdraining the soil to render swampy fields productive. Farmers and gardeners were urged to adopt more aggressive methods of turning the soil. An article in the 1846 *Prairie Farmer* magazine noted, "Our present mode of gardening, so far as is reduced to system, has been derived from our English neighbors, with whom the very mention of a plow in the garden is the rankest heresy." England's smaller properties required only the use of spades, but heavy-duty plows were the tool of choice on the prairie. Yet old habits died hard. "A great many people plow as if they were afraid of breaking through the earth and disturbing the Chinese. This kind of carefulness will not do in a garden," warned the *Prairie Farmer*.[6]

Gardeners struggled with tough prairie sod and the many variants of wet and dry prairies. The challenge to develop prairie-worthy plows consumed many local tinkerers and entrepreneurs. In 1833, John Lane, a blacksmith from Lockport, Illinois, claimed title to the first steel plow, although John Deere's Self-Polishing Plow, invented in 1837 in Grand Detour, Illinois, clearly got more recognition. Noting the importance of better planting and harvesting technology, Cyrus H. McCormick moved his famous reaper factory from Virginia to Chicago in 1847. Breaking up prairie soil became a booming business.

Winds of Change

Chicago's exposure to blustery winds posed additional problems for the gardener. John Wight bemoaned the perils of fierce winds for successful gardens, and his 1850s lament rings a familiar bell with Chicago gardeners even today.

> But oh the spring! We may as well say we have none. We make numerous attempts at one annually; but every notch we get forward has its offset of a slip or two backward, till somehow or other it gets to be summer. These northern gales frown down on us, reeking with the breath of polar bears and icebergs, for days together, blasting our tender plants, and our hopes together . . . First, come two days of south wind,—the mercury rapidly going up to 75° and 85° Fahrenheit, calling forth the tender leaves prematurely,—the ground yet as cold as November. Then, short as the crook of your elbow, are six days of north wind,—the mercury falling to 38° and 45°. Oh, horror! Now look to your starting vegetation. But this is not all. The north wind brings a rain. The water comes in an avalanche, and the fine soil drinks and holds it with the eagerness of a toper. Now come two days more of south wind and heat. Your poor tree stands with its feet in the ice, and its head in the fire. If it, or its ancestors, have been crossed with the northern seal, and the salamander both, it may live.[7]

Chicagoans had to find other ways to cope with the inescapable wind. Unlike New England, no sleepy hollows or valleys protected plants or people from the gales. Groves of trees were few and far between. Windbreaks made of evergreens became the solution of choice, with belts of arborvitae or Osage orange popular landscape elements. The Midwestern landscape was transformed from an endless, undulating plain to checkerboard farms bound by walls of trees and evergreen hedges.

Level the Playing Field

Chicago's topography, its utter flatness, was both foreign and frightening to visitors. Comfortable with the rolling countryside of the South or the mountains of New England, emigrants to Chicago were unnerved by the vastness of the plains. Pioneer Juliette Kinzie, riding on horseback in the 1850s near what is today the western suburb of Riverside, had an unobstructed view for ten miles to the Kinzie homestead on Lake Michigan. "We could look across the extended plain, and on its farthest verge were visible two tall trees, which my husband pointed out to me as the planting of his own hand, when a boy. Already they had become so lofty as to serve as landmarks, and they were constantly in view as we traveled the beaten road."[8] This flat terrain had a profound effect on landscape design. The fledgling town of Chicago was a prime candidate for the surveyor's typical grid system. Civil engineer James Thompson platted the town in military squares in 1830, and this pattern grew geometrically to outlying suburbs. Without mountains or craggy coasts to intervene, streets and properties were drawn in neat, regular boxes. Today, finding evidence of horticultural talent and creativity is easy, given this checkerboard land. Simply look for a curve that breaks out of the box. Neighborhoods including Lake Forest, Riverside, Clarendon Hills, Highland Park, and Norwood Park, all laid out by landscaping professionals, are readily identified by their unique curvilinear streets. Curves in Chicago are neon signs blinking "gardener lives here." Witness the efforts of ardent early horticulturist the Reverend Simon James Humphrey, who insisted that his street in Oak Park be curved. Today, Elizabeth Court remains the only curved street in otherwise gridlocked suburban Oak Park.

Flat ground necessitated innovations in garden technology and design. Water did not drain properly, and much of the Chicago area was marshy. Many Chicagoans made their fortunes in developing tiles (i.e. Hinsdale's William D. Gates), or in creating pumps (i.e. Oak Park's H. W. Austin) to drain the swamps.

In New England, water lines often originated on hillsides and flowed downward to fuel ornamental garden fountains. In Chicago, artificial hills had to be created, or pumping mechanisms installed. Windmills were also used to pump water to irrigate the plains (see plate 1). Indeed, instead of celebrating Chicago's unique level plain, early landscape designers such as John Blair, William Le Baron Jenney, and Frank Calvert, and later designers such as Frederick Kanst literally made mountains out of molehills in their designs for public and private spaces. Lakes and ponds were scooped out of prairie soil, and mounds and hillocks created. The incongruity of artificial mountains on flat prairie land escaped the impressionable public, who sought trendy, gardenesque features.

Barbed-wire fencing, while utilitarian, was another important Chicago-area technology because the flat plains provided no natural boundaries separating livestock from garden fields. The scarcity of wood on the treeless plain also created a market for the new barbed-wire fencing. Once utilitarian, the ubiquitous garden fence became a strong element in garden design, even when livestock no longer roamed the plains.

Perhaps the most overlooked effect of Chicago's flat surface was the democratization of garden design. Unlike New Yorkers, who could hide their privileged sanctuaries in the hills and hollows of the Hudson River Valley, Chicagoans displayed their gardens right out in the open. Any farmer taking the plank roads to Chicago could see the fine city gardens of William B. Ogden, George Snow, William B. Egan, and others. Garden trends and taste initiated by Chicago's wealthy were very much in the public domain, and were rapidly assimilated by the working class.

Treasuring Trees

In an era where trees were highly valued for fuel, shelter, shade, fruit, and windbreaks, the prairie's lack of trees was a significant concern for new settlers. Most pioneers staked their claims near precious groves, often located eastward of rivers and streams, beyond the threat of westward-blowing prairie fire. Today, many outlying Chicago suburbs have a "Grove" suffix, reflecting the overriding importance of these early oases. Research into suitable trees for the prairie was an ongoing concern for every Chicago homeowner. In 1868, *Prairie Farmer* editors recalled the treeless prairie in its early days: "Twenty-seven years ago no one could have dreamed so wild a dream as is now realized. It was then doubted whether clover could be grown, and orcharding was not thought of. If forest growth had been natural to the prairie, why had it not been clothed like Ohio and Indiana with the thick leaved forest? Of course, the soil was at fault or we should have timber, ergo: fruit trees cannot be grown."[9]

Proper tree culture was a subject of intense debate in many western agricultural papers. Growing "western" trees and plants became a matter of civic pride, as well as common sense. Western nurserymen railed against the false claims of eastern "tree peddlers" whose unscrupulous sales tactics besmirched the profession. As late as the 1870s, Illinois nurserymen complained about eastern tree salesmen selling roses grafted on willow stock at high rates. This 1879 diatribe from the Horticultural Society of Northern Illinois is typical: "There was considerable discussion upon the subject of the nuisance of the omnipresence of the tree-peddlers and many instances recounted of their rascally cheating . . . Some members thought there should be a law to prevent such outrages; others contended that licenses should be demanded for tree-peddling, with heavy penalties attached; and that no one should be allowed to sell without a certificate guarantee from some responsible nurseryman."[10] The solution, as agreed at this convention of nurserymen, was to educate the public about using locally grown trees. Nurserymen established thriving businesses growing trees suitable for the Midwest climate, such as evergreens from Robert Douglas or David Hill and P. S. Peterson's trees and shrubs, and these businesses were staunchly endorsed by local newspapers.

FROM FRONTIER FORTS TO WORLD'S FAIRS

Chicago's horticultural progress is best examined in tandem with the city's key historical milestones. The city's incorporation in 1837, the great Chicago Fire of 1871, and the world's fairs of 1893 and 1933 each marked dramatic evolutionary periods in Chicago's horticulture.

Fort Dearborn and First Florists, 1803–36

We have no remnants of Chicago's very first gardens, those of the nomadic Potawatomi, Chippewa, Ottawa, and other Native American nations who lived here in the early 1800s and prior. But when Fort Dearborn was built in 1803 to protect the strategic harbor of this new territory, growing vegetables was a matter of survival. Provisions from the federal government arrived sporadically by slow boat, so it behooved the military to raise their own produce. From the moment they set foot on shore, soldiers at Chicago's Fort Dearborn began digging in the fields. A map of the fort drawn in 1808 by Captain John Whistler shows a large officers' garden and several smaller gardens. The drawing, reproduced in many history books, indicates a simple geometric garden design, with paths separating symmetrical beds. Around the edges of Whistler's sketch, the woods surrounding the fort are depicted much more loosely, indicating perhaps the untamed nature of the native forests.

The original fort burned and was rebuilt in 1816, and by the 1830s included well-tended fruit trees. The banks of the Chicago River that bordered the fort were "occupied by the root-houses of the garrison." One historian recounted the challenges of gardening at the fort:

> It was impossible for the garrison, consisting of from seventy to ninety men, to subsist upon the grain raised in the country, although much of their time was devoted to agricultural pursuits. The difficulties which the agriculturist meets with here are numerous; they arise from the shallowness of the soil, from its humidity, and from its exposure to the cold and damp winds ... The grain is frequently destroyed by swarms of insects. There are also a number of destructive birds of which it was impossible for the garrison to avoid the baneful influence, except by keeping, as was practiced at Fort Dearborn, a party of soldiers constantly engaged

in shooting at the crows and blackbirds . . . But, even with all these exertions the maize seldom has time to ripen, owing to the shortness and coldness of the season.[11]

Fort residents wrote about the gardens to friends, and even exchanged produce. Soldier Daniel Baker confided to a Detroit correspondent in June 1817 that he had "been mostly occupied in making a garden and other necessary arrangements for living." In November 1818, Baker shipped some of his harvest to his friend with an explanatory note describing "a bushel of cranberries gathered within a few days past, and I believe of a better quality than those generally brought in by the Indians."[12]

Some settlers established homes outside of the fort, although many families shared temporary quarters within the garrison walls. For the most part, homes and gardens were simple and functional. Charles Lee's farm, dubbed Hardscrabble, near today's Bridgeport, was typically utilitarian. The John Kinzie home across from the fort, however, was deemed a mansion and had the grandest garden of the times. The image of this simple frame home is instantly recognizable by four tall Lombardy poplar trees and a picket fence in the front yard. Although some accounts question the presence of the poplars, the trees seem credible as a pioneer homesteader's fast-growing windbreak.

Behind the Kinzie home were vegetable gardens and outbuildings, honey locust, and cottonwood trees. In his *Reminiscences of Early Chicago*, author Edwin

While certainly simplistic by today's standards, the gardens at the John Kinzie home were considered substantial for the era. A mature tree, most likely a cottonwood, grows behind the house, near the fenced-in vegetable gardens and outbuildings. The Lombardy poplar trees in front might have served to break the southwest winds. Courtesy of Chicago Public Library, Special Collections.

Gale romanticized the honey locusts of the fort era: "These locusts were marvels of grace and beauty, and I have always rejoiced that on one cold, cheerless autumn day, about 1841, … I gathered a bushel basket full of pods from these trees, the beans of which my father planted on his farm at Galewood, now within the western limits of the city, where quite a number of vigorous trees are still standing."[13] Gale transplanted two of these trees to his own home in Oak Park in 1870, and prized them "as souvenirs of Fort Dearborn and my early school-boy days." Who can say how many Oak Park homes share these legacy locust trees today?

Most gardening in the Fort Dearborn era was for subsistence, but the garden of Elijah D. Harmon, surgeon for the fort in the early 1830s, was an exception. Like many physicians of the 1800s, Harmon was interested in gardening for its medicinal as well as ornamental potential. Harmon bought land south of the fort, on Lake Michigan. Juliette Kinzie wrote, "When we chose the path across the prairie towards the south, we generally passed a new-comer, Dr. Harmon, superintending the construction of a sod fence, at a spot he had chosen, near the shore of the lake. In this inclosure [*sic*] he occupied himself, as the season advanced, in planting fruit-stones of all descriptions, to make ready a garden and orchard for future enjoyment. We usually stopped to have a little chat. The two favorite themes of the Doctor were horticulture, and the certain future importance of Chicago."[14]

Horticulture and the future of Chicago were about to take a giant step forward. On August 12, 1833, the Town of Chicago was incorporated with a population of 350. On September 26 of that year, the Treaty of Chicago was signed. Native American tribes ceded their land around the new town, and were given a three-year period to vacate the area. With land sales flourishing and the threat of Indian attack removed, settlers began to consider Chicago a permanent home. Beautifying their grounds was the next logical step, with wealthy Chicagoans creating elaborate landscapes and middle-class residents gradually planting ornamental flowers in addition to vegetables.

Edgar Sanders, one of Chicago's earliest and most influential nurserymen, credits Englishman Samuel Brooks with starting the first florist business in Chicago in the 1830s.[15] Brooks was one of many immigrants to imprint European gardening customs on Chicago's soil. Writing in 1884, Sanders said, "Fifty years ago it was customary for gardeners to save all the seed they could before emigrating, and doubtless many new, fine fruit of supposed chance origin that have made this country famous, were the product of this seed-saving from the choicest in European gardens."[16] Sanders recounted that Brooks nearly filled a small schooner with his supplies of tools, plants, and seed. He arrived first in New York with an entourage of wife and nine children but, in 1833, moved to Chicago, lured by rosy circulars advertising "land $1.25 per acre, climate mild, peaches flourished, cattle needed no shelter." Sanders described the unfortunate reality Brooks discovered instead: "Instead of the mild climate he found a very harsh one as compared with London. This man, direct from the acknowledged center of the civilized world, dropped down on a spot forbidding in appearance, and direct from nature's workshop, and started a Nursery and Florist business. Imagine the change for such a man, bringing his piano along, the first one here, as a supposed necessity for his family."[17] Despite the culture shock, like many immigrants, Samuel Brooks was energetic and eager to start business. In the three years between the city's incorporation as a town and incorporation as a city on March 4, 1837, the population had multiplied over ten times, to more than 4,000. Even as the city grew by leaps and bounds, homeowners wanted a little bit of natural beauty indoors and out. Brooks, and later nurserymen who followed in his footsteps, established thriving businesses. It was to be the start of organized horticultural enterprise in Chicago.

Garden City Era, 1837–70

Explanations abound as to how Chicago earned the nickname Garden City. Civic leaders waxed eloquent about the city's idyllic features, and described a town filled with promise. More realistically, some historians say that national financial crises of 1837 and 1857 caused Chicagoans to grow gardens out of sheer economic necessity. In his 1874 book, *Chicago and Its Suburbs*, Everett Chamberlin noted this flurry of landscaping activity after the panic of 1837. Rather than move to bigger homes, "men began to whitewash their fences, plant for fruits, and for shade, clean up their yards, plank their walks, and lay out their grounds both for convenience of culture and for beauty . . . Thus, the place became the 'Garden City.'"[18]

John Kennicott suggested that this luxuriant growth was illusory, since homeowners planted fast-growing but inelegant and impermanent trees: "Who does not know, that in the battery of heroic names, Chicago boasts that of The Garden City of the West! Have horticulturists nothing to do with this professional assumption? Will [Chicago] retain the title unquestioned, if she continues much longer to plant cotton-woods instead of maples, and half as many locusts as elms? And truth to say, I fear the great abundance of these cheap deciduous trees constitutes her chief title to the name of a Garden City."[19] The true etymology of Chicago's sobriquet may remain unproved, yet it is clear that the three decades between the city's incorporation and the Chicago Fire of 1871 brought enormous change and growth to Chicago's gardens. Horticultural societies were formed for education and networking. Railroads brought the world to Chicago, and Chicago's fruits to the world. More nurserymen came to Chicago and, following Samuel Brooks's lead, planted orchards and fields of ornamental flowers. These were consumed by the newly affluent class who began to landscape the grounds of their tasteful mansions. The Garden City era was a time of discovery, expansion, and most importantly, communication among horticulturists.

In 1841, the first issue of the Chicago-based *Prairie Farmer* rolled off the presses. At first a monthly magazine, the *Prairie Farmer* soon became a regular weekly newspaper that was widely distributed throughout the Midwest. Founded by John S. Wright and J. Ambrose Wight, both of the Chicago area, the newspaper became the voice of the western farmer and gardener. Correspondents from around the Midwest exchanged ideas for improved husbandry, horticulture, and farming. Editorial and writing responsibilities often were assumed by prominent Chicago-area horticulturists, including Dr. John A. Kennicott, Edgar Sanders, Jonathan Periam, Mathias L. Dunlap, and Orange Judd. The *Prairie Farmer* was to the Midwest what several more-established horticultural papers were to the East. Good relationships were enjoyed among these publishers, and subscriber discounts could be had through a mutual exchange program. The *Horticulturist*, *Warder's Review*, the *Florist* from Philadelphia, Luther Tucker's *Country Gentleman* from Albany, and the *Rural New Yorker* were just some of the papers with whom *Prairie Farmer* editors traded.

Cheerleader, arbiter of horticultural debate, town clarion, the *Prairie Farmer* was much more than a newspaper. Despite several changes in editorial leadership, the *Prairie Farmer* provided continuous coverage on horticultural and agricultural issues for decades. The editors penned these prophetic words in 1843: "It is our delight to look forward a few years, anticipating what is to be the condition of the industrious farmers of the West ... The farmers are to be the lords of the land. At the East, the merchant, lawyer, doctor &C. fill the chief places; but here it is to be the farmer. He is to be the man of wealth, influence and independence—the *gentleman*."[20]

Chicago's earliest nurserymen frequently had backgrounds of culture and education. This background was often sublimated in the hard work of establishing a pioneer nursery but shone through in their civic improvements and articulate publications. John Kennicott, Mathias Dunlap, and Lewis Ellsworth were among the first nurserymen of the 1840s. Kennicott, the Old Doctor, started his nursery called the Grove, in what is now Glenview, Illinois, in 1842. One of several Kennicott brothers, Dr. John was often seen riding through the tall prairie grass, making house calls on patients and botanizing en route. He frequently picked up an unusual prairie plant or tree fruit on his travels and brought it back to his garden at the Grove. A contemporary of nationally known horticulturist Andrew J. Downing, Kennicott often wrote for Downing's magazine, the *Horticulturist*.

Mathias Dunlap, who used a pen name, Old Rural, when writing for the *Chicago Tribune* or other papers, started his farm in Cook County and later moved to Champaign, Illinois. He became one of many important personages linking the newly created university, now the University of Illinois, with Chicago. Lewis Ellsworth began his nursery in Naperville in the 1840s, as a small business. By 1853, his fifteen-page catalog listed well over one hundred kinds of summer, autumn, and winter apples, over fifty types of pears, thirty varieties of cherries, and a limited selection of peaches, plums, apricots, quinces, grapes, and other small fruits like raspberries and gooseberries. He also offered a handful of ornamental trees and shrubs, evergreen trees, and roses. Old-fashioned garden flowers such as pinks, crocuses, gladiolus, tulips, and lilies were also for sale.

Nurserymen such as Ellsworth, Dunlap, and Kennicott were prominent leaders in their respective towns. In his 1950 retrospective article for *American Nurseryman* magazine, author Frank Heinl noted, "The pioneer nurserymen of Illinois were more than mere tree growers . . . all had a burning desire to build something, a school, a church, a business or a society of some sort."[21] Heinl reported that Kennicott and Dunlap were at the forefront of the movement to create the University of Illinois. Kennicott, Dunlap, and Ellsworth were also outspoken abolitionists and, like many early horticulturists, were leaders in local or state government.

Following the formidable footsteps of these pioneers, the 1850s brought a new breed of nurseryman. Edgar Sanders, a prodigious garden writer as well as nurseryman, came to Chicago from England in 1857, and described this next wave of gardeners: "The Second Era [of nurserymen] came after railroads, cheap postal rates, and other Improvements. The representative man of this new class . . . had accumulated a small capital, either as private gardener or nursery hand, usually the former, and mainly natives of foreign countries. Neither capital nor native Americans had yet taken kindly to this business. To be successful in it a man must be 'handy,' so he can build his own greenhouse, if not his own shanty and he must combine the nurseryman, truck-grower, florist, bricklayer, carpenter, glazier, teamster and man of all work."[22] These Renaissance men of horticulture began growing plants on a larger scale, to satisfy the demand of Chicago's growing population. Farmers and city dwellers alike wanted trees for shade and windbreak, or fruit trees to supply the dinner table. Nurserymen became known for their specialties: Robert Douglas of Waukegan was known for his evergreens, P. S. Peterson was the man to call for transplanting large trees, Swain Nelson of (present-day) Glenview had a good sense of garden design.

Leading nurserymen constantly experimented with varieties, and created a new palette of plants for the prairies. With their newfound specialties and plant introductions, Chicago nurserymen shared expertise through the emerging horticultural societies. Like their East Coast brethren, nurserymen, farmers, and skilled gardeners in Chicago banded together in professional societies. The key purpose of the societies was information exchange, accomplished through conventions, fairs, and published meeting notes. The lobbying power of these groups was significant; resolutions and recommendations were often sent to state and national legislative groups.

The *Prairie Farmer* served as a popular forum for announcements and news of fellow gardeners and nurserymen. As early as 1843, the *Prairie Farmer* sounded a call for the establishment of a "Farmer's Club," and even offered its offices as a meeting spot. The first agenda topics focused on prairie breaking, recommended grasses, and ideas for promoting agricultural improvement. Between 1840 and 1870, a number of similar societies sprang up. One of the first groups formed was the Chicago Horticultural Society (CHS). Created in 1847, its charter members included professional nurserymen such as John Kennicott as well as "wealthy amateurs" W. B. Egan and William B. Ogden. The so-called wealthy amateurs were themselves interested in horticulture, but employed gardeners to manage their grounds.[23] The CHS held a number of competitions for cut flowers and other greenhouse specimens. The CHS disbanded but was reconstituted in 1890 just before the Columbian Exposition. It exists today and, among other responsibilities, manages the Chicago Botanic Garden in Glencoe, Illinois. Other early societies included the Chicago Gardeners' Society (ca. 1859), Northwestern Fruit Growers Association (1851), Cook County Agricultural and Horticultural Society (1866), and Illinois State Horticultural Society (1856).

Group membership frequently waxed and waned when different clubs with kindred interests merged. It was common for leading horticulturists to belong to many, if not all, of the groups. These decades saw the most spirited and productive associations of local horticulturists. With urgent questions still unanswered on fruit and flower culture, plowing techniques, and entomology, networking with fellow gardeners was the best way to get information. The societies had lofty goals, including the establishment of model farms, exhibition halls, and agricultural schools. Creative ways of obtaining funds to support the society often went beyond mere dues collection. In 1843, for example, the Illinois State Agricultural Society secretary sought support from Governor Thomas Ford to use monies from state-collected bounties on wolf scalps. A later proposed transaction involved purchase of a racetrack to be used as a horticultural fairground in season.

Early society membership also included patrons of horticulture, the so-called wealthy amateurs. This proved to be a mutually beneficial relationship wherein patrons provided meeting space and other amenities, while nursery professionals shared tips for improved gardening. Horticulture patrons also helped provide objectivity among professional groups, which were sometimes splintered by competing interests. In the early days, clear distinctions were not made among florists, nurserymen, farmers, orchardists, market gardeners, or the yet-to-be-defined profession of landscape architecture. These disciplines often overlapped and caused turf battles and confusion. Horticulture patrons (i.e. wealthy customers) were able to bring focus to the various societies' goals. As *Prairie Farmer* editors opined, "It is far better an amateur with leisure and devotedness to the cause of horticulture should be at the head of the society than one in trade. There are always little matter [*sic*] occurring that cause unpleasant feelings."[24] Popular topics among society members often included fruit culture and pest control. As the societies matured, their membership became increasingly interested in the business aspects of marketing of orchard and farm produce. Standing committees for the Horticultural Society of Northern Illinois, for example, focused exclusively on fair marketing standards, and shipping and packaging best practices. As distribution options expanded and railroads flourished, Chicago's geographical position as a transportation hub became even more relevant to gardeners.

Railroads and New Markets

The emergence of railroads changed gardening in Chicago forever. With rail transport, the economics of agriculture favored the farmer, with a nation of prospective buyers as close as the nearest train depot. Shortly after the 1848 maiden voyage of the *Pioneer*, a steam-engine train that ran from Chicago to Oak Park, *Prairie Farmer* reader James T. Gifford penned this "Ode of Illinois Farmer to Steam Car."

> Fair lovely lawns which are spread in the West
> With their rich fertile soil drew me forth from the east . . .
> Soon of the produce my fields brought to hand,
> Soft were the roads o'er the rich mucky land . . .
> The strong band of union the car's iron rail
> Which unites Garden City with lovely Fox Vale.[25]

Railroads also meant stiffer competition for Chicago gardeners from southern growers who were now able to ship produce well in advance of Chicago's harvest schedules. Chicago gardeners hastened to develop new glasshouse and hotbed techniques that would let them get a jump on the northern growing season. Their choice of crops also changed to reflect these new market conditions, as noted by nurseryman Jonathan Periam in the 1880s: "Years ago every gardener was anxious to beat his neighbor in getting early potatoes, green corn, tomatoes, egg plant, etc. The Chicago gardener pays little attention to extra early vegetables, except lettuce, raised entirely under glass in the winter. Why? The South comes into competition in everything that can be transported, hence there is no money in such vegetables. But the cultivation of plants too delicate to bear transportation, like lettuce, has been largely extended."[26]

In addition to changing the produce mix, and expanding business for market gardeners, railroads made possible the commuter suburb. Suburbs burgeoned

SOUTH WATER STREET, CHICAGO.
A SECTION OF THE BUSIEST WHOLESALE MARKET IN THE WORLD.

1914, Max Rigot.

With improved rail transportation, and with Chicago's central
location, the city became a major
marketplace. Open-air markets
such as those at Haymarket
Square and South Water Street
flourished for wholesale or retail
purchases from market gardeners. From the author's collection.

in the last quarter of the nineteenth century, beginning in the 1860s with the
nascent railroad. Dramatic improvements in railroads, nurseries, and gardening societies certainly paved the way for horticultural growth in Chicago. But
it was a more gradual and subtle shift in mindset among Chicago gardeners in
the 1860s through the 1880s that brought true horticultural innovation. After
decades of emulating foreign plant choices and cultural techniques, Chicagoans
finally embraced their own climate and gardening. The shift came slowly, aided
and abetted by Mother Nature, with the inevitable decline of nonnative plants.
Nurseryman Robert Douglas recalled one pivotal year when horticulturists were
dramatically faced with the destructive power of nature: "Then came the winter
of 1855–56, known to this day as the hard winter, and swept almost everything.
The spring of 1856 found the horticulturists of the Northwest about as the 10th
of last October [the day after the 1871 fire] found the citizens of Chicago; only
that the people of Chicago knew of what material to build a new city, but we did
not know of what material to build new orchards."[27] After this terrible winter,
with major losses in crops, trees, and plants, Chicago gardeners started looking

seriously at plants more suited for this region. Douglas continued, "From this time forward it was not enough to know that a fruit was of good size and flavor, a fine thrifty grower, and an early and abundant bearer. The question before all others was, is it hardy? Not, is it hardy with you at the East? But is it hardy in the Northwest?" More experimentation was pursued with native plants such as Osage orange, hardy apples, and winter-tolerant evergreens. Horticulturists started to challenge the eastern-dominated status quo, and to recommend planting techniques specific for the Chicago area. Typical was this 1855 writing from the ever prescient John Kennicott, who managed to poke fun at eastern horticulturists while commenting on the propagation habits of strawberries. Responding to an argument that strawberry runners always produce plants true to the parent, Kennicott drew this analogy: "This may be all so in New York, where plants, politics and men all obey the laws of conservatism, but out here in Illinois, where men and strawberries cut loose from conventional usages, things are different. Here, runners, although 'rooting branches,' do not always follow the law of their parentage; but spring up into plants on their own hook, cut loose from the old fathers and mothers, and go ahead themselves entirely, bearing just such blossoms as they choose."[28]

The influential Edgar Sanders, himself an English immigrant, also began tailoring his horticultural recommendations to local conditions. Writing about conservation issues, Sanders noted in 1864, "In this, as with many other things, it may not be necessary to follow strictly in the footsteps of England, but that the country as a whole . . . would be benefited, they and their children, by a more extensive planting of evergreens and other trees, as is sure as that we are fast using up our resources in wood and timber."[29]

Public Parks and Private Gardens

Cognizant of national trends, Chicago was precocious in developing landscaped cemeteries and public parks, and ultimately created a park system to rival those of the most gentrified cities in the world. In Victorian America, cemeteries served not only as places to mourn the dead, but as pastoral strolling grounds. Rosehill Cemetery on Chicago's north side was founded in 1859 and initially landscaped by William Saunders, an esteemed landscaper from Philadelphia. Thomas B. Bryan, a leading Chicago industrialist and horticulture patron, led the effort to create Graceland Cemetery, landscaped by a host of professionals including locals Swain Nelson and later O. C. Simonds.

Predating the public park system, early Chicago had a small bit of greenery in the downtown area, particularly around the Court House in the center of town. Many called this green space unimpressive, as evidenced by these resolutions from the Chicago Gardeners' Society in 1860: "*Resolved*, That the condition of the public parks and the Court House Square in this city, is a disgrace and a

These views show an early attempt to landscape Courthouse Square with trees. Ca. 1860s. From the author's collection.

reproach upon the public spirit and good taste of our citizens. *Resolved*, That the Executive Committee of this Society are hereby directed to take active measures to improve and ornament the Court House Square, provided it shall be found practicable so to do."[30] During the next year or so, John C. Ure, president of the society, endeavored to improve the grounds with plantings of trees and walkways. Attempts to improve any public green space, such as Courthouse Square, met with passionate debate among citizens who felt entitled to comment on "their" property. Growing public sentiment for better health and clean air also spurred the growth of public parks in Chicago. In 1864, a city ordinance transferred sixty acres of the Chicago Cemetery to establish Lincoln Park. One year later, $10,000 was secured for improvements, and a plan by landscape gardener Swain Nelson was approved. An act of the state legislature in 1869 clarified boundaries of three park districts in Chicago, on the north, west, and south sides. Plans for the parks were made by leading landscapers of the time, including Frederick Law Olmsted (South Parks) and William Le Baron Jenney (West Parks). The parks served not only to improve the green space in Chicago, but also to set a new standard for private home gardens. In writing about Union Park, a small, early park on the near west side, Chamberlin says, "The beautiful grass plats are studded with trees, fountains, rustic seats and arbors . . . It is a favorite haunt of promenaders and driving parties."[31] Homes surrounding Union Park were landscaped equally well.

Although developers and property owners recognized the symbiotic relationship between public green space and private property, high turnover in landownership precluded investment in most private landscaping. In the 1840s, real estate speculation was so rampant that properties were traded within months, even weeks. Land was not viewed as an asset to be held, cultivated, and con-

Like many early small parks, Union Park was created to enhance the property values of nearby homes. The landscape design shown in these early views favored rustic bridges and manmade rockwork. 1883. From *The New Chicago Album* (New York: Wittemann Brothers, 1883).

ceivably passed down to future generations, but as a commodity to be quickly resold. Improvements such as landscaping were not economically feasible. John Wright commented in 1849, "In the matter of ornamenting grounds, and shading streets, those who care for it work at disadvantage. The magic ideas connected with such phrases as 'corner lots,' 'rents,' 'double in value,' 'going up,' are too much for 'trees,' 'shrubbery,' and 'roses.' Sometimes the two sets of ideas can be yoked together, and made to pull the same way, and then 'a smashing business' is done."[32] Nonetheless, some wealthy landowners commissioned extensive landscaping with hired professionals. The 1844 city directory listed six men with an occupation of gardener, including one Thomas Kelly, gardener to Mayor William B. Ogden.[33] As the *Prairie Farmer* explained, "Thomas Kelly . . . like most of those of his occupation in this city, acquired his art across the water."[34] Indeed, many of Chicago's first landscapers were head gardeners of estates in Ireland or England. They brought with them the techniques and plants native to the United Kingdom. At the Ogden residence, on Chicago's north side, Kelly tended a small vineyard, along with plums, currants, and raspberries. The stars of the flower garden were impatiens and lady's slippers. But his collection of evergreens—pine, larch, balsam fir, and arborvitae—struggled, possibly due to his inexperience with local growing conditions.

Ogden, Kennicott, Egan, and a handful of other leading figures were among the earliest to beautify their home estates with flowers, a movement that became stronger in the 1850s as more local nurseries sprang up. People began to see their homes as permanent, not as waystations on the road to the next frontier stop. Popular literature, and discussions within horticultural societies, began promoting gardens as necessary for the moral well-being of children and women. Planting gardens thus became a way to tame the frontier, declare oneself cultured, and even instill morality in youth. As the fervor and avaricious aftermath of the 1849 gold rush struck, Illinois horticulturist Mathias Dunlap warned his colleagues in an 1858 Illinois State Horticultural Society convention: "If you do not wish your grown up sons and daughters to stray off into far off lands, when the scramble for gold, for power, or for place, stirs up all the worst passions of our nature . . . make your home attractive and set them the example of rational and refined enjoyment, teach the love of flowers, the appreciation of the beautiful in nature."[35] Further discussion at this meeting raised the notion that frontier children were more impetuous than their New England counterparts. To curb roving spirits and protect them from temptation, gardens were cited as a solution. "The effect of gardening upon the young would be salutary. Children at the West have a great deal of peculiarly American independence, much of the free and easy, not so much controlled by the conventionalities of society as at the East . . . If we would tame down this spirit, we must make our homes beautiful and attractive."[36]

Pseudoscientific observations were made on the calming effect of gardens:

> [John] Kennicott said he had often carried flowers through the streets to see what effect would be produced upon those who saw them. It did not take many such experiments to convince him that most persons have a love for flowers and that with children this natural taste amounts to a passion. Let any one carry a handful of flowers slowly by a group of school children and see how quickly their eyes will sparkle . . . Who can doubt the importance of fostering this natural love of the beautiful, so marked a feature in the disposition of children.[37]

Kennicott further opined that gardens promoted well-being and health, particularly for women. In the Victorian era, women avoided the outdoors because tanned skin and calloused hands signified a lack of breeding and wealth. Instead, Kennicott urged the press to write more about the benefits of gardening and noted, "The influence of gardening upon health is generally overlooked. Our females need more fresh air and exercise to check the physical degeneracy that is certainly fastening upon the American people."[38] With the popular press recommending gardens for health, culture, and moral improvements, the demand for gardens and gardeners began to increase. In an 1861 *Prairie Farmer* article titled "Gardeners Wanted," the editors noted, "In the 'old country' there are more than can be employed . . . young men who have acquired sufficient knowledge, emigrate to countries where their services will be needed . . . Well educated practical gardeners are needed in this country."[39]

Fires, Fairs, and Suburbs, 1871–93

The best-laid plans of Chicago's forefathers, from elaborate schemata of parks and boulevards to grand drawings for private homes, came to a burning halt with the great Chicago Fire of 1871. Many nurserymen, like other businesses, were displaced by the fire, to say nothing of the countless homes destroyed. The entire city's physical layout and demographics were remapped through new building codes and an exodus to the suburbs. New building codes requiring brick construction made suburban homes a more economical choice, and that, combined with easier train commutes, set the stage for a boom in suburban development. Homeowners, displaced by the Chicago Fire, flocked to newly developed suburbs. No longer summer cottages, homes in outlying areas became year-round residences. Nationally, this also coincided with the so-called golden age of seed and plant catalogs (from the 1880s to the early 1900s) where elaborate chromolithographs tempted home gardeners with promises of beautiful grounds. Taking cues from the extensive landscaping in city and suburban parks, homeowners upgraded their own gardens to include the latest floral fashions, ranging from geometric carpet bedding to simple dooryard gardens.

Real estate speculators formed "improvement companies" and purchased tracts of land outside the city limits. The healthful benefits of nature and greenery became a mantra in the popular press, and the green space of suburban

Perhaps presaging Chicago's enduring horticultural prowess, Ezra McCagg's greenhouse was improbably one of the few remaining structures after the great Chicago Fire of 1871. From the author's collection.

design was a popular lure. By commissioning nationally recognized landscape architects, such as Frederick Law Olmsted in Riverside, town developers significantly elevated the role of greenery and gardens. These suburbs commanded high prices. Woodlawn, near Washington Park (then South Park), began selling lots in 1866 at $160 per acre, and by 1874 lots were selling at $4,000 to $5,000 per acre.[40] About Highland Park, Everett Chamberlin wrote, "The price paid by the [Improvement] Company, six years ago, for the whole tract embraced in the town plat was only $40 per acre. Within the past few weeks, they disposed of 85 acres of the poorest land at the round figure of $1,000 per acre."[41]

Chamberlin highlighted the importance of trees and greenery in many newly formed suburbs. Of La Grange, he noted, "one thousand shade trees of various kinds have been out about the lots and streets, besides a large number of fine fruit trees planted by [founder] Mr. Cossett." Established in 1871, the Beck's Park subdivision in Englewood (at 67th and Halsted) included a park with artificial lakes and fountains and some 15,000 newly planted shade and ornamental trees. Incorporating parks into a suburb's design also enhanced property values of homes nearby. In Englewood, for example, John Raber's home fronted eastward toward the South Parks. Raber's home, purchased in 1862 for $17,000, was worth $75,000 in 1874. Similarly, in Washington Heights (now the Beverly community), Chamberlin noted the value of the newly created Prospect Park: "This contains six acres, with serpentine walks and grass plats, evergreens, flower-beds and ornamental shrubbery. It makes property in its vicinity extremely desirable."[42]

In a choice location near the South Parks, John Raber's home included miniature ponds with goldfish, winding walkways, trimmed hedges, and beds of flowers. Ca. 1874. From Chamberlin, *Chicago and Its Suburbs*.

The exodus to the suburbs was just one of the marked impacts on horticulture after the Chicago Fire. Horticultural societies continued to flourish and became even more dominant in national affairs. At a meeting in Chicago on June 14, 1876, a group formed to become charter members of the American Association of Nurserymen, Florists, and Seedsmen. This group, hosted by Chicagoan Edgar Sanders, ultimately became the American Association of Nurserymen.

Chicago had always been a host city for garden society meetings and horticultural fairs, and this only increased after the fire. In the 1840s, the CHS had sponsored a number of floral exhibitions. In the 1850s, various agricultural societies, both national and statewide, held their fairs in Chicago. The Great Sanitary Fair of 1865, landscaped by John Blair, drew crowds to the lakefront. But after the Chicago Fire, it became even more important to demonstrate to the outside world that Chicago was back in business. The 1870s and 1880s saw a number of regional fairs, including various interstate exhibitions. But it was the Columbian Exposition of 1893 that really put Chicago back on the map. Landscaped by Frederick Law Olmsted, the 1893 World's Fair set a new standard for architecture and horticulture. Jackson Park was transformed from a barren wasteland to a beautiful oasis of green space and lagoons. The Horticulture Palace, designed

The Horticulture Palace housed exciting displays of new plants and fruits during the Columbian Exposition (1893). Designed by William Le Baron Jenney, the Palace was prominently sited across from the Wooded Island, Frederick Law Olmsted's masterpiece of naturalistic landscaping. Courtesy of Chicago Public Library, Special Collections.

by William Le Baron Jenney, was filled with products of floriculture, pomology, and viticulture from around the world. Leading nurserymen and seed companies from the United States and foreign countries vied for positions on the grounds to display their new flower introductions. A world horticultural congress was held, with participants from around the globe.

By the late 1880s, Chicago's role as a central marketplace clearly affected the prospects of local gardeners and nurserymen. In his retrospective on the status of local horticulture, Edgar Sanders observed that strawberries, peaches, and even celery, once big crops in Chicago, were no longer grown: "Take celery . . . where I come from, north of Chicago, the land was found just suitable. The Germans, mostly, kept increasing their patches, until there were two or three hundred acres . . . In a little while a sort of wheezy cry came that Kalamazoo was growing celery. 'Ah!' the South Water street [Chicago produce market] men said, 'can't beat Lake View celery' but it kept coming in, and better and better. Prices drop . . . The celery is down, high prices spoiled their [the nurserymen's] appetite, not a few pulled up stakes and hied to California."[43] Florists, with new glasshouse technologies, had better prospects, according to Sanders: "Why it is positively magical, the transformation then and now in this department. There were just three greenhouses at that time of fifty feet each. Now a person is apt to say, so many acres in glass when speaking of many great florists with steam engines of a hundred and fifty horse power to heat them. Then only an old fashioned flue was used . . . If our city is short [of cut flowers], they are bundled up in ice if in summer, cotton-batten if in winter and sent off thousands of miles to decorate my ladies' chamber."[44] Although the two decades between the Chicago Fire and the Columbian Exposition raised Chicago's profile as a major horticultural center, the task remained for Chicagoans to cultivate a unique, regional style in the gardening world.

The Jungle and the Prairie, 1894–1933

At the end of the nineteenth century, the flood of immigrants to Chicago, coupled with the explosive growth of factories and industrialized jobs, made for desperate overcrowding in city tenements and simmering social unrest. Upton Sinclair painted the ugly side of this picture in his 1906 novel, *The Jungle*, about Chicago's meatpacking stockyards. But Chicagoans were energetically pursuing solutions to these social ills through settlement houses such as Hull-House, philanthropic programs, and public-awareness campaigns of the Progressive Era. Conservation fervor also flourished at this time as a way to preserve the unsullied innocence in Nature. Chicagoans were major players in this new national environmentalist movement. Groups such as the Prairie Club of Chicago, led by influential Chicagoans including architect Dwight Perkins and landscape architect Jens Jensen, explored the countryside around Chicago. Through their efforts and kindred groups such as the Wildflower Preservation Society, endangered green spaces, such as the Cook County Forest Preserves and the Indiana National Dunes, were preserved.

Jens Jensen, a Danish landscape architect who began his career in Chicago in the West Parks system, was a charter member of the Prairie Club. In the 1880s while working in Union Park, Jensen created "The American Garden," featuring native plants. This garden was a tremendous hit with the public and much easier to maintain than the exotic plants favored by Victorians. At the turn of the century, while Frank Lloyd Wright began his school of prairie architecture, Jensen and his colleagues launched a parallel movement in landscape architecture. Like landscape gardener O. C. Simonds before him, the charismatic Jensen embraced the prairie environment, using native plantings and emulating the look of lazy prairie rivers and rockwork in his designs. Chicago gardeners finally had a champion of regional flora.

Jensen's work was in great demand, both for public properties (Columbus Park and Garfield Park, among others) and private residences. Many of Chicago's elite asked for Jensen landscapes. Following the country-estate movement popular in America in the early 1900s, Chicago's leading citizens purchased large tracts of land in outlying suburbs such as Lake Forest and Highland Park. Jensen created naturalistic landscapes using native trees, water features, and striated limestone reminiscent of prairie bluffs.

Jens Jensen, a charismatic Danish immigrant, is recognized as one of Chicago's preeminent landscape architects who celebrated the region's native plants and landscape. Courtesy of Sterling Morton Library and Archives, Morton Arboretum, Lisle, Illinois.

Of course, not everyone in Chicago could afford Jensen's talent. Even though the 1893 World's Fair heralded the City Beautiful movement in urban planning and landscaping, industrialization took its toll on natural greenswards, replacing them with crowded slums. Chicago's urban tenements and deplorable working conditions were not conducive to ornamental gardens. Gardens were unknown to most poor immigrants, as activist Jane Addams wrote:

> We were also early impressed with the curious isolation of many of the immigrants; an Italian woman once expressed her pleasure in the red roses that she saw at one of our receptions in surprise that they had been "brought so fresh all the way from Italy." She would not believe for an instant that they had been grown in America. She said that she had lived in Chicago for six years and had never seen any roses, whereas in Italy she had seen them every summer in great profusion. During all that time, of course, the woman had lived within ten blocks of a florist's window; she had not been more than a five-cent car ride away from the public parks; but she had never dreamed of faring forth for herself, and no one had taken her.[45]

In this era, the Progressive social movement encouraged the creation of small neighborhood parks to provide green space for the working poor. A strong supporter of the movement, Jens Jensen built smaller parks for Chicago's immigrants, even as he designed estates for the rich. Despite the extreme poverty in the tenements, efforts were made by philanthropists, activists like the Prairie Club, and designers like Jensen to ensure that all Chicagoans had access to nature and greenery.

World War I and the subsequent onset of the Great Depression limited recreational gardening activity during the 1920s. But by the 1930s, Chicago was ready for another international exposition. In the depths of the Great Depression, Chicago returned to its lakefront to host its second world's fair, the 1933 Century of Progress Exposition. Experiments with colorful lighting and the design of ultramodern, windowless buildings emphasized outdoor display gardens. Midwest sense and sensibility toned down the dramatic modern garden style, which debuted in France, and the influence of the prairie school of landscape architecture was apparent in several naturalistic displays. Home gardens were also featured in the outdoor horticultural area, as well as displays from major garden societies. By showcasing Chicago's gardening talent, and featuring the prairie style of landscape architecture, Chicago had finally gained the recognition of the horticultural world.

The 1930s also saw the maturation of Chicago's horticultural publications. Chicago had become home to more agricultural journals than any other city in the world by the late 1800s.[46] Subscribers throughout the Midwest enjoyed the

The horticultural grounds at the 1933 Century of Progress World's Fair included vignettes of gardens from around the world. From the author's collection.

Prairie Farmer, one of Chicago's earliest agricultural periodicals. The *Orange Judd Farmer*, published by former New Yorker Orange Judd, relocated to Chicago and became a rival publication in the 1880s. The *Farmer's Review* featured more agriculture-related articles. With correspondents from around the nation, Chicago produced its own biweekly *Gardening* magazine published by William Falconer in the 1890s. *The House Beautiful*, a turn-of-the-nineteenth-century monthly magazine published by Herbert S. Stone in Chicago, described the ideal house and gardens. Chicago's media influence was not limited to the printed word—local horticulturists gave lectures at other cities' professional meetings, and as radio became popular, various stations broadcast gardening talk shows hosted by local garden club members or nursery representatives.

By the 1930s, the shift in Chicago's horticulture from the prosaic to the aesthetic was clearly embodied in the local publication of the new *Landscape Architect*. Produced by Chicago landscape architect F. Cushing Smith, the magazine highlighted the marriage of architecture and landscapes, seemingly a universal sign that Chicago's horticulture had indeed reached a pinnacle of maturity.

The century of gardening described in this book begins and ends at the same spot along the shores of Lake Michigan. The 1933–34 horticultural exhibition at the Century of Progress World's Fair was held a stone's throw away from the early gardens within the palisades of Fort Dearborn, but there was a world of difference between the fledgling cornfields of the fort and the international garden displays of the fair. Decade by decade, Chicago grew into its flowery sobriquet of the Garden City.

TWO

GROWERS AND SOWERS

Planters of the Plains and Parkways

Chicago's gardens are as much about people as about plants. Pioneer Chicagoans were risk takers who shared a sense of adventure and a willingness to learn new things. Their entrepreneurial spirit motivated them to explore the natural world, invent agricultural tools, and improve horticulture. Chicago's early gardeners, through individual experimentation and group discoveries, not only reshaped their own physical world, but also influenced the nation's horticultural and scientific progress.

Unlike early real estate speculators who only experienced Chicago as absentee landlords, Chicago's first gardeners and naturalists staked their reputations and livelihoods on mastering the natural world. They left no nest eggs behind them on the East Coast, nor were their land claims hedged with broader investment portfolios. Their first challenge was to understand and interpret Chicago's regional flora, and then to classify, cultivate, and make it pay. For scientists, Chicago's natural world offered an uncharted territory of new plants. Their objective was to document and study new species. Nurserymen, florists, and market gardeners, however, were more interested in finding easily grown plants to meet burgeoning demand. Both science and commercial interests often overlapped, with mutually beneficial results.

Chicago's women gardeners brought yet another new perspective to the field of horticulture. Chicago women's clubs were among the most numerous and influential in the country during the last quarter of the nineteenth century. Most clubs included a horticulture or forestry subcommittee, which was often focused on conservation matters, or a gardening group that targeted civic improvement. Chicago-based women's garden clubs, often requiring sponsored memberships, included some of the region's most wealthy and outspoken ladies. The women's impact on gardening, while often constrained by prevailing male-dominated mores, was soon felt in areas such as city planning, social improvements, and landscape architecture.

Special-interest groups developed as a time-tested way of sharing information among horticulturists. Horticultural societies, based on East Coast and British models, were a mainstay throughout Chicago's history. Societies were formed at many organization levels, from city to county to state. Chicago-area members often participated in multiple societies, including those back East, such as the Massachusetts Horticultural Society and Pennsylvania Horticultural Society. By keeping ties with their eastern brethren, Chicago horticulturists not only were apprised of the latest news, but also exchanged valuable information about the new frontier. As individuals, and collectively in professional societies, scientists, nurserymen, florists, and kindred spirits forged indelible marks in Chicago's soil and soul.

THE SCIENCE OF HORTICULTURE: FROM PLANT HUNTERS TO ECOLOGISTS

Amateur and professional botanists and scientists were intrigued by the wealth of plant diversity in Chicago-area prairies, wetlands, and forests. Many species had never been seen before. Initial field trips were conducted by individuals such as George Vasey, who led explorations throughout Illinois's countryside. Other adventurers included Robert Kennicott, who began exploring his own backyard at the Grove, and then branched out to Illinois and ultimately to Alaskan frontiers. Both men eventually made national headlines through their work with the Smithsonian Institution. Chicago's scientific research grew from individual pursuits such as these to cooperative programs sponsored by scientific societies and newly established museums. In 1857, the Chicago Academy of Sciences became Chicago's first museum, cofounded by a group of physicians, leading society men, and naturalist Robert Kennicott. The Illinois Natural History Society was founded in 1858 (and exists today as the Illinois Natural History Survey), and the Illinois State Microscopical Society was founded in 1868. With the strength of such organizations behind them, Chicago's botanists progressed from plant hunters like George Vasey to the Father of North American Ecology, Henry C. Cowles.

The Chicago Academy of Sciences embraced a broad scope of interests, including botany as a major focus. In 1860, the academy resolved to hold five monthly field meetings in different parts of the state. Their inaugural trip, from Chicago over three hundred miles south to Cobden, Illinois, was attended by one hundred people and sponsored by the Illinois Central Railroad. The academy's transactions proclaimed: "We venture to assert that nowhere in this country has there occurred an excursion of such a purpose, that can compare with this."[1] Professional and amateur scientists enjoyed the day-long ride, arriving in

FIELD-TRIP FLOWERS

This list, as documented by Illinois naturalists on an early railroad tour of Illinois, presents a fascinating flora. The sequence of plants unfolded as observed from north to south across five degrees of latitude in Illinois. It offers a picture of the flowers across the entire breadth of Illinois in 1860. Here is the list in the original order, with the original nomenclature and capitalization.[1]

Spiderwort (*Tradescantia Virginica*)
Wild Orange Red Lily (*Lilium Philadelphicum*)
Phlox (*P. Divaricata, P. aristata, P. bifida*)
Tall Coreopsis (*Coreopsis tripteris*)
Tickseed (*Coreopsis palmata*)
Cone Flower (*Rudbeckia hirta* and *R. triloba*)
Grass Pink (*Calopogon pulchellus*)
Black Sampson
Moccasin Flower (*Cypripedium spectabile*)
White Lady's Slipper (*Cypripedium candidum*)
Milk Vetch (*Astragalus distortus*)
Horse Weed (*Erigeron Canadense*)
Robins' Plantain (*Erigeron bellidifolium*)
Queen of the Prairie (*Spirea lobata*)
Vine Bark (*Spirea opulifolia*)
Meadow Sweet (*Spirea salicofolia*)
Wind Flower (*Anemone Pennsylvanica*)
Wild Lupine (*Lupinus perennis*)
Butterfly Weed (*Asclepias tuberose*)
Hairy Puccoon (*Lithospermum hirtum*)
Hairy Waterleaf (*Hydrophyllum appendiculatum*)
Painted Cup (*Castilleia coccinea*)
Watershield (*Brasenia peltata*)

Sweet Scented Water Lily (*Nymphae oderata*)
Yellow Pond Lily (*Nuphar advens*)
Silky Cornel
Kinnikinnik (*Cornus sericea*)
Plantain (*Plantago Virginica*)
Phlox (*Phlox acuminatum*)
Horsetail Scouring Rush (*Equisetum robustum*)
Yarrow (*Achillea millefolium*)
Flowering Spurge (*Euphorbia corollata*)
Long Fruited Anemone (*Anemone cylindrical*)
Meadow Rue (*Thalictrum Cornuti*)
Low Bindweed (*Calystegia spithamaea*)
Early Wild Rose (*Rosa blanda*)
Cowbane (*Archemora rigida*)
Blazing Star (*Liatris pycnostachya* and *L. spicata*)
Cone Flower (*Rudbeckia subtomentosa*)
Prairie Dock (*Silphium terebinthinaceum*)
Rosin weed–Compass Plant (*S. lacineatum*)
Sneezewort (*Helenium autumnale*)
False Indigo (*Amorpha fruticosa*)
Lead Plant (*A. canescens*)
Pasture Thistle (*Cirsium pumilum*)
Goat's Rue–Catgut (*Tephrosia Virginiana*)

1. "The First Field Meeting of the Chicago Academy of Sciences," *Prairie Farmer*, July 5, 1860, 1.

Cobden at 9:00 p.m. En route, "did the train stop at a station, or at a watering place to catch breath or take a drink, entomologists and botanists with vials and cases, were ready and active in the prosecution of their favorite studies." Watching through the windows of the train, botanists documented plants in the order observed in the countryside.

The growth of Chicago's scientific institutions paralleled those of other major Midwest cities. Unfortunately, even as the Civil War slowed museum growth in most states, the Chicago Academy of Sciences was dealt an even more severe blow with the city's 1871 fire. The academy's building at Wabash and Van Buren streets was completely destroyed, along with over $200,000 of collected specimens. Fortunately, some specimens survived in private herbaria and remain today in a number of Chicago institutions. Ultimately relocated to a new building in Lincoln Park, the Chicago Academy of Sciences published the *Bulletin*, a series of essays on natural history. In 1891, the *Bulletin* contained a 168-page essay by William K. Higley and Charles S. Raddin, titled "The Flora of Cook County and a Part of Lake County, Indiana." Higley was a Northwestern University professor of botany and served as the academy's secretary from 1897 to 1906. Except for a smaller list by Henry H. Babcock, Higley and Raddin's essay was the first extensive listing of regional flora for Chicago.

In addition to list making and professional exchanges, some early botanists bridged the gap between the academic and lay world. Henry H. Babcock, for example, was a founding member of the Chicago Botanical Society in the 1860s, and hosted meetings for the group's scientific membership at his private Chicago Academy on 18th Street.[2] He actively exchanged seeds and botanical information with countries throughout the world, sharing his findings with his peers and the general populace alike. One of Babcock's greatest ambitions, unfortunately unfulfilled, was the launch of a Chicago botanical garden. With a few like-minded citizens, Babcock proposed a world-class botanical garden within the South Parks system. In June 1875, he sent a circular to botanical gardens and private individuals around the world, asking cooperation in creating this garden. By December of that year, he had received over 3,200 packages of seeds and bulbs, 775 species and varieties of living plants, over 1,000 herbarium specimens, and volumes of books from gardens or collectors in Prague, Pisa, Melbourne, Zurich, Calcutta, Montpellier, Amsterdam, and cities in the United States. Ninety-five species of native plants were collected and planted, and seeds of almost 200 native plants were collected.[3]

H. W. S. Cleveland, then landscape architect for Chicago's South Parks, developed a long-range plan for the grounds, incorporating Babcock's botanical garden. In the interim, two and one-half acres were prepared for the plants, and four plant-houses, located in the southeast corner of the northwestern areas of the South Parks, were constructed to grow tender plants. Cleveland remarked on Babcock's dedication and knowledge, "He [Babcock] found time to inform himself thoroughly of all the latest improvements in the construction and heating of plant houses, and a building was erected under his direction for the purpose, which might serve as a model of economy of form and adaptation to its object."[4]

Scientists and physicians were especially interested in this botanical garden, which was planned to be one of the largest in America. In 1876, the *Chicago Medical Journal* suggested that in the United States, only Harvard's botanical garden was comparable. The journal opined, "It is well known that the Government gardens at Washington subserve chiefly the purpose of providing bouquets for members of Congress, while that at St. Louis—small, popular, and showy—is scarcely adapted to the study of botanical science."[5] The *Pharmacist* noted, "To pharmacists and physicians the Botanical Gardens should be of no ordinary interest ... The study of botany has at all times been deemed of great value and importance to students of medicine and pharmacy. Through neglect of this study in the schools our medical men are in lamentable ignorance of the value of many indigenous plants as remedial agents. To such an extent is this true, that we owe to accidental causes, or to the traditions of the 'noble red man,' the imperfect knowledge we have of native medicinal plants and their uses."[6]

Unfortunately, the South Parks Board, in a vote on July 21, 1877, suspended operations of the botanical garden, citing overall cost-cutting measures. Two specimens of each variety of all plants were to have been saved for potential future use in a later botanical garden.[7] This plant ark did not happen in the four short years remaining of Babcock's life. Scientists mourned the lost opportunity, as in the writings from this 1891 Babcock biographer on the fate of the botanical garden: "Here were the labor and the hope of years brought to naught; scientific treasures of inestimable value were wasted; faith with contributors was violated; the cause of science in Chicago was disgraced."[8]

Chicago's universities were pivotal at this time in formalizing botanical education. What was once the passionate pursuit of adventurous individuals such as Kennicott and Vasey became a codified program of learning. In 1901, Charles B. Atwell, professor of botany at Northwestern University, planned an on-campus demonstration garden of every native plant in Cook County. The University of Chicago, which opened in 1890, retained preeminent botanists of the day, such as John Merle Coulter, who was appointed to head the Department of Botany in 1894. Coulter was a leading American botanist who had founded the *Botanical Gazette* nearly twenty years prior. Under Coulter's leadership, in 1896, the university built the Hull Biological Laboratories, a four-building complex that included a botany pond, rooftop greenhouse, laboratories, and classrooms. In the early decades of the twentieth century, University of Chicago botanists brought science not only to college students but to the general population

An elaborate greenhouse was built atop the University of Chicago's Botany Building. American Environmental Photographs Collection, AEP Image Number AEP-ILP72, Department of Special Collections, University of Chicago Library.

through countryside field trips, and to high school curricula. H. C. Cowles led walking trips to ecologically rich areas including the Indiana Dunes with groups such as the Prairie Club. His protégé, H. S. Pepoon, conducted botany classes at Lake View High School. Coulter's *Botanical Gazette* was the leading journal of American botany, and the University of Chicago's Department of Botany was ranked first among its academic peers.[9]

MAKING PLANTS PAY: THE BUSINESS OF HORTICULTURE

Scientists and botanists may have thrilled at discovering native plants, but earning a living through horticulture was an uncertain undertaking in early Chicago. Growing conditions were experimental, nursery stock was hard to obtain, and customers were fickle in their loyalty toward upstart western nurseries. Those who chose to grow plants as a livelihood needed self-confidence and, because few books referenced western horticulture (which, at that time, could mean anything west of the Appalachians), a great network of colleagues to share information. Professional associations and societies were critical links for Chicago's horticulturists to expand their knowledge and establish lobbying power.

Distinctions among nurserymen, florists, and seedsmen were blurred in early years, and professional groups often combined these disciplines. Generally, the early interpretation of the trades meant that nurserymen grew stock, typically trees and shrubs, on their own grounds; florists specialized in flowers, sometimes growing them under glass or in the field and sometimes retailing cut or potted plants; and seedsmen, as the name implies, produced and packaged seeds. Before specialization, individual gardeners often pursued all three trades. While there may have been overlap and contention among these three groups, all shared common interests. Governmental regulations, railroad tariffs, and marketing practices were of vital interest to all horticultural trades, and caused individual practitioners to seek solidarity in professional associations. Local horticulture societies were the first to host fairs and flower shows where tips and techniques could be freely traded among exhibitors and visitors.

The CHS, said to be the second oldest society in Illinois,[10] included this mix of civic leaders and working nurserymen. John Kinzie, Chicago's great pioneer, was president of the group, and other city leaders like Mayor William B. Ogden and entrepreneur W. L. Newberry were also members. John Kennicott, another founding member, was one of Chicago's pioneer horticulturists, and is representative of the type of Renaissance men who explored all aspects of Chicago's flora.

Few names were more respected in Chicago's early horticultural history than that of John Kennicott. A physician, Kennicott was called the Old Doctor, to distinguish him from other Kennicott kin who were also doctors. A native New Yorker, Kennicott brought his family to Cook County in 1836, leaving a teaching career in New Orleans. At thirty-six years old, he staked his claim in the Grove, a woodsy section of what is now suburban Glenview. Forty of Kennicott's 230 acres were devoted to the nursery. According to one of many eulogies, "His first work after arriving in Illinois; after providing a rude shelter for his wife and child, was the planting of fruit trees and flowering shrubs and plants about his home, and the creation of a private nursery, with which to supply his own wants and those of his neighbors."[11] Kennicott wrote for a wide variety of horticultural periodicals, and experimented with plants not only for medical purposes but also for commerce.

Nurserymen: Growing Plants on the Prairie

Early nurserymen such as John Kennicott, those who grew and sold trees for a living, had a particularly trying job. As one early grower wrote in the 1850s, "The nature and requirements for a tree are less understood than any animal or thing with which we come so frequently in contact."[12] This writer, a nurseryman himself and a customer of eastern nurseries, described his disappointment with fruit trees untrue to their purported strain, poorly packed material, and a "much be-puffed flowering plant; which we not only planted but distributed on their authority. That plant is now an intolerable weed, sprawling through the soil and making itself odious." The writer opined that western nurserymen had demonstrated better integrity and customer service. What they needed to make their businesses profitable, however, was more knowledge of plant culture.

Kennicott's writings, and particularly his leadership in horticultural societies, greatly expanded Chicago nurserymen's knowledge of local growing conditions. In the 1850s, the CHS temporarily disbanded, to lie quiescent in fits and starts until its rebirth in 1890.[13] Kennicott helped organize the Cook County Agricultural and Horticultural Society in 1856 and supported the Chicago Gardeners' Society, whose membership largely consisted of professional nurserymen.[14]

Chicago horticulturists were also quite influential in regional societies. The state of Illinois, ranging across three U.S. Department of Agriculture hardiness zones, provides a diversity of growing conditions. Chicago's early horticulturists benefited from lessons learned in the southern part of the state where early French settlement predated Chicago. State societies date back to 1819 with the formation of the Illinois Agricultural Association. But it was in the 1850s, with the impetus of railroad and further westward expansion, that some of the most influential and long-lasting professional groups were formed. In each, Chicago individuals played prominent roles.

In 1853, the Illinois Agricultural Association became the Illinois State Agricultural Society, which in 1871 evolved into the Illinois Department of Agriculture. The focus of this group, per its 1860 constitution, included the "promotion of Agriculture, Horticulture, Manufactures, Mechanic, and Household Arts." Organizationally, the group comprised county delegates, and county fairs complemented annual state fairs held throughout Illinois. Topics of interest to this group ranged from horse breeding to drainage to bread making to orcharding. Early Chicagoans participated in the society and, in fact, won several awards at the 1861 state fair held in Chicago. Nurseryman Robert Douglas won second place for the finest arrangement of nursery-grown fruit trees and shrubs, and businessman Charles E. Peck of Winnetka won acclaim for the best transplanted forest trees.

The Agricultural Society had a broad mission emphasizing farm crops and animal husbandry, yet two other groups with even more specific horticulture missions were formed in the 1850s. The Northwestern Fruit Growers Association (formed in 1851) included officers John Kennicott and Arthur Bryant Sr. (from Princeton, Illinois), and such Chicago-area members as Mathias Dunlap, Robert Douglas, Lewis Ellsworth, and John A. Wight. The fruit growers studied root- and stock-grafting and best apple types, and held exhibitions of the same. Lists of recommended apples were periodically distributed, and exhibitions and annual conventions also met throughout the state and Midwest. The group attracted non-Illinois residents, including luminaries such as John A. Warder and Charles Downing. In 1857, a joint meeting was held in Alton, Illinois, between the newly formed Illinois State Horticultural Society and the Northwestern Fruit Growers Association at which the two merged into the Illinois State Horticultural Society. This group, still active today, was a prominent voice not only in horticultural circles, but also in social aspects such as landscaping of public parks, the industrial education movement, and conservation.

An interesting but seemingly ill-fated group formed in 1860 with the creation of the Northwestern Agricultural Society. The *Prairie Farmer* reported that this newly formed society, an endeavor spearheaded by John Kennicott, W. B. Egan, and others, purchased the Garden City Race Course in Brighton, six miles south of the city. Capital stock of $60,000 was raised through the society, with thirty members of the society subscribing $22,000 as of February 1860. This site was chosen in anticipation of the state fair and other events, but some accounts intimated that it was a real estate boondoggle. The *Prairie Farmer* noted that the "objects of this organization, as set forth in its Constitution are *rather* imposing, viz: The promotion, protection and encouragement of agriculture, and its kindred arts and science; the assertion of the importance of a fitting recognition of this great primary interest, so closely allied to the success, not only of Chicago, but of our whole country; the creation of an Agricultural Bureau at Washington; the founding of an Agricultural College, Museum, Library and Reading Room, in Chicago; the establishment of annual fairs and exhibitions, and the improvement of a permanent park, pleasure and fair grounds."[15]

Prairie Farmer editors applauded the lofty goals of the nascent Northwestern Agricultural Society but were not favorably impressed with the location. The site of the first fair caused great controversy, with many believing that the Garden City Race Course was poorly chosen. Nonetheless, this first fair of the Illinois State Horticultural Society was hosted in Chicago's Brighton area. Held from September 8 to 13, 1861, in conjunction with other kindred societies (the Illinois State Agricultural Society and Chicago Mechanics' Institute) as part of the Illinois State Fair, the horticulture department gained recognition, although the overall fair was not successful. Undeterred, then president of the Illinois State Horticultural Society John Kennicott posted an open invitation in local newspapers to host the society's annual meeting in Chicago on December 3, 1861.

Writing to "all the Brotherhood of Rural Art and Science, West and East, and all who wish to learn the secrets of Horticulture, or are willing to teach them," Kennicott exhorted kindred spirits to Chicago's Bryan Hall for four days of "subjects of great moment to the initiated, and of equal interest to the tactical amateur and unadvised beginner in the nursery, lawn and garden, in the orchard and on the farm."[16] In what was to become a frequent collaboration, railroad tycoons from the Galena and Chicago Union; Chicago and Northwestern; Alton and Chicago; and Chicago, Burlington, and Quincy railroads promised free or reduced railroad fare for all society members attending the meeting.

As railroads improved, Chicago horticulturists were more connected with their brethren within the State of Illinois. In 1860, C. Thurston Chase of Chicago headed a committee that proposed dividing the state into seven fruit districts. Geographical subgroups were thus formed, such as the Horticultural Society of Northern Illinois, which embraced areas west of Chicago including Princeton, Rockford, and Mount Carroll, south including Kankakee, and north including Lake County. Most of Chicago's leading horticulturists belonged to the northern Illinois society. Notes from various society subcommittees give insight as to pressing issues of the day.

For example, in 1872, there were eight subcommittees of the Horticultural Society of Northern Illinois, but by 1877 the interests had expanded to sixteen special subcommittees. Of these groups, six were focused on the culture and proper selection of fruits and vegetables. Two focused on the increasingly important areas of marketing and keeping fruit. Two continued the study of soils—prairie clay and loam evidently still posing agricultural issues—and two were concerned with the types of trees best suited for the area. Typically comprising three members, a representative from Chicago or a nearby town was on each committee (except grapes and entomology) as follows:

Cultivation of Apple Orchards: E. H. Skinner, Rockford; S. G. Minkler, Oswego
Gathering and Keeping Apples: S. M. Slade, Elgin
Small-Fruits, Embracing all Classes of Berries: Lewis Ellsworth, Naperville
Plums, Culture and Varieties: A. Bryant Jr., Princeton
Vegetable Gardening: George S. Haskell, Rockford; E. C. Hatheway [sic], Ottawa
Greenhouse and Window Plants and Bulbs: H. W. Williams, Batavia; H. C.
    Vaughan [sic], Chicago; Frank Ludloe, Naperville; B. O'Neil, Elgin
History and Progress of Horticulture in Northern Illinois: H. D. Emery, Chicago
Ornithology: C. W. Douglas, Waukegan
Landscape Gardening: Bryant [sic], Princeton; H. S. A. Cleveland [sic], Chicago
Vegetable Physiology: Edgar Sanders, Lake View; Bryon D. Halstead, Chicago
Geology and Soils: E. H. Beebe, Geneva
Marketing Fruits: H. D. Emery, Chicago
Cherries, Culture and Varieties: S. M. Slade, Elgin
Adaptation of Varieties to Soils and Locations: D. C. Scofield, Elgin
Timber Planting for Economic Uses: Robert Douglas, Waukegan; D. C. Scofield,
    Elgin
Culture of Evergreens: Dr. W. A. Pratt, Elgin

J. C. VAUGHAN AND THE SECOND CROP OF NURSERYMEN

J. C. Vaughan, and later his sons, epitomized the second wave of Chicago nurserymen who would capitalize on technology and professional networking to build their business. The Vaughan Company strategically located their Western Springs operations next to the Chicago, Burlington, and Quincy Railroad tracks to expedite shipping. They used "modern" trucks to haul stock between their field operations and Chicago downtown warehouses.

Like many nurserymen, Vaughan located near a railroad and used it for interstate deliveries. He had a fleet of open-air trucks to haul plant material locally. Courtesy of Western Springs Historical Society.

J. C. Vaughan began selling nursery stock in Chicago in 1870, and opened Vaughan's Seed Store in 1876. By 1889, Vaughan had 7,500 feet of glasshouses in Western Springs in a nursery and greenhouse business that would expand to over 35,000 feet of glass in 1926. In this western Chicago suburb, Vaughan's trial gardens, with cannas, peonies, chrysanthemums, and more, were legendary in beauty and extent. With branches in New York and downtown Chicago, Vaughan also started a catalog business that became international in scope.

Vaughan's Western Springs location became famous for its landscaping displays. Rustic bridges crossed lily ponds, and tree-lined driveways led to a large nursery office. A boardinghouse was established on the grounds to house the workers. As well as growing his own stock, Vaughan bought seeds and stock from around the country and the world. Vaughan's "international pansy mixture," for example, won awards at world's fairs in 1893, 1898, and 1901, and resulted from "searching for the best in every land," according to the 1910 catalog. Vaughan's "12 Best Sweet Peas," also advertised in that catalog, were bought from English hybridizers and then grown by California growers. Through Vaughan's trial gardens, the company was able to determine how well plants thrived in the Chicago region, and whether it was more economical to buy or grow locally. Vaughan liberally shared his plant experience with his catalog customers.

Vaughan brought about many plant introductions, often naming them with local attributes. In the 1890s, Vaughan introduced the 'Bridgeport Chicago Drumhead' cabbage, in honor of Chicago's Irish Bridgeport community. The "Chicago giant self-blanching celery" was a white celery that could be sown as early as February. In 1906, Vaughan introduced an award-winning light scarlet gladiolus named after fellow Chicagoan, Mrs. Francis King. Capitalizing on brand recognition in New York and Chicago, Vaughan marketed both the "Chicago Parks" and "Central Park" lawn seed. In 1903, at a meeting of the American Florists and Ornamental Horticulturists, J. C. Vaughan was presented with a diamond ring and declared the "Dean of Chicago horticulturists."[1]

The company exhibited extensively in flower shows, locally, nationally, and internationally. Throughout the decades, the Vaughan name figured prominently in world's fairs including Chicago's Columbian Exposition of 1893 and the Century of Progress (COP) fair of 1933–34. J. C. Vaughan was chairman for the World's Fair Flower Show Association, which coincided with the St. Louis World's Fair of 1904. Following his father's footsteps, Leonard H. Vaughan was on the executive committee of many of Chicago's garden and flower shows, and had an extensive exhibit in the COP World's Fair in 1933–34.

1. "Flowers to Be Supreme," *Chicago Daily Tribune*, August 22, 1903, 4.

In 1874, the Illinois State Horticultural Society once again reorganized into three horticultural societies representing three growing districts: northern Illinois, central Illinois, and southern Illinois. Northern Illinois included Cook, DuPage, Kane, Lake, McHenry, Kankakee, and seventeen other counties.

Chicago horticulturists not only were influential on state and regional committees, but they also played an integral role in forming professional associations at the national level. Many national organizations were started at Chicago-based meetings, or often included Chicagoans among their charter members. The powerful American Association of Nurserymen, Florists, and Seedsmen, for example, had its beginnings in the Chicago area. At the January 26, 1876, Horticultural Society of Northern Illinois meeting in Crystal Lake, Illinois, members recognized the need for a more broadly based professional association to address common problems. In his seminal history of U.S. nursery-industry associations, *A Century of Service*, Richard White credits this meeting as a crucial catalyst for national organizations. This pivotal gathering marked the start of the American Association of Nurserymen, Florists, and Seedsmen, and Edgar Sanders, Chicago's preeminent florist, was elected its first president.

On June 14, 1876, the group held its inaugural convention in Chicago's newly constructed Grand Pacific Hotel on La Salle and Jackson and at the exposition building. This inaugural meeting included an exhibition of horticultural products

Chicago hosted many regional and national horticultural meetings, such as this growers' convention at Vaughan's nursery grounds in Western Springs. The building shown here was part of the nursery and at times served as a boardinghouse for workmen. Courtesy of Western Springs Historical Society.

and equipment, and, at the invitation of the South Park Commissioners, a tour of South Park, where H. W. S. Cleveland was implementing a landscape design by Frederick Law Olmsted. Included in this first meeting were discussions of fruit-grading standards, transportation, and postal rates.[17] In its coverage of the meeting, the *Chicago Tribune* concluded, "The West is now producing nursery products enough to supply the present demand of the whole country whereas it was formerly the case that it depended on the East and it is therefore of the utmost importance to Western interests to secure Chicago as the central point for the display and sale of all products that come from the nurseries and their auxiliaries."[18]

Despite the *Tribune's* admittedly biased endorsement of Chicago, controversy arose over where the association's annual conventions should be held. Competition among cities became so fierce that in 1897, a motion was passed by the group that meetings "should not be located either east or west of the Chicago meridian for more than two years in succession."[19] Chicago's central location was seen as a compromise. Between the years 1876 and 1933, the association hosted its annual meeting in Chicago eighteen times—far eclipsing the next most popular city, Detroit, which hosted the meeting four times in the same period. As a frequent host city, Chicago further established itself as a leader in the horticulture industry.

Chicago's nurserymen influenced prevailing business practices with their pioneering "protective associations," aimed at minimizing business credit risk. By sharing customer credit information, nurseries could avoid dealings with disreputable businesses. Richard White reports that one of the earliest of these protective groups was organized in the west in 1871, and a writer signed simply "Chicago" urged *Gardener's Monthly* subscribers that eastern nurserymen should follow suit. Locally, Jonathan Periam and other Chicagoans such as J. C. Vaughan, Edgar Sanders, P. S. Peterson, and W. D. Allen formed the Nurserymen's Tree Planters Protective Association of Greater Chicago and Vicinity in 1876 to accomplish similar goals as well as exchange horticultural information.[20] The group also established standards for tree sizes, and warrantees for replacement trees.

Chicago nurserymen were leaders in trade and technology innovation. For example, Plant Quarantine No. 37, enacted by the U.S. Department of Agriculture on June 1, 1919, caused serious disruption to the American nursery industry. Under this quarantine, an embargo was placed on foreign imports of nursery stock and grafts. This left many nurserymen, who had become dependent on outside stock purchases, short of stock and lacking the expertise to grow it themselves. Arthur H. Hill of the D. Hill nursery in Dundee proposed an organization to pool collective information on plant growing. At a meeting on June 16, 1919, in Chicago, the American Association of Propagating Nurserymen was formed,

SCHOOLS AND SOCIETIES

The importance of horticultural societies and associations began to wane in deference to the rising importance of educational institutions. When the University of Illinois was established in 1867 through the Land Grant Act, its academic staff began to take stronger leadership roles in the societies' activities. Professors from the university were often asked to give presentations. Life members of the Illinois State Horticultural Society and frequent lecturers included professor of horticulture Joseph C. Blair (member since 1899), University of Illinois vice president and professor T. J. Burrill (1870), professor C. S. Crandall (1902), professor H. B. Dorner (1911), instructor of olericulture C. E. Durst (1911), and professor of landscape gardening Wilhelm Miller (1912). Many of these professors had strong connections with Chicago: C. E. Durst lived in the city, and Wilhelm Miller designed Chicago-area estates and gave frequent lectures to the society about the work of landscape artists O. C. Simonds and Jens Jensen.

Horticultural experiment stations, operated by society members in cooperation with the university, began to replace reports of individual farmers and horticulturists. Often, leading horticultural society members assumed this additional responsibility: Experiment Station Number 1, for the Northern District of the Illinois State Horticultural Society, was located in Princeton and operated by the Bryant family. The society also maintained advisory committees to the university's department of horticulture and to the state entomologist.

One of Illinois's most influential early entomologists, Stephen A. Forbes, was not only a popular speaker to the society, but one whose certification program became of utmost importance to Illinois nurseries. Forbes, raised in northern Illinois, studied at Chicago's Rush College for a year, but ultimately turned to the study of biology and natural science in southern Illinois. After a particularly harsh plague of San Jose scale in Illinois in the 1890s, legislation was introduced in 1899 requiring the state entomologist to inspect nurseries for infestation and pest management. The inspection job, which Forbes described as "originally unwelcome, often delicate, and sometimes disagreeable,"[1] was fraught with difficulty and changing rules. Forbes's opinion on a nursery's health, however, soon became much more than academic. If he deemed a nursery injurious to others (because of communicable infestations), he was empowered to order the offending stock destroyed, if no remedial action was taken. In 1912, 145 nurseries were certified by the state entomologist, including 47 in the Chicago area (see the table in this box).

Of the total nurseries certified in the state in 1912, ten of the seventeen nurseries with fifty or more acres were in the Chicago area. Four of the six Illinois nurseries with one hundred or more acres were in the Chicago area. Many of the nurseries in the list were family affairs, with sons learning the trade from the fathers. Multigeneration or multifamily nurseries included Bryants, Klehms, Hills, Petersons, Vaughans, Douglases, and Nelsons. Information was passed down from generation to generation, with family members often joining the same horticultural societies. As the societies deemphasized fraternal exchanges of information in favor of more academic lectures from the university, memberships in the society began to wane, and the chain of family society memberships was broken (plate 3).

1. Stephen A. Forbes, "What Should the State Require of a Negligent Owner of a Dangerous Orchard?" *Transactions of the Illinois State Horticultural Society*, n.s., 45 (1911): 77.

Certified Chicago-area nurseries as of December 10, 1912

Name	Location	Acres	Certificate
American Horticultural Co.	Glenview	18	124
Aurora Nursery Co.	Aurora	50	25
Austin Nursery Co., A. B.	Downers Grove	35	28
Bahr, Fritz	Highland Park	3	100
Barnard Co., W.W.	Chicago	15	104
Beaudry Nursery Co.	Chicago	60	97
Bryant and Son, Arthur	Princeton	200	2
Buckbee, H. W.	Rockford	15	15
Buckland, J. V.	Ringwood	10	30
Carbary, E. W.	Elgin	2¼	115
Clavey, Fred D.	Deerfield	15	53
Cotta, H. R.	Rockford	½	67
Douglas, C. W.	Waukegan	1	52
Douglas' Sons, R.	Waukegan	60	47
Dyniewicz, Wladyslaw	Chicago	30	6
Franzen, Frank O.	Chicago	0.05	41
Freeman, John H.	Evanston	1	71
Graceland Cemetery Co.	Chicago	0.05	37
Harms, R. J.	Kenilworth	4	40
Hartwell and Son, J. L.	Dixon	2	139
Haxton, Fred	Chicago	0.125	110
Hill Nursery Co., D.	Dundee	150	21
Johnson, A. G.	Chicago	0.625	114
Joliet Nurseries	Joliet	40	129
Kabo, Mrs. Margaret C.	Joliet	1	121
Kadlee Nursery Co., Frank	Evanston	4	72
King, James	Elmhurst	40	43
Klehm and Sons, John	Arlington Heights and Moline	97	48
Knox, Robert	Winnetka	1	102
Leesley Brothers	Chicago	80	88
Marson and Son, C. W.	Chicago	43	119
Maywood Nursery Co.	Maywood	85	99
Nelson and Sons Co., Swain	Glenview	165	61
Palmgren, Charles A.	Glenview	14	63
Peterson Nursery	Chicago	185	66
Pfund, C.	Oak Park	10	136
Pfund, Wm.	Elmhurst	4	34
Pierce and Son, H. B.	Antioch	13	74
Porter and Son	Chicago	19½	107
Pottenger, J. W.	Kankakee	4	9
Ringler Rose Co.	Chicago	1½	89
Rowland, John H.	Elgin	0.167	27
Schroeder and Sons, F. O.	Morton Grove	8	123
Spurlock, M. J.	Park Ridge	0.125	106
Vaughan's Seed Store	Western Springs	80	57
Von Oven, F. W.	Naperville	40	42
Wittbold, Otto	Chicago	40	113

Source: Report of chief horticultural inspector of State of Illinois, *Transactions of the Illinois State Horticultural Society*, 1912, 43–46.

with Hill as president and F. W. Von Oven of Naperville Nurseries as secretary-treasurer.[21] The group was active through 1934, and had a peak membership of 134 firms in 1929.

Chicago nurserymen were among the first to recognize the business benefits of organized publicity and marketing. Parlaying his knack for promotion, in 1907, Vaughan became general chairman of the National Council of Horticulture, a joint lobbying and publicity effort of the Association of American Nurserymen, Society of American Florists, and American Seed Trade Association. Never shy of experimenting with new advertising methods, Vaughan sponsored radio shows throughout the 1920s and 1930s, with commentary on garden design. Bringing the garden-design cycle full circle, "Ma Perkin" of Vaughan's Seed Store was a daily radio personality. The apocryphal Perkin sold her seed packets for an "Old fashioned garden" through the mail, with a "handwritten" note containing seed advice (plate 2).

The American Association of Nurserymen, Florists, and Seedsmen evolved over the years. In 1883, the American Seed Trade Association was formed, and one year later, the Society of American Florists was created to address more specialized interests of these groups. With two major constituents defecting, the group changed its name simply to the Association of American Nurserymen in 1887. Now known as the American Nursery and Landscape Association, it has more than two thousand member firms throughout the United States. The Illinois State Nurserymen's Association was formed in 1925 and included many Chicago-area residents among its officers.[22]

John A. Klehm's home was surrounded by orchards of apples and cherries. Although fairly plain, the home still boasted floral planters. Ca. 1860s. Courtesy of Arnold J. Klehm.

Nurserymen became involved in city beautification efforts that coincided with the Progressive Era and widespread adoption of Arbor Day programs between 1900 and the 1920s. As homeowners became more knowledgeable about gardening, nurserymen often gave lectures on proper tree-planting methods. Occasionally, efforts to promote city beautification were perceived as self-serving, with additional tree plantings seen as greater revenues for the nurserymen. Combating this image, the Illinois State Nurserymen's Association adopted the motto "Public First and Nurserymen Second" at their annual convention in 1923. In reviving the slogan "Plant a Tree," so popular during previous Arbor Days, the nurserymen planned a nationwide movement under the direction of J. A. Young of Aurora, Illinois, to plant trees to beautify communities across America. Fifty-six national horticultural and civic organizations endorsed this Chicago-born idea.[23]

Family nurseries not only helped pass along knowledge from generation to generation, but customers and neighbors obtained cultural and design ideas from viewing the grounds. Frequently the most attractive landscaping in the neighborhood surrounded a nurseryman's home—on the principle that his own grounds should reflect his talents. Of course, there were a number of situations where a nurseryman's property lacked evidence of solid design, but more likely this would have been a case where the nurseryman prided himself on stock versus design expertise.

Nurseries that did not diversify, and relied instead on sales of tree and shrub stock, often faced significant business challenges. As their customer base shifted from farmers to suburbanites, demand for trees lessened. Suburbanites had

Nurserymen's grounds were often destinations where customers could obtain growing information, or simply browse among the plants and enjoy the ambiance. This nursery, marked "Pool's garden," could possibly be that of Isaac A. Poole, an early gardener and CHS exhibitor in the late 1860s. From the author's collection.

smaller properties to grow trees, did not need rows of them as much for wind-breaks, were less knowledgeable about tree care, and were less likely to invest in the time and money required to grow worthwhile trees. Hence programs such as "Plant a Tree," while seemingly self-serving for the nurseryman, helped inspire the new homeowner with the possibilities for home beautification. Even as Chicago's position as a great transportation hub helped local growers ship their stock, competitors' stock from around the nation also became more readily available. Edgar Sanders observed as early as the 1880s that of the more than one thousand listed Illinois nurserymen, only seventy-eight were dedicated solely to the nursery business; others had branched into the more lucrative florist trade. Sanders noted that whereas Chicago's nurserymen in the 1880s suffered from the double-edged sword of the railway, the florist gained ground.

Florists Flourish

Florists, those who sold retail plants and cut flowers, might seem a luxury on the prairie, but such was the love of flowers among early Chicagoans that florist shops could be found in the city as early as the 1840s.[24] Technology was a great boon to the florist business. Steam inventions fueled not only the railroads to transport flowers, but also heating technology within the greenhouse. The railroad greatly helped the retail florist. As a center of transportation, Chicago became a hub for florist organizations and publications. When the Society of American Florists (SAF) was formed in Chicago in 1884, its original purpose was to provide hail insurance for members, because hail was a prominent threat to crops, and glasshouses in particular. Starting with twenty-one charter members, the group expanded to four hundred members within a year. In the 1930s, the organization moved its headquarters to Chicago. The *Florists' Review* magazine originated in Chicago in 1897, and was published there for ninety years.

It is not surprising that the florist industry formed its own society. The florist business in Chicago, like the nation, grew from one-man shops with a jack-of-all-trades proprietor, to sophisticated mass-production operations. Fritz Bahr, long-time florist of Highland Park, recalled the years before the 1880s when the SAF was formed: "In most cases it was just a matter of existing. There was good stock grown, but for so little money that there was little chance of profit or hope of settling bills once a month. The average florist and the gardener were on a level with the coachman."[25] As the business became more complex, new methods of shipping and marketing became unique challenges for the florists.

Store-bought flowers were a luxury in frontier Chicago, purchased only for special occasions. Typical is this transaction between businessman Benjamin Taylor and early florist Edgar Sanders. On June 1, 1858, Taylor wrote to Sanders asking if he would "be so kind as to prepare a bouquet suitable to present to Mrs.

SAMUEL BROOKS, FATHER OF THE CHRYSANTHEMUM

The story of Chicago's early florists is best exemplified by that of Samuel Brooks. Brooks was one of the city's earliest florists, arriving in 1833. Born in London in 1793, Brooks apprenticed under professional florists in England, and formed his own business in his midtwenties. Part of his success in London was his willingness to sponsor expeditions to China and other exotic countries to collect new and rare plants. Azaleas, Norfolk Island pines, and many new chrysanthemums, then a novelty in western Europe, became part of his stock. His work with chrysanthemums earned him the accolade "Father of the Chrysanthemum" in *Gardeners' Magazine* in 1890.[1]

At forty years of age, he immigrated to the United States with his nine children, his wife, and a governess. Leaving his family in Buffalo, New York, he set out to explore the business potential in Canada. According to one account, the scarcity of customers and amount of work needed to clear the Canadian forests dissuaded Brooks from settling there. Instead he heard glowing reports of Chicago, including "that the country was an open Prairie, no trees to fell, and he could get all he wanted at a dollar and a quarter an acre, that cattle needed no shelter in winter . . . that the climate was mild and trees flourished." Attracted by this description of nirvana, Brooks sent most of his stock and equipment by ship through the Great Lakes. He loaded two prairie schooners with the remainder of his belongings and began the two-month journey from Buffalo to Chicago, "studying the flora on his way [as] though almost a new country."[2]

What this educated man thought when he arrived in Chicago in 1833 is anyone's guess. According to one report, a wild bear had just been killed outside of the city proper, and seven thousand Indians still congregated there. Brooks's necessities, packed in the schooner, included a piano[3] and eight-day clock—hardly the stuff needed to fight off wild bears or trade with Indians. Brooks soon found the case for Chicago's horticulture to be overstated. The ground was frozen even in October, and the winter of 1833–34 was particularly severe. He relocated to a farm on the Des Plaines River and, with characteristic civic duty, served as postmaster and opened his house for religious worship, as there were no nearby churches. Using his knowledge of plants, he is said to have served as a physician in this uninhabited area, as well as a justice of the peace. After some seven years of farm life, Brooks returned to the city and his chosen florist profession. He built Chicago's first commercial greenhouse on Adams Street between Clark and Dearborn streets. As the city of Chicago began to expand around him, Brooks moved his residence and business to the South Side, then called Cleaverville, and ultimately to Hyde Park.

Like many early horticulturists, Brooks was also an amateur botanist. He wrote one of the first records of local flora in an 1847 *Prairie Farmer*. On a foray into the Calumet region near Chicago, Brooks, accompanied by John Goode of Chicago's Sheffield Nursery, documented unusual plants that might make good florist and garden flowers. Among his findings were "a new species of Cacalia, with nettled leaves, and large heads of pure white flowers" which he deemed "a plant well deserving a place in our gardens, for its showy appearance when most other flowers are on the decline."[4] He also gathered "large quantities of a beautiful annual Polygala, with heads of flowers resembling the Globe Amaranthus," and while acknowledging it was common in moist areas, noted that he'd never seen as fine specimens. Two species of *Hypericum* and three or four orchidaceous plants were also of interest to him. He found a patch of *Myrica gale* (which he called sweet gale or bog myrtle), which he'd been seeking for awhile, and opined that it deserved a place in the garden because of its unusually shaped leaves, neat habit, and highly aromatic appeal.

1. Edgar Sanders, History of Samuel Brooks, manuscript, Edgar Sanders Collection, Chicago History Museum.
2. Ibid.
3. Pianos were a sign of culture in Chicago, not only of music appreciation, but also of the expense and trouble involved in bringing them across the prairie. Many claim title to bringing the first piano to Chicago.
4. Samuel Brooks, "Plants near Chicago," *Prairie Farmer*, December 1847, 369.

Taylor on the Anniversary of our wedding day."[26] The bouquet was to be sent to Taylor's office, which, from his letterhead, appears to have been the *Daily Evening Journal*, at 1:00 Saturday afternoon. Taylor further specified that the bouquet be "so packed that it can be carried *sixty* [emphasis in original] miles by railroad to Indiana" where his spouse awaited.

Such was the job of a successful florist in 1850s Chicago. Only a handful of florists braved the task of building commercial greenhouses to survive Chicago winters, obtaining seed or propagate flowers, and packaging them successfully to withstand a sixty-mile shipment by train. Early florists, according to Bahr, grew everything they retailed on their grounds. As transportation and competition from other states increased, florists soon expanded their supply to include purchased stock from other growers. Chicago's open-air markets on Water and Randolph streets offered florists the opportunity to purchase wholesale that which could not be grown more cheaply by oneself, or to sell wholesale, via railroad, that which could not be sold locally.

During the 1880s, as postfire Chicago resumed its former busy pace, the florist business picked up. "Probably very few persons have any idea how much money is spent during the social season in Chicago . . . in the way of floral decoration," reported the *Chicago Tribune* in 1881. "The business has grown to large proportions during the last five or six years, and if the statement of a well-known local florist is to be accepted . . . the average annual expenditure for flowers in Chicago is in the neighborhood of the very respectable sum of $175,000." Social functions proved the most popular for cut flowers, with weddings, anniversaries, and funerals "conducted with as near an approach as possible to the average horticultural show."[27] As the city's wealth continued to grow, so did the disposable income earmarked for flowers. The *Tribune* noted just four years later, "Fashionable society uses more and more flowers every year . . . Young women delight to be presented with more bouquets than they can carry, and the floral decoration of the home is a very large item in the bill of expenses."[28] Certain holidays, such as Decoration Day (now Memorial Day) and Christmas, might tap the resources not only of the city's florists but of neighboring suburbs for miles around.

Even as national trade organizations grew, the local Chicago Florists' Club was organized in 1886. Edgar Sanders and Willis N. Rudd were among the prominent officers of the club. This group held meetings twice monthly in the Masonic Temple at State and Randolph, and cohosted meetings with kindred national or state-level groups whenever their conventions were in Chicago. In 1887, one year after organizing, Chicago florists were sufficiently prepared to host the third annual convention of the SAF. Over a thousand delegates from across the country were expected, and Chicago's florists planned a grand affair. After a three-day business meeting in the First Regiment Armory, convention attendees were given a grand tour of the South Parks. In addition to reviewing the grounds and greenhouses, the Chicago florists took on Philadelphia florists in a much-ballyhooed baseball game (final score: 18 to 4, Chicago). A farewell dinner was held in the armory, which was decorated with extensive floral arrangements of roses and gladiolus. The Chicago florists escorted their guests to the Grand Pacific train station accompanied by the music of the Second Regiment marching band.[29]

In the late 1890s, Chicago florists were among the early adopters of new technology and marketing ideas. Florist delivery via telegraph to distant American cities and even Europe was announced in the *Chicago Tribune* in 1897. Aiming for competitive advantage, florists began packaging their arrangements in attractive boxes, almost as colorful as the contents themselves. One local writer noted, "Plain white boxes are almost unknown at the fashionable flower shop, the bunches of roses, tulips, daffodils . . . going to their recipients in long and square receptacles of pale cream, blue, green or pink tinted pasteboard, covered all over with designs of straying blooms, and strapped by broad ribbons of some harmonious hue which on top is tied into a great flat bow."[30]

At the turn of the century, competition intensified among florists, and especially against itinerant street vendors. Gone were the days when John Kennicott could wander the streets of Chicago, bouquets in hand. Retail florists with established storefronts protested vigorously against flower peddlers who sold cut flowers and boutonnieres on major streets in the Loop. In 1899, after Mayor Carter Harrison Jr. approved an order restricting this trade, proponents of flower vendors cried foul, noting that the vendors offered a cheaper alternative. One proponent asked, "Who has not seen a shop girl or tired-looking woman, a grimy laborer or newsboy buying for five or ten cents enough flowers to brighten some humble home or bring the message of the fields and sunshine into a cheerless sick room. So long as the fruit and vegetable peddlers are with us, let the flower vendor stay too."[31]

The battle against flower vendors continued throughout the next few years, with petitions to the city council to remove vendors in 1901. Flower vendors, many of them recent immigrants, became suspects in incidents of theft from private gardens. A 1901 *Chicago Tribune* headline declared "War on Flower Thieves," noting "police and citizens believe that men who steal blooms are agents of downtown vendors."[32] In this instance, private and public gardens in Hyde Park and Austin were targeted, and "police orders were sent out . . . commanding all patrolmen to be on the lookout for men who are stealing the blooms, and citizens have promised the Inspector to keep close watch on their yards." Flower filching had become a major offense, just as apple stealing had incensed homeowners in previous years.

The street vendor wars were just one instance of local professional florists banding together. In this case, many florists contributed to a lobbying fund to promote legislation against street sales. Florists and gardeners also were divided on the role of labor unions. With its Haymarket riots and Pullman strikes of the late 1800s, Chicago was a hotbed of organized labor activity. In 1903, the gardeners and florists union went on strike in the West Parks, noting that with more than $600,000 worth of plants under their care, "nonunion men cannot be secured to give these plants the care they require" and that "nearly every trained florist in the United States is a member of their union."[33] Just prior to that, in August 1903, six hundred delegates of the SAF, now called the Society of American Florists and Ornamental Horticulturists, met for a dinner in Chicago where the mood was "anxious over the labor outlook," according to the *Chicago Tribune*. Noting that "gardeners and hothouse employees in Chicago are being unionized and plans are on foot to form a national body," the SAF had invited F. W. Job, a noted employers' advocate, to speak. Delegates from other cities appeared less interested in Job's remarks, according to the *Tribune*: "With appetites appeased, [they] forgot the menace of organized labor in the fear that they would be late for the theater." Still, the paper noted, "the advent of a union among the florists is considered so momentous in Chicago that fifty local dealers met previous to the dinner and organized an employers' association to deal with the workers."[34]

By the early 1900s, sophisticated greenhouses were integral components of the commercial florist trade. John C. Moninger Company, established in 1868, operated one of the nation's largest greenhouse manufacturing plants in Chicago. The Moninger Company built custom greenhouses for private and commercial concerns throughout the country. In 1908, Illinois ranked first among other states in area covered by glasshouses. The Illinois State Horticultural Society reported that while smaller florists continued to thrive, many operations were of considerable extent. "A large percent of these are in the neighborhood of Chicago, and several firms have each over a million square feet of glass."[35] In the same year, the SAF again selected Chicago for an annual meeting, prompting the *Chicago Tribune* to claim, "With this selection [as host city] goes the compliment of conceding to Chicago the distinction of being the leader of all American cities in the raising of flowers . . . A city that many years ago took the name of the Garden City, and which displays the motto *Urbs in Horto* on its municipal seal, is certainly entitled to recognition by those who are concerned or interested in horticulture."[36]

In the 1920s, florists had so capitalized on Chicago's position as a market center that Wabash Avenue between Randolph and Lake streets was unofficially designated Florists' Row. Dominated by the powerhouse firms of E. C. Amling, the Chicago Flower Growers' Association, and A. L. Randall Company, whose combined business was said to represent three-fifths of the Chicago florist trade, Florists' Row was convenient to both the open-air markets and train terminals. In the 1930s, florists and greenhouse horticulturists became even more prominent in orchestrating large flower shows. Along with the burgeoning garden clubs, by 1930 the April flower show held at the Chicago Stadium was declared the largest in the nation, finally eclipsing that of rival New York City.[37]

TOP This Chicago florist shop shows how refrigerated coolers helped expand the selection of stock available. Ca. 1910. From the author's collection.

BOTTOM Great efforts were taken to transport fresh flowers from florist to recipient. While some city dwellers enjoyed the luxury of walking to a corner florist, others relied on local cartage. Although Florists' Transworld Delivery (FTD) was not founded until 1910, flowers could be ordered by telegraph or even written notes, necessitating deliveries by horse from wholesaler or retailer. This view of Forest Glen Floral Company on West Madison Street was taken in December 1890, a busy holiday season for florists. From the author's collection.

It was florists such as Paul Battey of Glencoe, Garfield Park horticulturist Frank K. Balthis, and Garfield conservatory horticulturist August Koch who in 1934, along with various garden club officials, launched the Midwest Horticultural Society, a home and city beautification group that supplanted the CHS, whose activities were beginning to wane.

Networking, through joining professional societies, and a willingness to adopt new technologies and business practices were critical to florists' success. As Highland Park florist Fritz Bahr wrote in his 1922 *Commercial Floriculture* book, "I am convinced that the future of the florist business rests entirely in our own hands. In no other part of the world is there a better chance to develop this industry than here, or a better market... it is a matter of producing to supply the demand; and by means of cooperative work through the S.A.F. and the F.T.D.,

EDGAR SANDERS, "NESTOR OF CHICAGO FLORISTS"

One of the most enduring and influential leaders in Chicago's horticultural world, nurseryman and florist Edgar Sanders arrived in Chicago by way of Albany, New York, in 1857. The English-born Sanders learned his trade as an undergardener, then as head gardener on some of England's finest estates. While in Albany, he tended the grounds of Luther Tucker, the original publisher of Andrew J. Downing's *Horticulturist*. At about that time, Tucker started *Country Gentleman* magazine, where Sanders launched his own prolific American writing career. A regular columnist for the *Prairie Farmer*, Sanders also wrote for other Chicago-based papers, including the *American Florist*, the *Chicago Times-Herald*, the *Chicago Daily News*, and *Good Form*, a household journal. Suggesting that Sanders write under a pen name, in 1892, James S. Judd solicited Sanders's input for the *Prairie Farmer's* rival, Chicago-based *Orange Judd Farmer*.[1] A member of many national committees and societies, including the Massachusetts Horticultural Society, Sanders was also a popular correspondent for New York periodicals, including *Popular Gardening and Fruit Growing*, *American Agriculturist*, *American Gardening*, *Farm and Fireside*, and the *Florist's Exchange*.

In 1905, the British-based *Gardeners' Chronicle* dubbed Sanders the "Dean of Chicago Horticulture," and partially claimed credit for his success, noting, "We often have occasion to note the frequency which men of British origin have come to the front in American horticulture. We suppose in time American-born cultivators will naturally displace them, but even so many of them will be of British extraction. In any case, it is pleasant to read of the success of our countrymen. Among them none is more respected than Mr. Sanders."[2] Early in his career, Sanders had written for the *Chronicle* and for the *Gardener's* magazine of London.

He was a charter member of many horticultural organizations, including local groups such as the CHS, the Chicago Gardeners' Society, the Cook County Agricultural and Horticultural Society, and the Chicago Florists' Club. He took leading roles in many state societies and in national groups such as the American Association of Nurserymen, Florists, and Seedsmen. His correspondents and acquaintances were themselves legendary, including L. H. Bailey, Peter Henderson, Thomas Meehan, and publisher A. T. De La Mare.

and by means of continuous advertising to educate the public, each year will be made a better one than that which preceded it."[38] Until air travel became more commonplace, thus facilitating plant shipments from South America and other tropical climates, the floral trade thrived in Chicago.

Chicago Horticulture Finds the Market

Chicago, the nation's hub, was a natural marketplace for farm and garden produce. As the transportation capital of the country, and with extensive trade exchange infrastructure, garden produce was bought and sold in Chicago on a massive scale. At first by boat, then plank road, then rail and truck, farmers brought their crops to the city. A new type of horticulturist, market gardeners,

These national and international connections helped keep Sanders abreast of the latest trends in horticulture and landscaping. His columns ranged from specific advice on plant culture to reports of private gardens he had visited. His advice was hands-on and practical, arising from personal experience at his residence in Lake View, where he built his first fifty-foot greenhouse in spring 1857. He purchased his home, at 1639 Belmont, when land cost about $700 per acre. W. W. Boyington, a leading architect in Chicago, drew the plans for Sanders's house, a modest gothic cottage. Sanders made some modifications, including a profusion of sunny windows "for having spent the previous eight years with no south windows, we were determined for the future to make up the deficiency by having at least one south window in all the important rooms."[3]

Ruth Hall, a fellow correspondent for the *Prairie Farmer*, visited Sanders's Lake View greenhouse in 1863, asking and receiving permission to "look closer at the beautiful flowers smiling through the glass." Hall described rose-dyed azaleas, hyacinths "from the royal purple to the purest snow," white double violets, heliotrope, and mignonette blooming in Sanders's greenhouse. There were also "right regal camellias" and "old-fashioned Polyanthus, which I remember as bordering so many cottage gardens," as well as "solemn Box, once used as the quaint and grim ornament of the dead." In short, Hall recommended a trip to Sanders's facilities as a "very passable Eden for February," and urged all readers to visit.[4]

Like many of his fellow nurserymen, before the days of professional landscape design, Sanders served as a de facto design expert, laying out the grounds of several private estates and of Calvary Cemetery. Sanders opened a flower store in downtown Chicago in the Sherman House at 58 Clark Street in 1867. Four years later, like many of his colleagues, he suffered losses from the 1871 fire. He stayed in business until partially incapacitated by a stroke in 1902. Upon his death in 1907, various eulogies referred to him as the "Grand Old Man" of Chicago, "Papa Sanders," and the "Nestor of Chicago Florists." His pallbearers included colleagues J. C. Vaughan and George Wittbold.

1. James S. Judd to Edgar Sanders, November 23, 1892, Edgar Sanders Collection, Chicago History Museum.
2. *Gardeners' Chronicle*, June 24, 1905.
3. *Prairie Farmer*, August 18, 1866.
4. Ruth Hall, "A Trip to Lake View," *Prairie Farmer*, April 11, 1863, 231.

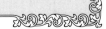

hauled cartloads and later truckloads to the famous Haymarket, Water Street, and Randolph Street open-air markets. The brisk trade in fresh produce mirrored the frantic activity in the grain pits at the Chicago Board of Trade. Chicago's market conditions were reported and avidly followed in local and national newspapers. Changes in marketing technologies and consumer buying patterns were often seen first in Chicago, and then radiated out to the country. Similarly, Chicago gardeners had the opportunity to view firsthand what their competition in other parts of the country were growing, and were often able to modify production accordingly.

For market gardeners near Chicago in the latter half of the nineteenth century, fruits and vegetables were the mainstay (versus grains, which could be stored longer and sold via the Chicago Board of Trade). Of course, identifying and cultivating stock that withstood the rigors of transportation across bumpy plank roads and, later, jostling steam-engine trains, was a top priority.

In the 1860s, the *Prairie Farmer* presaged Chicago's market success: "No city in our country will be better supplied with all kinds of fruit of the choicest and most excellent quality and at low rates than will Chicago in a few years. Connected as it is by railroads with the early producing regions South, from which fruit can be transported here almost as fresh as when produced . . . our supply will be inexhaustible."[39] Yet with unsophisticated refrigeration, inadequate storage facilities, and rudimentary transportation, the life of an early market gardener was difficult. Pioneer market gardener Jonathan Periam, who was often recognized as the first to market, cultivated fifty acres including early cabbage, tomato, cauliflower, eggplant, and sweet potatoes. Like most market gardeners, Periam's fortunes depended on the vagaries of weather. The hard frost in spring of 1862 destroyed one thousand newly transplanted tomatoes and ruined several acres of beans and corn. Showing the resiliency of Chicago's early growers, Periam immediately replanted the beans, ready for another day at Chicago's market.

Great inroads (literally including an improvement in transportation) were made in the 1870s. Only a year after the Chicago Fire, Jonathan Periam commented that the fruit markets throughout major cities in the Midwest, including Chicago, Milwaukee, and Kansas City, were filled to overflowing. So successful were the markets in a major hub such as Chicago, that produce became available even to the edges of the western frontier: Periam observed miners in the Rocky Mountains enjoying fruits from the Midwest, and "for the first time I saw an apple in the hand of an Indian in Wyoming Territory."[40]

Market gardeners of the late 1800s endured long days. Charles Klehm, a second-generation nurseryman from Arlington Heights, recalled the early days of market gardening, circa 1885: "When I was about 16 years of age, my father made me haul the cherries to Chicago. It was a drive of about twenty-five miles

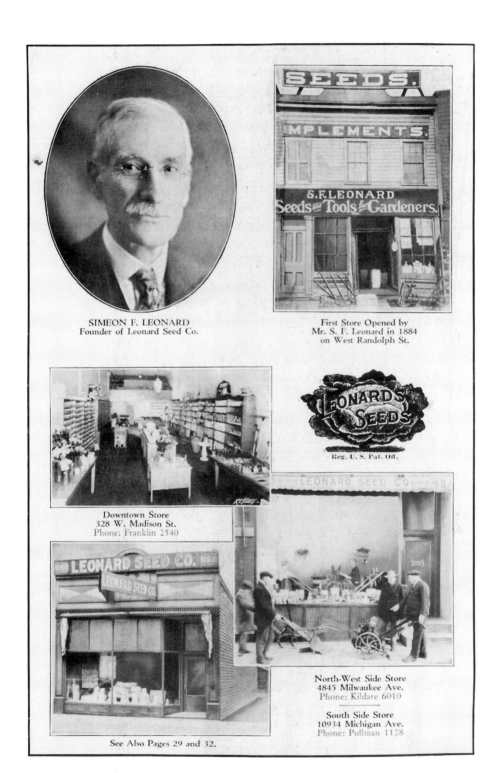

SIMEON F. LEONARD
Founder of Leonard Seed Co.

First Store Opened by
Mr. S. F. Leonard in 1884
on West Randolph St.

SEEDS.

IMPLEMENTS.

S. F. LEONARD
Seeds and Tools for Gardeners.

LEONARD'S SEEDS

Reg. U. S. Pat. Off.

Downtown Store
328 W. Madison St.
Phone: Franklin 2540

LEONARD SEED CO.

North-West Side Store
4845 Milwaukee Ave.
Phone: Kildare 6010

South Side Store
10934 Michigan Ave.
Phone: Pullman 1128

See Also Pages 29 and 32.

Leonard Seed Company was a specialized company founded in 1884 that catered to market gardeners. Its original store at Randolph Street Market was at the hub of the market garden trade, while branch offices around Chicago served outlying gardeners. From the author's collection.

each way, and it would mean starting at 12 o'clock at night to get to South Water Street by 6 o'clock in the morning, then putting the team in the livery stable until 11 or 12 o'clock and getting home about 4 or 5 o'clock and loading again for the next trip. This went on until the cherries and currants were gone. Now [circa 1924], the trip takes about 1¼ to 1½ hours each way."[41]

Major issues in the early days of market gardening included packaging, storage, and distribution of produce, particularly fruit. Individual errors, or chicanery on the part of one's competitors, could readily render a business unprofitable. Chicagoan W. H. Schuyler commented in 1875, "The fruit grower, probably, experiences greater loss in money value through negligent, ignorant and fraudulent marketing of fruit, than through want of knowledge of the laws that govern its culture. This fact has been forcibly impressed upon me for the last two or three years by daily personal observation of the Chicago markets."[42] With no generally accepted standards governing packaging and shipping, markets were flooded with inferior fruit, barrels of various sizes, and the proverbial bad apples hidden amid an upper layer of good. In addition to providing leadership in national packaging and shipping laws, early Chicagoans formed the Northwestern

Some greenhouses were mom-and-pop businesses. Here, members of the Breiter family were all engaged in the market gardening business. From the author's collection.

Fruit Growers' Distributing Association in January 1875. Members of the guild were sworn to honest packaging practices, and the group promised better dissemination of market supply and demand information.

Chicago produce markets were true examples of American laissez-faire. In the late 1880s, the principal market was on South Water Street, where sellers maintained storefronts. At the North Avenue market and the Haymarket, produce growers sold directly from their wagons. A *Chicago Herald* reporter described the scene:

> The Haymarket is one of the institutions of the city. There is a marketmaster appointed, whose duty it is to see that certain hours of "opening" and "closing" are observed, and to collect a tax of every team located there for the purpose of selling produce ... The buying is done almost entirely by grocers and peddlers ... Some of them are on the scene at 2 o'clock in the morning, and buy a load of the early comers. This they frequently sell in time to get back at 10 o'clock for another load, which they go more or less musically crying through the streets. Prices, of course, vary, and are determined by demand and supply, and also by quality.[43]

The Peter Nepper greenhouses in Rogers Park were located somewhat north of the main glasshouse district in Lake View. The family lived in the farmhouse at right. Ca. 1900. Reproduced by permission of Rogers Park/West Ridge Historical Society.

Representative prices for that time at the Haymarket were cabbage, 60–85 cents per dozen; onions, three bunches for 25 cents; and tomatoes, 10 cents per dozen.

While local Chicago market gardeners thronged to the Haymarket and other markets, so did produce growers from nearby states. Often Michigan fruit growers sent their produce via ship. Their ordeal was further complicated by seemingly minor logistics such as the opening and closing of Chicago's bridges. If the bridge timing was off schedule, shippers might miss the critical time of day to sell. Petitions were sent to Chicago's city council to improve adherence to bridge-closure schedules. Railroads, of course, brought in produce from around the country. Chicago's market gardeners altered their mix of crops depending on the supply and prices seen from competitors.

The next mode of transportation to affect market gardeners was the horseless carriage, and with it dawned the age of the truck gardener. In the first decades of the twentieth century, market gardeners retired their workhorses and hauled goods to market in the newfangled trucks. August Geweke of Des Plaines, a leader in the truck-gardening movement, observed, "Success in truck gardening is open to all. The work is simple. Anybody can do it. It is known as the last resort for people in our great city of Chicago to help themselves out if they are hard up and have nothing else they can do." Geweke went on to note, however, that the key difference between a successful and unsuccessful truck farmer was planning and crop specialization. Gone were the days of Jonathan Periam, who could dabble in a variety of plants; it was now the government's job to test plant varieties. Geweke noted, "The common truck gardener can not carry on much experimental work. This would prove too expensive. He should know that such work is being done for him by the State and by the United States Government."[44]

Geweke served an unprecedented multi-decade term as president of the Cook County Farmers' and Truck Gardeners' Association. This group, formed in 1902, began with a group of truck farmers in the Jefferson Park community who lobbied against proposed licensing on produce. Throughout the years, the association worked for better roads, equitable land taxes, and better marketing facilities. One of their first campaigns, in 1902, was to lobby for more space at the Randolph Street market. The group petitioned Chicago's city council to approve measures to widen the street, and to condemn certain land parcels to provide space for the enlarged market.

By about 1910, truck gardening in the Chicago area was deemed profitable work, even as farmers transitioned from horse-drawn carts to trucks, or "power wagons" as they were then called. Typical is the experience of John Burmeister, a twenty-five-year-old German-born tenant farmer renting from James McGawn of Mont Clare (part of the town of Jefferson on Chicago's north side). Burmeister's daily routine sounds very similar to that of Charles Klehm, despite the intervening decades. Four times each week Burmeister left his home at midnight, traveling south on Grand Avenue to the West Randolph market—about an hour's drive. He typically sold $50 to $110 of goods before hitching his team for the trip home by noon the following day.

This was considered a good income, and reflected Burmeister's prowess as a salesman. Chicago's produce markets were a microcosm of the polyglot city itself, and successful truck farmers needed the skills of a diplomat. The *Chicago Tribune* explained, "The best soil tiller in Cook County would break up and go out of business unless he also is a past master in selling his stuff after he backs his wagon to the curb in West Randolph street. He must talk to a dozen persons at once, most of them speaking different languages ... and keep in a good humor with everybody."[45]

The interests of the truck gardeners and those of the many groups involved in Chicago's city beautification plans converged on the issue of road improvements. In 1913, there were about thirty country roads leading into Chicago and about one thousand miles of country roads in Cook County, all of them deemed in poor condition by Henry Hyde, a leader in Chicago's city beautification movement. Hyde asserted that, while Illinois ranked first in the nation's agriculture, it ranked twenty-fourth in percentage of improved roads. The bad roads prevented truck farmers from making timely deliveries to the city, and food was wasted. According to Hyde, "within the last year, a good many tons of root crops, onions, carrots, parsnips, beets, [and] turnips which were raised on farms not more than twenty miles from the city hall, have been dumped into holes and gulleys."[46] In Hyde's interview with Geweke, the latter proposed distributed markets in addition to central Randolph Street. He also opined that if there were a commitment to road maintenance, the farmers union would form a cooperative to purchase and operate motor trucks to haul more produce on a more consistent basis.

The number of registered motor trucks in Chicago nearly doubled between 1911 and 1912,[47] reflecting a nationwide interest in the "power wagon" potential. Henry Farrington, editor of the Chicago-based *Power Wagon* publication and the "greatest living authority on motor trucks" according to one *Chicago Tribune* writer,[48] wrote frequently of the benefits of motor trucks. In 1915, he began a newspaper article with the sensational prediction that "the motor truck will save the world from famine if the great European war lasts as long as Lord Kitchener thinks it will."[49] Farrington reasoned that the economics of trucking would finally win over the most diehard horseman. He even proposed as a (presumably facetious) alternative, "Horses are not bad eating."

World War I boosted the use of trucks and of market gardens in general. August Geweke, who was chairman of the war-garden produce committee of the Illinois Council of Defense, promoted war gardens as a patriotic and economical measure for Chicago-area residents. The Chicago Board of Education offered a controversial solution to the labor shortage that truck gardeners faced during the war years—allowing high school–aged boys to work on the farm. Geweke, in reported accounts, opined that farm labor should not come under the heading of factory work for purpose of child labor laws, and that with the extra assistance, truck farms could produce three times the crops. "If we didn't believe that farm work was beneficial—far better than letting the boys and girls run loose in the streets—we wouldn't even suggest any reform," Geweke said.[50]

The automotive industry spawned another important variant of the truck farm during the war years. As the automobile itself became more available to the average Chicago citizen, farmers created roadside markets at the edge of their farms. Leading the way were the rural school gardens in Cook County. In 1917, under a new program by the county superintendent of schools, students tilling about five thousand school gardens in the county began posting blackboards listing available fresh produce at the roadside of their farms. Motorists, enjoying a weekend drive along country roads, were attracted to the signs, and stopped for fresh vegetables and fruit. Farmers also progressed from posting signs at the road to bringing the produce stands right to the driver's door. The burgeoning automobile industry encouraged the trend by establishing a national "Farm-to-Table" week in November. Chicago motorists were urged to drive to outlying roadside markets to beat the high cost of living.

Both roadside markets and truck gardens flourished in the 1920s and 1930s. The University of Illinois offered extension courses, in cooperation with the Cook County farm bureau and Cook County Farmers' and Truck Gardeners' Association. This school was an outgrowth of the university's experiment station begun in the early 1920s in Cook County. Two-day workshops were held in South Holland and on Chicago's north side in 1925. Topics included fertilizers and insect control, especially against striped cucumber beetles, onion maggots, and potato leafhoppers.

In 1931, the *Tribune* printed a retrospective on the trucking industry comparing current conditions to those of previous decades. Whereas horse-drawn trucks once carted produce from up to eighteen miles away, motor trucks now hauled to Chicago markets from three hundred miles away. Randolph Street and South Water Street markets remained open year-round, compared to seasonal hours in earlier years. Consumers could expect better quality, improved grading and packaging, and lower prices because of increased competition among truck farmers. Year-round availability of vegetables such as lettuce, beans, and carrots became possible only in the 1920s.

Market, or roadside, gardens also prospered thanks to the advent of automotive technology in the 1920s. The Chicago area was host to a number of very active motor clubs—from the venerable Chicago Automobile Club and Chicago Motor Club, to smaller groups such as the La Grange Motor Club and Calumet

Motor Club. These groups were simpatico with farmers who lobbied for better roads. The clubs promoted weekend excursions that traveled along country roads bordering roadside markets (plate 4).

TOP The Joliet State Prison maintained its garden grounds with inmate help. From the author's collection.

MIDDLE St. Mary's Mission House, about ten miles north of Chicago, was founded in 1909 as the first U.S. seminary whose purpose was training priest and brotherhood candidates for foreign missions. One of the main courses for the seminarians was farming, so that they were well equipped to serve the populations in undeveloped countries around the world. From the author's collection.

BOTTOM The Franciscan Monastery was founded in 1888 on Chicago's south side. Produce from the garden was shared with the community it served. From the author's collection.

JONATHAN PERIAM: AN EARLY MARKET GARDENER

Jonathan Periam's long horticultural career spanned many fields, but he was particularly influential in shaping Chicago's market gardens. Periam came to Cook County from New England as a teenager in 1838. His energetic, positive approach to innovation, coupled with tried-and-true backup crops, contributed to his success. Periam wrote of his horticultural success and failures in the "Kitchen Garden" column for the *Prairie Farmer*, and served as editor of that magazine from 1879 to 1884. He authored several agriculture books, including the 1,029-page tome *Home and Farm Manual* (New York: N. D. Thompson and Company, 1884).

Periam's influence spread far beyond Chicago. In the preface to the *Home and Farm Manual*, the publishers wrote, "The author needs no introduction to the reading public. In the capacity of author and journalist his constituency has for more than a quarter of a century been the whole American

EN ROUTE TO THE MARKET

As transportation and other technologies improved at the turn of the twentieth century, Chicago felt the brunt of market imperfections. Governmental regulations simply could not keep up with the conflicting pressures on all of the players in the market chain. From grower to shipper to marketer to retailer to consumer, everyone seemed to have an ax to grind. Since Chicago was the most visible marketplace, all the flaws of the nationwide system were highlighted there. In 1913, at the annual meeting of the Illinois State Horticultural Society, discussions focused on marketing practices. Food prices were generally rising across the nation at this time, and market inefficiencies were largely blamed. Speeches were given by individuals representing different sectors of the marketing network, many of them from Chicago.

Chicagoan Mrs. J. C. Bley, president of the Pure Food Commission, gave a speech entitled "The Consumer." Bley and her group discovered that, due to market inefficiencies "millions of barrels and boxes of apples were stored in the United States . . . and there was a possibility . . . of large quantities being thrown away in the spring."[1] Bley's group invited commission men and representatives of retail grocers' associations in Chicago to organize an apple sale. She reported that the sale was a success, with several thousand barrels and boxes sold at prices amenable to both consumer and retailer. Among Bley's proposed solutions was elimination of the middleman by direct sales from producers to consumers through coop groups. She also questioned why most apples consumed in Chicago were from out-of-state—with higher prices. She proposed better marketing of local produce, and an Illinois apple show—an idea that would come to fruition in 1918.

Also present at this same 1913 horticultural meeting were representatives of the transportation industry: agents from the railroad and Wells Fargo. J. C. Clair, industrial commissioner of the Illinois Central Railroad in Chicago, explained some of the tribulations on the distribution side of the market. He emphasized that the railroad had established several demonstration farms along its lines, suggesting that farmer education was an important service provided by the railroad. He also cited figures supporting his claim that railroad costs were a fraction of the total cost to consumers. L. J. Troja of Chicago, an agent for Wells Fargo, noted that his company had recently become a distributor of farm produce. Troja observed that lack of standards among farmers in preparing shipments (e.g., in grading and boxing produce) caused major problems. Fluctuating prices—while produce was en route—caused disagree-

people—his name everywhere a household word. Nor is it confined to the limits of our own country; former books, the product of his pen, have reached the phenomenal sale of 50,000 copies beyond the confines of the American continent."

Like his horticultural brethren, Periam was active in local and state horticultural societies. A life member of the Illinois State Horticultural Society, Periam was said to be the oldest living member at the time of his death in 1911. When the University of Illinois was chartered in 1867 as the Illinois Industrial University, Periam served as head farmer and superintendent, Practical Agriculture, from 1867 to 1869.[1]

1. Franklin Scott, *The Semi-Centennial Alumni Record of the University of Illinois* (Champaign: University of Illinois, 1918), 927.

---

ments among producers and buyers. Troja opined that Wells Fargo was improving conditions by finding markets for producers, and offering posted price lists.[2]

Adding to the dissent was John Denny, whose occupation and speech title, "The Commission Man," exposed him to controversy. Indeed, Denny expressed his vulnerability in his opening remarks: "I will endeavor to plead my own case, as well as that of my fellow jobbers. You will notice that in appearance, I do not look much different from the rest of you . . . you have probably expected to see me with 'horns and hoofs,' breathing 'fire and brimstone,' . . . as would cause the children on sight of me to run to their mothers for protection."[3] Indeed, the commission men, or jobbers, and middlemen, were popularly and universally reviled as necessary evils. Denny asserted that jobbers received only a fraction of the final sales price, and contrary to popular opinion, they worked long days and assumed great financial risk. All parts of the marketing network were culpable for high prices, he opined, and the country as a whole had gone "reform mad," which, with newly imposed regulations, added higher costs.

Last to offer his point of view was A. G. Hambrock of Chicago, the secretary for the Retail Merchants Association of Illinois. Hambrock laid the fault of high prices squarely on the consumer. In his view, consumers demanding high-priced services, such as last-minute telephone delivery orders, buying on credit, and generally living too high a lifestyle, caused retailers to pass along price increases. Hambrock also noted that new legislation such as stricter sanitary packaging laws increased produce prices.[4]

That many of these industry associations and spokesmen hailed from Chicago is not surprising, given the high volume traded on South Water Street. As the nation's most celebrated marketplace, market garden issues were magnified and reported around the country. Nearly everyone at that 1913 Horticultural Society meeting had a horror story to tell of the Chicago markets. Reforms and new technologies helped alleviate some of the problems, but even today, in our worldwide markets, the comment from A. G. Hambrock may ring true: "There are many reasons for the high cost of living and it has been often said that the high cost of living is the cost of living high."

1. Mrs. J. C. Bley, "The Consumer," *Transactions of the Illinois State Horticultural Society*, n.s., 47 (1913): 117–22.
2. L. J. Troja, "The Express Company," *Transactions of the Illinois State Horticultural Society*, n.s., 47 (1913): 126–36.
3. John Denny, "The Commission Man," *Transactions of the Illinois State Horticultural Society*, n.s., 47 (1913): 136–48.
4. A. G. Hambrock, "The Retailer," *Transactions of the Illinois State Horticultural Society*, n.s., 47 (1913): 148–56.

Thanks to improved glasshouse technologies, Chicago-area gardeners were able to bring full lines of produce out as early as June 1, and to offer wider varieties. Nonetheless, garden writer Frank Ridgeway noted that "the first home grown vegetables of the season are considered a treat, but they are no longer welcomed with exclamations as they were a few years ago. Vegetables look more or less the same on the city green grocers' counter, whether they are grown in local gardens or in gardens two thousand miles away."[51] Until large agricultural companies took over small family farms, market and truck gardens flourished in Chicago, with their heyday in the Jazz Age with peaks during the Depression and war years.

Other specialized forms of market gardens took hold in the Chicago area in the Progressive Era of the early twentieth century. Inmates of orphanages and prisons were often pressed into service in gardens as a form of vocational training and to raise food for the table and market. Young boys, sometimes labeled juvenile delinquents, or else those who were indigent or orphans, were often sent to benevolent institutions where farming was taught as a suitable trade. In the 1890s, St. Mary's Training School of Feehanville, Illinois (now Maryville Academy in Des Plaines), housed nearly seven hundred boys, many of whom farmed the nearby fields. (Archbishop Feehan himself was said to have an extensive conservatory at his residence in Chicago.) At about the same time, the George Farm in Glenwood, part of the Illinois Industrial Training School for Boys (now the Glenwood School for Boys and Girls), also offered agricultural hands-on experience. The Industrial School for Girls, initially housed at the old Soldier's Home on the lakefront in south Evanston, included among its curricula skills needed for kitchen gardening. Any surplus produce from these institutions was marketed to the public as a source of income and a means to teach economics.

Prisons, even major jails such as that of Cook County, soon recognized gardening as a useful occupation for their inmates. In 1900, prison official Bernard Prasil began what was to be a forty-year career of gardening at the Cook County jail (then called the House of Correction). When he first arrived at 26th and California, he found only three flower beds outdoors, and "inside the walls of the institution there wasn't a blade of grass." During his tenure, Prasil supervised more than twenty thousand inmates in creating forty-seven flower beds, some thirty feet in diameter. "Men who had been thieves and had committed manslaughter looked after the 15,000 geraniums, cannas, salvias, petunias, and vines with tender care."[52]

LADIES OF THE WEST: CHICAGO WOMEN SHOULDER THE PLOW

"The first troubles that a new gardener, or young gardener, or old gardener will have, is the woman question." The woman question was indeed under scrutiny when truck gardener August Geweke spoke these words in 1915. Women's suffrage was soon to be a battle won, and Chicago's women, from social settlement leader Jane Addams to temperance reformer Frances Willard, were influential players. In horticultural circles, women had been making enormous contributions, although recognition was not readily awarded by their male colleagues. Geweke held a common provincial view that paradoxically acknowledged both women's importance and subordinate role: "There is hardly any business that is so closely connected with woman's work, or the partnership and cooperation so necessary, as in gardening. As gardening is carried on by most people in the state, especially around Chicago, the woman has a great deal to do . . . if the farmer chooses to go into the garden business, he will either have to possess a wife or else get one."[53]

Women's Pens Mightier Than the Plow

Chicago's women gardeners were never pampered hothouse flowers. Whether plucked from comfortable surroundings back east, or part of a hardscrabble husband's pioneer lifestyle in the west, Chicago women relied on flowers and gardens to make the prairie their home. Even while tied to strict Victorian mores, Chicago ladies discovered outlets in horticultural pursuits. Their voices were first heard and supported in the *Prairie Farmer*, as in an 1843 letter from "a fair correspondent, who signs herself 'Lizzy.'" Lizzy was one of the first to recognize the beauty of prairie flowers, noting that "language is too tame to tell the love I bear them." She encouraged "farmer's wives and daughters" to transplant wildflowers around the home. Her letter concluded, "Sigh not for your eastern homes, though they were blest . . . learn to love these here, and learn to love and cultivate the flowers, and to be happy here and hereafter."[54] Lizzy recognized that making peace with native flora would help make the new frontier a home.

Women themselves questioned the propriety of their involvement in the *Prairie Farmer* but were heartily welcomed by the editorial staff. An anonymous female correspondent wrote in 1846 that, while her husband was not fond of farming, she and her five children loved reading the *Prairie Farmer*. She opened her letter with the comment, "I don't suppose that it is fashionable for ladies to send their names as subscribers for your paper, neither do I know that it is altogether out of place for them to do so."[55] Editors assured her that there was nothing unladylike about subscribing to their paper. In fact, in future issues, the newspaper catered to feminine interests with well-meaning (but gender-biased) columns such as "Diary of a Housekeeper," "The Housekeeper's Improvement Club," and "At the Tea Table."

The 1850s brought the considerable talents of Frances D. Gage, who sometimes wrote under the pen name Aunt Fanny, to the literary scene. Gage first attracted attention in the horticultural world when she dared to challenge reviews of the Cooper apple at a pomological congress. Her essay in the *Ohio Cultivator*, reprinted in the inaugural issue of the *Western Horticultural Review*, included her comment, "But I believe in woman's rights, and surely the house-wife ought to know as well as her husband what apple is the most palatable and useful in all ways; and I move, ladies, that we get up a congress of women, and pass resolutions."[56] Invited by the editors to continue writing, Gage modestly discounted her own horticultural prowess, but her eloquent descriptions of childhood wildflower hunts and landscaping around her one-acre lot contradicted her self-effacing denial.

Gage's writings offered realistic views of Midwestern women gardeners. In 1853, Gage traded her country home in Ohio for a city lot in St. Louis. Writing of the move in the *Ohio Cultivator*, Gage debunked the myth that a city woman's life was any easier than that of her country sister. Rising before dawn, Gage went to the market and bought lettuce, radishes, pie plant, greens, onions, and asparagus at five cents a bunch "and a bunch is about as big as a piece of chalk," she noted disparagingly. Nonetheless, she reveled in her 40-by-20-foot backyard, where she hoped to plant pansies, verbenas, roses, heliotropes, and mignonette. "And I bet a sixpence that I will have as many flowers in my little pocket greenery as half of you cultivate on your big farms and grand gardens."[57] Like many women of the era, Gage moved westward with her husband, and immediately set about planting a garden to make the new place feel like home.

Gage often penned her letters from Chicago in the 1860s. Her brother, John Gage of north suburban Gage's Lake, was an outspoken proponent of horticultural education for both men and women. In 1850, he wrote, "In horticulture . . . there are many things that a woman can perform as well at least as a man. She can learn and understand botany, select and arrange flowers in an ornamental garden; she can graft on buds of flowers, shrubs or fruit trees . . . and do many other things better than the men; and should, therefore, participate in the benefits of an education."[58] Frances Gage's essays from Illinois reflected the liberal thinking that seemed to have run in her family. Her 1860 *Prairie Farmer* article "Dress for Women" ridiculed current ladies' fashions. "I have seen many a woman carry more flowers on her head than could be plucked from her garden or door yard in a season," Gage wrote of extravagant hats.[59] She urged women to wear sensible clothes and get outdoors for exercise and health. She crisscrossed the State of Illinois lobbying for women's rights. Pleas for women's rights wove through many of Gage's horticultural essays.

One of the last essays Gage wrote for the *Prairie Farmer* was titled "Western Women." She went toe to toe with another author who haughtily implied that western women were "not like New York ladies." Indeed, said Gage, "Our Western women are—unlike New York women—sensible; wear thick shoes in winter and paint roses on their cheeks over the cooking stove, or with a five mile race over the prairies on a breezy day to see an old friend." Her concluding salvo rang: "Western women support the men, while New York men support the ladies."[60] Gage, who appears to have moved back East after the 1870s, later published a book on temperance, *Steps Upward*.

Joining the ladies "At the Tea Table," the *Prairie Farmer's* condescendingly titled women's column, was Ruth Hall, who took up the pen shortly after Frances Gage. Hall wrote on a wide range of topics, including fashion, the work of country wives, and current events. Like Gage, Hall minimized her horticultural prowess—but in her case, she may have been correct. In an 1863 article, Hall described her first attempt at making a garden at her new home about a mile outside of a large city (presumably Chicago). Her husband thought the project foolish, so with the advice of a "country bred neighbor," Hall began a kitchen garden on her own. Hired help plowed under the virgin prairie soil, and then Hall selected her crops. She chose carrots "represented as looking well in a bed," and beets "formed a pretty contrast." Corn came next, "for the sake of its poetic connection." She covered a fence with morning glories and beans, and tucked four-o'clocks and lady's slippers in pocket gardens. Altogether it appeared that Hall planned her garden for beauty rather than for harvest.

She was quite surprised by the amount of space that squash vines took up, and positively horrified by the snails and the "most disgusting looking bugs" found in the garden. Still, she said, "I was really very happy in my garden." Unfortunately, reflecting the misconceptions of the times, "then the cholera began, and green things were deadly poison [her husband] said." The horses, cow, and sheep were let into the garden gate, and Hall's garden was demolished. "Thus Eve and I suffered by a garden; and, having tried the experiment once or twice with similar results ... I gave up farming without a sigh."[61]

Despite her own brown thumb, Hall admired horticulture and wrote frequently of flower shows and fairs. She painted a picture of Chicago's antebellum spring city gardens as "the donning of new grass dresses by the miniature lawns in the neighborhood of the avenues, the bursting of the lilac bushes into leaf ... the glorification of parlor windows with Hyacinths." Hall thanked those "wealthy enough to provide such pleasures" and, tongue in cheek, chastised florist Edgar Sanders for offering such tempting wares. "Even the very poor cannot escape his fascination," Hall wrote of Sanders, "for I saw an English woman invest ten cents on a Cowslip root, [when her] boot evidently needed a patch, yet the infatuated female looked as happy with that dear reminder of home." [62]

With husbands and sons gone to fight, the Civil War years offered women new gardening opportunities. One writer who signed herself Mrs. Observer asked, "Does it not seem that this war is on purpose to call forth all the powers that have so long lain dormant in the female element of our race?" She described the efforts of a McHenry County family who, beginning with a half bushel of flax, ultimately made a thriving business of "weaving eighty-nine and a half yards of cloth by hand wheels and loom." The ladies in this family worked alongside the men, some of whom had been wounded in the war. Thus, with flax and females, Mrs. Observer challenged, "Let us all put our shoulders to the wheel, and learn [sic] the rebels that cotton is not King, nor will we have him reign over us."[63]

Chicago's women garden writers continued to flourish, particularly at the turn of the century. Inspired by women such as England's master gardener and author Gertrude Jekyll, Chicago women added a Midwestern voice to the growing chorus of women garden writers. Louisa Yeomans King, writing as Mrs. Francis King, began her first garden in Elmhurst, Illinois, before moving to Michigan. She maintained ties in the Chicago area even as she rose to national prominence. Lena May McCauley wrote *The Joy of Gardens*, as well as a regular newspaper column on gardening. Schoolteacher Louella Chapin wrote *Round About Chicago*, a celebration of Chicago's natural areas such as Palos Park and the Sag. Chicago's wealthier women gardeners were also busy writing. Harriet Hammond McCormick wrote *Landscape Art, Past and Present* about her own Lake Forest garden, Walden. Frances Kinsley Hutchinson wrote about the makings of her garden, Wychwood, at her summer home in Lake Geneva, landscaped by the Olmsted firm. Kate Brewster, an amateur gardener in posh Lake Forest, ironically wrote *The Little Garden for Little Money*, a book in the Little Garden series compiled by her friend Louisa King. Writing continued to give voice to women gardeners even as their exploits in the field increased.

Getting Down to Business

Mary Felke, aka Bouquet Mary or Flower Mary, was perhaps Chicago's first female florist. Arriving in Chicago in 1854 at age fourteen, she successfully parlayed streetside flower sales into a proper greenhouse business in the 1860s. Edgar Sanders recalled her as being "among the old settlers in the florist world," "of that solid, hardy class of German women," who initially worked as a "peddler of bouquets made almost exclusively of prairie flowers." Felke picked wildflowers where the Deering plant ultimately stood in the 1890s. Felke and her husband bought a lot on Mohawk Street near Centre Street, then considered far out in the country, and built a greenhouse in 1862. Her business thrived until 1871, when the Chicago Fire destroyed her property. But Sanders reported that in spring of 1872, Felke was back in business. Sanders wrote of a visit to Bouquet Mary's greenhouses in 1884, noting that the business had now expanded to cover three 75-by-150-foot lots with five connected greenhouses. Sanders looked for Felke herself, and "sure enough she was there, rotund and jolly, giving instructions to her help, which is still composed of members of her own

family."[64] Felke grew cut and pot roses, ivies, cinerarias, calceolarias, callas, geraniums, and fuchsias. She supplied her wares to many of the major hotels, such
as the Tremont House and Grand Pacific Hotel. By the 1890s, Felke had sold her
Chicago greenhouses and opened a new florist business in Wilmette. When this
business, too, burned in 1898, Felke was determined to rebuild. Her daughter,
Mrs. Anton Than, also owned a florist business in Chicago and helped Felke
reestablish her shop.[65]

Another important woman of the times was Elizabeth Emerson Atwater.
Atwater was born into a wealthy family in Vermont and moved with her husband to Chicago in 1856. Educated at the Troy Seminary, a well-regarded ladies'
school, Atwater used her botany background to prepare scientifically precise collections of flowers from around Chicago and from her travels. She is credited
with the discovery of the moss plant, *Bryum atwateriae*, which she found at the
foot of Yosemite Falls in California.[66] Elizabeth Atwater was described as being
the forerunner of the golden age of botanical collecting, according to the *Chicago Naturalist*.[67] Her botanical collection, thirty boxes containing three thousand
specimens, happily survived the Chicago Fire and is in the archives of Chicago's
Peggy Notebaert Nature Museum.

When the Land Grant Act was signed and the industrial schools opened for
study of agriculture and allied arts, Illinois was among the first schools to admit
women. Women first enrolled at the Illinois Industrial University (now the University of Illinois) in 1870, just two years after it opened. Yet despite equalities in
state-sponsored education, horticultural societies were still male dominated. In
1860, Jane C. Overton signed her letter to the *Prairie Farmer*, "Yours most outra-
geously indignant and anti-molifiable." Overton was outraged because women
were not allowed to judge at the state fair. "We should have half the voice at least
in awarding prizes in the horticultural department," she said. "If my husband
were a member of that Board, he would go with holes in his stockings the next
six months, I tell you. It is well for him he does not endorse the action of this
stately, dignified, anti-woman's rights Executive Committee of the Illinois Agricultural Society."[68]

The 1874 *Transactions of the Illinois State Horticultural Society* reprinted "as fair
criticism" an essay from the *Chicago Tribune*, titled "Mrs. Sam Jones goes to the
Meeting of the Illinois State Horticultural Society." The fictionalized Mrs. Jones
joined her husband at the society meeting in Champaign. She observed that
among the sixty members, there were only three or four women. "There were
so few women that I was almost sorry that I had attended," she commented. "I
looked over the programme, and could not find the name of a single woman on
the list who was to take part in the proceedings; and, during the whole of the
three days' session, not one of us females was invited to say a word."[69] Nonetheless, Mrs. Jones had quite a bit to say in this essay, including her theories on

location of the Industrial University: "What could possess the Trustees to locate the building so far out of the way . . . A citizen told Sam that it was to please a real estate ring . . . If that is true, these men should be made to move every brick." And on bud variation and Darwinian theory, Jones said: "I never felt so much like asking questions or making a speech in my life, and, had I been a man, I should have gone in . . . This is a queer doctrine for the great State Horticultural Society, to say the least . . . If the Society would admit the women, their natural curiosity would lead them to investigate some of these matters, and set them right before the people."

Never lacking opinions, until about the 1890s Chicago's women gardeners nonetheless had to hide behind pen names, or their husbands' occupations while attending horticultural meetings. Even when asked to speak at a horticultural society meeting, women were expected to provide some form of "proper" entertainment. In 1879, for example, after her very articulate speech to the Horticultural Society of Northern Illinois, Miss Minnie Meyers of Yorkville rendered a piano solo. Mrs. M. J. Cutler of Kankakee, whose 1889 topic for the society was "The Garden in the House," wrote her speech entirely as an allegory with fictional characters, Madge and Jerry. Apparently, without poetry or pianos, women's speeches were not taken seriously. While women were honorary members of the society, not until 1902 do we see a female lifetime member of the Illinois State Horticultural Society. The attitude toward women in the society may be inferred from this passage, taken from the 1879 *Transactions of the Illinois State Horticultural Society*.[70]

> DR. SCHROEDER: ". . . In Europe the women and children do most of the [market gardening] work . . . American women don't do enough work, especially in the garden . . . The farmer's wife should help the husband . . ."
>
> MR. MURTFELDT: "That's the reason why I am going back to Europe. I can get rid of this work myself and make the women do it." (Laughter.)
>
> MR. MINKLER: "How would it do, Doctor, for the men to help the women do the housework?"
>
> DR. SCHROEDER: "Yes, I do it." (Laughter.)

Yet progress was being made on other fronts. The *Prairie Farmer* observed in 1888 that women were engaged in the finer work of horticulture. Still, while the magazine reported that it was not uncommon for women to cultivate large gardens for "health and amusement," only a very few ventured into the production greenhouse business. "Methods of propagation by budding and grafting, the starting of difficult seeds, the laws by which the increase of many plants is governed, and general botanical knowledge, are sealed books to many women,"[71] said the *Prairie Farmer*.

In the early 1890s, Edgar Sanders reported on a variety of horticulture-related occupations. Remarking on women as florists, Sanders said, "There is not
a single part of the business that has not...been forced on her as a drudge in one
part of the world or the other, and having tasted of its bitters let her also, if she
can, enjoy the sweets that are connected with the present field of florists' work"[72]
(plate 5). In 1892, Sanders wrote of "Women as Tree Agents" and "Women as
Farmers." He observed that ladies made excellent door-to-door sales agents, as
they found opportunities where men found closed doors. Women farmers were
hard to find, he said, only because "it is only within the last three decades, even
in America, that business pursuits have been fairly opened to women."[73]

In 1893, Chicago hosted the Columbian Exposition, and Chicago's women,
as epitomized by powerful socialite Mrs. Potter Palmer, were scrutinized under
the national spotlight. Their insistence on having a voice in world's fair matters,
their employment of women in design and construction of the Women's Building, and their overall sophistication in matters of culture made an impression
on the world. While Chicago celebrated American landscape architect Frederick Law Olmsted in landscaping the overall world's fair grounds, Francophile
Bertha Palmer insisted on using a French landscape architect for the Women's
Building grounds. In her 1893 *Munsey's Magazine* article, "The Women of the
World's Fair City," author Mrs. M. P. Handy reported on the intrepid nature and
sophistication of Chicago's women. Large numbers of successful businesswom-
en flourished in Chicago, Handy reported, and there were scores of women who
"discuss municipal improvements, public education, and charities, with fully as
much zeal and with as thorough an understanding as is displayed by their husbands and brothers."[74]

"It is a great mistake to suppose that the Chicago woman is provincial," Handy cautioned. "In the first place she is generally from somewhere else, and is
only an adopted daughter of Chicago; in the second, if she belongs to the better
class, she has probably had a finishing course at some well known school or college in the East, and is quite as well educated and as polished as her sisters of
any other American city." Chicago, as a frontier town, afforded new opportunities to women who might otherwise have needed an aristocratic heritage in the
East. Handy noted that Chicago did not "insist upon grandfathers" or other
pedigrees, and any woman with talent as an author, poet, inventor, or "a 'mission' and the eloquence wherewith to proclaim it, then even though you may be
as poor as a churchmouse, yet, like that mouse, you may move in good Chicago
society."[75]

Despite these seemingly limitless opportunities, women were only gradually
entering the florist or nursery trades on a large scale. In 1893, Martha L. Rayne,
an Oak Park journalist, published the book *What Can a Woman Do?* Gardening
was prominent among the nearly thirty suggested business occupations open

to women. She began her essay, "It is a matter of surprise that so few American women attempt to earn a living in this way, and that a work that is both pleasant and profitable should be left almost entirely to the foreign born population." Rayne predicted that gardening would be more profitable than being a seamstress or "shop-girl," and that gardening was eminently suited to the nurturing side of women.

One year later, Ellen Bryant Freeman of Princeton gave a similar speech highlighting labor issues and urging women to take on healthful occupations like gardening instead of sweatshop jobs in the factories. She asked rhetorically,

> Is the objection raised that women have not the physical strength to perform outdoor labor; that these occupations are too severe for them? Does any one suppose that it requires more strength and endurance to cultivate the soil with the labor-saving implements now in use, out in the fresh air and life giving sunshine, than it does to stand all day over the wash tub, or ironing table, in a hot, steaming, stifling atmosphere? Let us be magnanimous for once, and leave laundry work to the men . . . and if we must work hard let us do something that shall not only develop our muscle . . . and give us sounder health, more buoyant spirits.[76]

Edgar Sanders, who had himself reported on women entering various green industries, attended Freeman's presentation, and observed that women were making strides as retail florists and greenhouse owners. However, he seconded Freeman's caveat that "a good deal of work connected with gardening was not quite so rosy as the pleasanter part might lead one to fancy."[77]

By 1900, Sarah Randolph Frazier of Yorkville noted that there were "many successful seed women all over the land."[78] She suggested flower growing as "an industry which ought to appeal to women with country homes, to whom no other business is open." Raising violets was a profitable undertaking, she opined. Growing roses, and extracting their essential oils for perfume, was another growth industry she recommended. But the odds were still against female florists. Charlotte Megchelsen began her florist business on Chicago's south side. Her first business venture failed as customers went to the larger florist nearby. Undaunted, she worked two jobs to enable her to hire another woman to work at the shop. With only $50 in starting capital, by 1910 her business still thrived, although she remained the only female florist shop proprietor on the South Side.[79]

Female landscape gardeners and landscape architects emerged in Chicago in the early 1900s. Frances Copley Seavey practiced as a landscape gardener and was committee chair of the Improvement Association Department of the Park, Cemetery, and Landscape Gardening Association. Some of her early writings espousing thoughtful garden design appeared in the Domestic Science column of

SEEDS · PLANTS · ROSES
1934

Six Beautiful Hardy
PERENNIALS
Easy to grow from seed
Pkt.
(1) Delphinium, Gold Medal Hy- 15c
brids.
(2) Delphinium, Chinese Blue But-.08c
terfly
(3) Painted Daisies 10c
(4) Kansas Gay Feather 15c
(5) Anemone, De Caen. 15c
(6) Geum, Lady Stratheden 10c
Total Value 73c

SPECIAL OFFER
One full size seed
packet of each of 45c Post-
the six .... paid

NRA
WE DO OUR PART

SEEDS *grown by a*
WOMAN..

CHARLOTTE M. HAINES   *Rockford Illinois*

From the early 1900s through the 1930s, Charlotte Haines of Rockford parlayed her gender into a business asset with the slogan "Seeds Grown by a Woman." In the late 1890s, some nurserymen claimed women-owned nurseries a fraud—that they were actually owned by men but "fronted" by a woman for marketing purposes. Carrie Lippincott, of Lippincott's nursery in Minneapolis, asserted that she was indeed real, in her 1899 catalog. Charlotte Haines was also the true owner, not a fictionalized character. (Thanks to Jim Nau for this research.) From the author's collection.

the 1904 *Chicago Tribune*: "Every woman loves plants and flowers, and most of them realize their value . . . but until recently opportunities for education as to what constitutes good planting, from the artistic standpoint have been few and inaccessible. Now, however, a change has come about. On all sides one hears of improvement associations and rumors of more improvement associations of planting this and planting that." Seavey asserted, "Where to plant is of greater importance than what to plant. Plan must from the nature of things precede plant."[80] This motto governed the women's auxiliary of the American Park and Outdoor Art Association, of which she chaired the Chicago chapter.

More controversial in 1900 was the hiring of Annetta McCrea as the landscape gardener for Lincoln Park. McCrea's husband, a landscape gardener in Denver, had died seven years prior, and Annetta McCrea assumed his business. After living for awhile in Kalamazoo, Michigan, McCrea moved to Chicago about 1894. McCrea had definite plans on landscaping Lincoln Park. She aimed to fix labels noting botanical and common names for trees and shrubs, much like an arboretum. Striving for a look of "studied carelessness," she hoped to eliminate weedy trees such as cottonwood and elder and plant better ones, rectify poor pruning practices, increase the diversity of trees and shrubs, and add flowers. She wanted to engage the public in discussions about what they wanted in their parks. Superintendent Foster and landscape gardener Kanst of the South Parks were interested in McCrea's plans and decided to adopt some—particularly those of grouping shrubbery—in Jackson and Washington Parks. Superintendent Cooke of the West Parks said, "I am glad to see a woman have a chance to display her talent in the decoration of at least one of our parks . . . Women naturally have a keener sense of the artistic and it is proper that the landscape in our parks should show some signs of feminine delicacy."[81]

McCrea was not able to fulfill her vision. A scant six months after being hired, she was dismissed and became a cause célèbre. She protested that her termination was politically motivated because she refused to automatically grant contracts to "insider" nurseries. McCrea argued that her salary could not be a significant factor—that no other landscape architect in the country would accept her $100 monthly stipend. According to McCrea, she "took up the work because her home is in Chicago and she was willing to accept it at these terms to show the public what a woman could do as a landscape architect, and she hoped to make it simply a demonstration simply of the need of an architect in Lincoln Park."[82]

The *Chicago Tribune* wrote an editorial in defense of McCrea: "No sufficient reason has yet been stated for the preemptory removal of Mrs. McCrea . . . Political reasons do not warrant the removal of a competent employé, though she be a woman and hence unable to do political work."[83] Despite this endorsement,

In 1911, women collected signatures to support suffrage at a florist shop polling place at 4627 North Broadway. Reproduced by permission of Chicago History Museum, DN-0057979, Chicago Daily News Collection.

and even a plea to Governor John Tanner, McCrea was not reinstated. Not one to nurse her wounds, McCrea promptly secured the position of city consulting landscape architect in Marquette, Michigan, but retained her ties to Chicago. She lobbied extensively to the Buildings and Grounds Committee of the Chicago Board of Education for more school gardens. McCrea continued to pursue her chosen profession, persuading officials at the Chicago, Milwaukee, and St. Paul Railroad that gardens were needed along their rail lines. She created a position for herself there, and urged other women in public lectures that railroad gardening was an open field for women.

As the twentieth century continued, women became more outspoken in the political arena, and sometimes combined lobbying with business. In 1911, a woman florist became prominent in her lobbying efforts for women's voting rights. Nellie C. Moore, an ardent suffragist, opened her florist shop as a polling place during November. Working with Belle Squire, president of the No-Vote-No-Tax league, Moore asked each voter to sign a petition in favor of an amendment to the national constitution. Voters were given flowers upon leaving Moore's shop in the Twenty-fifth Ward.[84]

The Clubs, Shows, and Social Causes

As strong of spirit as individual female gardeners were, their real strength multiplied when women gathered in numbers. Women united in horticultural pursuits began to change the physical and social fabric of Chicago. Gardens and floriculture were seen as integral to humanity's moral and spiritual health, and who more appropriate to nurture the human spirit than women? Chicago women have a long history incorporating flowers into various social causes. One of the first major fairs held in Chicago, the Great Northwestern Sanitary Fair of 1865, was largely organized by women. A reprise of an earlier fair held to raise funds for the Civil War wounded, the 1865 fair, one of the first professionally landscaped events in Chicago, drew donations and support from around the world. According to the *Voice of the Fair*, a daily on-site publication, the fair "has given to our city an enviable fame in this country and Europe."[85] The *Voice* further noted, "The many kind, generous and patriotic ladies who have devoted so much of their time to helping in the Fair . . . deserve and receive the thanks of the managers of the Fair and the appreciative public." Although influential businessmen such as Thomas B. Bryan administered the fair, women across the nation had important roles, from raising funds to working on-site. Typical were women-sponsored church groups like the fair's Methodist department, where the ladies of Dixon, Illinois, raised $1,500 worth of vegetables, $900 in cash, and six hundred bushels of potatoes.

In Horticultural Hall, women staffed the popular, biblically themed "Jacob's well." This grottolike fountain surrounded by fir and cedar trees was attended by a costumed "Rebecca" and her attendants. Presaging the temperance movement that would take hold in Chicago, beverages that "will cheer but not inebriate" were offered here. The *Voice* opined that the popular Horticultural Hall, or something similar, should be retained as a permanent city feature, asking "if the city authorities will not take the hint—to give us something like this [Horticulture Hall] on a larger scale, where the young can resort to cultivate their taste for the beautiful. Our saloons and groggeries would be less frequented were better things more accessible." Flowers were a key element of the fair. Mrs. Asa Kennicott was in charge of the floral department, where "with rare taste [she] arranged thousands of bouquets in the most exquisite style," and "she has shorn her own magnificent garden of its burden of beauty and brought the offering daily to this shrine of patriotic devotion."[86] Other women, such as Miss Baily of Hyde Park and Mrs. Hoyne of Chicago, donated fresh flowers from their own gardens.

In the 1870s, Chicago women created the Chicago Flower Mission, similar to groups established in New York and Boston. Meeting on Wednesdays, this philanthropic group distributed cut flowers to hospitals, orphanages, the "Erring

Women's Home," tenement houses, and various "Sewing Women" groups (presumably underprivileged working women). Flowers were wrapped in brown paper and sent by train or horse and wagon from nearby suburban and city yards, and even from neighboring states. The 1877 annual report tallied contributions of 190 baskets, 236 boxes of flowers, and 878 bouquets. By 1884, the mission reported distributing 11,800 bouquets received from sixty cities and towns from Illinois, Wisconsin, Iowa, Indiana, Michigan, and Ohio. Considering the lack of refrigeration and great distances traveled, this was truly a logistic feat, managed through sheer organizational rigor.

The 1870s also saw the beginnings of the national Women's Christian Temperance Union, with Frances Willard of Evanston as president from 1879 to 1898. Under the auspices of the WCTU, women entered into real estate development in a widely publicized way. The Woman's Land Syndicate was formed in spring 1892 by Mrs. R. A. Emmons, business manager, who used the logo of the WCTU Temple building on her real estate letterhead. The syndicate bought $500,000 worth of lots in South Waukegan, an industrial development where sale or distribution of liquor was prohibited. In addition to temperance principles, landscaping and green space were key attractions, with marketing brochures advertising that "eighty acres of the most valuable land in the city has been dedicated as a public park, and thousands of dollars are to be expended out of private funds to beautify it."[87] While the development suffered financially in the economic downturn of 1893, the Woman's Land Syndicate brought nationwide attention to real estate developed and managed by women.

Gardening was a key component of other social movements as well. Mrs. J. B. Cobb of Waukegan, an avid gardener who won gardening awards as early as the 1860s in Illinois state fairs, operated a garden for underprivileged or "fallen" women in the 1860s. Wildwood, the former home of Colonel James Bowen on the city's far south side, also became a country retreat for working women in the late 1800s.

Flowers and gardens were emblems of culture and virtue, and farmer's wives often lobbied their busy husbands for ornamental and vegetable gardens for the farmstead table. As with the familiar case of the cobbler's shoeless children, farmers were frequently cited for unimproved landscapes. Horticultural periodicals continually chastised Illinois farmers for tending to their cash crops over their own backyards. Citing the spiritual welfare of farmer's children, do-gooders urged rural homeowners to spruce up their own gardens. From early *Prairie Farmer* writings to the beautification work of the 4H Club spearheaded by Chicagoan Myrtle Walgreen in the 1930s, farmers were constantly prodded to improve their landscaping. Rhetorically, the *Prairie Farmer* asked in 1845, "Reader . . . did you ever know a lady who cultivated flowers who was a slattern, and kept her house and family in dirt and disorder? We never did . . . *Home* should have flowers around it. They belong to and go to make *home*."[88] Mrs. Hillis of Dixon wrote in 1874, "Every mother has a right to a plot of well-prepared soil for herself and little ones to cultivate in floral treasures, and should enforce that right."[89]

In 1900, Sarah Frazier of Yorkville wrote, "Whether in the home, the garden or the school, we can voice our love of flowers by doing what we may to bring brightness into the lives of the children, the poor and the needy."[90] The 1900s Progressive movement was a catalyst that turned the written exhortations of prior decades to action. Nature was seen as a critical element in improving the life of the industrial poor, especially children. Plans for children's gardens were developed and published for every mother concerned about raising healthy, wholesome children. Chicago women such as Alice Hofer were influential members of the early national Playground Association, which fought to integrate outdoor play spaces into schools and neighborhoods. Jane Addams, with the help of her philanthropic friend Louise DeKoven Bowen, sent the children of Hull-House to the Bowen Country Club in Waukegan, a summer camp for underprivileged children.

Hull-House clients and others who lived in the tenement district obtained more garden experience through the work of women-led groups such as the Chicago City Gardens Association. It was organized by Hull-House leader Laura Dainty Pelham, who worked there for twenty-five years, until her death in 1924. In 1909, the association established 150 "small farms" on property loaned by International Harvester Company. The one-eighth-acre (150-by-37-foot) individual plots were leased at $1.50 each to needy individuals and families for growing vegetables for their own tables and small truck gardens. The typical yield was $25 per plot—then considered a fine profit. In this effort, the City Gardens Association formed partnerships with other benevolent societies, including United Charities, which offered to construct a playground, and the Chicago Women's Outdoor Art League, who offered to build a recreation house for the workers, designed by Jens Jensen. Women's clubs, particularly the West End Women's Club and the Garfield Park Women's Club, were actively involved in the effort. This benevolent endeavor, while helping newly arrived immigrants bootstrap to better living conditions, nonetheless presented conflicting views of women's progress: the *Chicago Tribune* reported, "As the women of course, for the most part, will be tillers of the soil, it will seem out there on a Monday morning, almost like the early days when women worked in the field."[91]

The project was quite successful, with more plots offered in subsequent years through the generosity of companies such as International Harvester and Western Electric. By 1915, the program had spread, so that a drawing was held for free plots at Hull-House, and other community gardens sprouted at 31st and South California, on Bellevue Place, and on fifty acres at Foster Avenue. [92]

The Chicago Women's Outdoor Art League exerted a strong influence on many of the City Beautiful projects launched in the early 1900s. In 1905, the league took a dramatic move to divorce itself from the national umbrella organization, the American Park and Outdoor Art Association (later the American Civic Association). The reasons the Chicago league separated from the national group largely reflected Chicago's independent and regional spirit. The group wanted its own identity, unique from its meaningless polysyllabic label as an auxiliary branch to the American association. Many members also objected to the scheme to plant fruit trees on Chicago's streets, an idea promoted by J. Horace McFarland, president of the American Civic Association in his presentation to the Chicago league in 1904. The planting—and high maintenance—of fruit trees in downtown Chicago was not feasible, according to many Chicago members.

As an independent local organization, much of the work of the Chicago Women's Outdoor Art League focused on beautifying public properties. It bought seeds for school children, and helped them plant schoolyards, parkways, parks, and settlement-house yards. Arbor Day became a cause célèbre for their tree-planting efforts. Later, the league promoted home beautification through gardening. Using the "bungalow" built on the city gardens near 31st Street and Marshall Avenue, the league promoted its surrounding grounds as a demonstration garden. Other bungalows—actually little comfort stations for gardeners to rest and store tools—were built and landscaped around the city. At 42nd Street and Chicago Avenue, the league erected a portable house and planted a model garden. Here, the *Chicago Tribune* reported, "a club whose members are all colored women has taken 40 of the plats, and the members are raising potatoes for the use of a home for Negro working girls."[93] The melting pot of the league's community gardens helped assimilate peoples from all cultures into Chicago.

Social work, including horticultural work, was often performed under the auspices of Chicago's formidable women's clubs. In 1893, the *Graphic* magazine observed that "among the unsuspected attributes of Chicago is the culture of her women." Key to this cultural development, according to the *Graphic*, was the extensive formation of women's clubs. When the (national) Federation of Women's Clubs held their meeting in Chicago in May 1892, Illinois reported more clubs than any other state, but, according to the *Graphic*, "this federated list does not even hint at the actual number of women's clubs in existence in Chicago."[94] The clubs were a way for women to exert influence in society, within the constraints of their gender-based roles. The Fortnightly (1876), Chicago's oldest women's club, featured women's suffrage as its original mission, but also focused on literary and intellectual pursuits. The Chicago Women's Club (1876), an outgrowth of the Fortnightly, was largely philanthropic, and, according to the *History of Women's Clubs in America*, formed an organizational model for women's clubs adopted by cities across the nation. Many women's clubs had subcommittees dedicated to floriculture, forestry, or landscaping. The clubs had powerful influence in local affairs. *Country Life in America* magazine reported, "As a result of a movement instituted by the Chicago's Woman [*sic*] Club, an ordinance to provide for 'the planting, preservation, control and culture of trees and shrubbery in the public streets' was adopted by the Chicago City Council. In May

TOP The University of Chicago botany program, among others, welcomed women students. Here May Theilgaard rows with H. C. Cowles about Big Bay, Madeline Island, Wisconsin. May Theilgaard Watts went on to become the Morton Arboretum's first naturalist, and wrote *Reading the Landscape of America*. Ca. 1916. American Environmental Photographs Collection, AEP Image Number AEP-WIS45, Department of Special Collections, University of Chicago Library.

LEFT Especially during the busy seasons, women took jobs at nurseries and greenhouses, as shown here at Vaughan's Nursery. Ca. 1917. Courtesy of Western Springs Historical Society.

1909, Mr. J. H. Prost was appointed forester, the first city official of the kind in the Middle West."[95]

The women's clubs strongly supported gardens for children and schools. As child labor laws were enacted, children found time on their hands in the summer—a condition of idleness that led to delinquency, in the minds of many citizens. The Chicago women's clubs took it upon themselves to develop vacation schools, among which gardening was featured. These vacation schools often focused on outdoor play, and were instrumental in the small parks and playgrounds movement that became so successful in Chicago. Based on the success of gardening at such schools as Chicago's Doolittle and Forrestville schools, the Chicago women's clubs wanted to make garden programs an integral part of education nationwide. "Thanks to the mothers the school children of Chicago may yet get a chance to make garden as part of their school work," the *Chicago Tribune* reported in 1910. "At least they will if the plans of the Illinois congress at the national convention ... are carried out."[96]

Gardens were seen both as restorative and educational for working women and underprivileged women. Curriculum at local orphanages and industrial schools included horticulture, and women's schools emphasized the topic as well. In 1912, suburban Park Ridge showcased its Industrial School for Girls, where residents learned to become truck farmers on the school's forty-acre farm. Gardens were also incorporated into the work environment of newly defined "women's jobs" such as that of telephone operators. The Chicago Telephone Company determined that a relaxing diversion during its female employees' breaks would include so-called fifteen-minute gardens. At its various exchange stations (e.g. Humboldt, Lake View, Stockyards, Canal, Wentworth, and Hyde Park offices), the company provided small lots for the "telephone girls" to work during rest periods and lunch. "Out to the flowers and shrubs fly the eager girl gardeners so soon as released from receiver and stool," gushed the *Chicago Tribune* in 1906.[97]

Women's clubs' subcommittees drove social and legislative change. In 1913, Mrs. J. C. Bley reported on the work of the Chicago Clean Food Club, which used consumer lobbying power to improve quality and prices of market goods. "As women are the spenders, so women must help solve the problem of cheaper, cleaner food," asserted Mrs. Bley. "We know that good food is rotting on the ground in your gardens and orchards because you cannot afford to gather and ship it." Through boycotts, publicity, and group cooperative purchases, the Chicago Clean Food Club helped put the spotlight on marketing and distribution issues for apples, for example, and in 1913 brought to the Illinois State Horticultural Society a comprehensive strategy to promote local apples.

Following the successful fight for the right to vote, women became more visible in leadership and previously male-dominated occupations. Women's groups gathered together to stage the first ever Woman's World's Fair in Chicago on April 18, 1925. With lectures and programs on women's issues, the fair, which became an annual event in later years, showcased a broad range of occupational opportunities for women. Exhibitors came from around the world, and the woman's fair concept was duplicated in cities around America and even in Switzerland.[98] Among the hundred or so women's occupations represented at the 1927 fair were conservationists, florists, flower growers, and gardeners. Flowers were an important decorative element of the fair, and Lillie Tonner Browne at 5832 West North Avenue was advertised as the "Official Florist for the Woman's World's Fair."

Interestingly, although the first women's club in America, according to the *History of Women's Clubs in America*, took its name from a botanical dictionary, garden clubs did not gain ground until about the turn of the century. Louisa King of Elmhurst was among the charter group who started the Women's Agricultural

Louisa King had achieved such national prominence by 1911 that this gladiolus was named for her, and was featured on the cover of Vaughan's catalog. Throughout her lifetime, many other plants were named for her. From the author's collection.

and Horticultural Association in America. Later renamed the Women's National Farm and Garden Association, the group elected King as its first president in 1914. Local garden clubs began to organize around the turn of the century. The North Shore Horticultural Society, incorporated September 8, 1905, included women from Chicago's affluent communities along Lake Michigan. One of the earliest clubs to organize, the North Shore Horticultural Society helped stage the first flower and garden show organized by women's clubs in 1927. That same year, twenty-nine clubs voted to form the Garden Club of Illinois. The Garden Club of Illinois, in turn, was one of thirteen charter members who organized the National Council of State Garden Clubs in 1929. Chicago hosted the first annual meeting of this council in 1930.

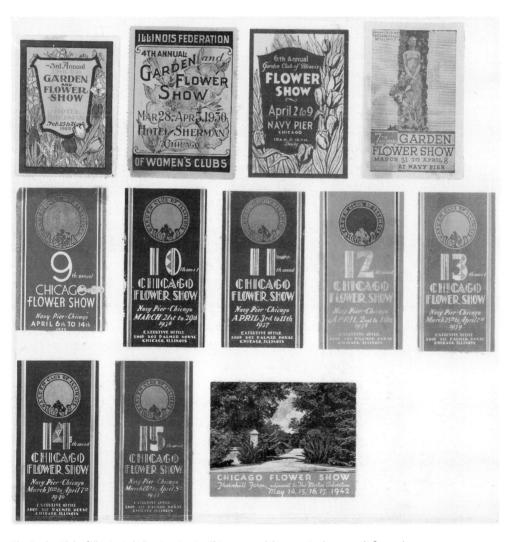

The Garden Club of Illinois, including twenty-nine Chicago-area clubs, organized many early flower shows. From the author's collection.

Tapping on their long history and expertise in organizing events, the Garden Club of Illinois sponsored a series of flower and garden shows from 1927 through the 1970s. The first three shows were held in Chicago's downtown Hotel Sherman, and operated by hotel manager John Servas. The list of early show patrons reads like a who's who of Chicago's society and included for various shows Mrs. J. Ogden Armour, Mrs. Walter Brewster, Mrs. Joseph Cudahy, Mrs. Frank O. Lowden, Mrs. Francis King, Mrs. Charles Walgreen, and Mrs. Charles L. Hutchinson. At the time of the 1929 Chicago Garden and Flower Show, plans were under way for Chicago to host a world's fair, and members of the Chicago World's Fair were part of the civic committee for this flower show.

John Servas, who'd gained experience in managing these local fairs, served as manager of the world's fair, and later parlayed his experience into managing other world's fairs, such as the 1939 New York World's Fair. Taking the 1929 Chicago Garden and Flower Show as an example, women were listed as patrons of the show, with management positions still largely taken by men. The Civic Committee included many significant individuals and institutions such as the Chicago Park District, University of Illinois, and Northwestern University, but not a single woman was represented. The Executive Committee had nine members, of which only two were women: Mrs. Frederick Fisher, in charge of the Garden Clubs Committee, and Mrs. Le Mean Mason, the director of advertising. The Awards Committee comprised only men, as did the twenty-seven-member advisory committee. In fact, all of the other subcommittees, from Layout and Design to the industry-related groups like the Retail Florists Committee and Pot Plant Growers Association of Illinois, were almost exclusively men.

After three years of male-dominated shows, the Garden Club of Illinois made a strategic decision. According to a club history, "in 1930, when Mrs. Karcher was elected president, she was determined that the Chicago Flower Show could and would be managed by the women of the garden clubs. The shows in previous years had been underwritten by the directors of the buildings in which they were given and the Garden Club of Illinois had no part in the business management of the show and no share in the proceeds."[99] In 1931, the first Chicago Flower and Garden Show managed by women was held at the Merchandise Mart. Subsequently held on Navy Pier from 1932 through 1941, the show garnered much acclaim as the only major show in the country planned entirely by women.[100] When Navy Pier was commandeered for defense purposes in 1942, Mrs. Joseph Cudahy and her brother, Sterling Morton, offered the Morton Arboretum as a venue to host the show.

In 1933, when Chicago hosted its second world's fair, the role of women in horticulture had clearly evolved. Kate Brewster was asked to head the horticultural committee of the fair in May 1932. The forthright Brewster accepted the position only on certain conditions: that the horticultural committee oversee all horticultural aspects of the fair, and that all exhibits conform to a unified plan developed by said committee. "There are many conflicting interests in the horticultural field and unless the committee is in a position to dictate terms to prospective exhibitors, the exhibit may easily degenerate into the usual 'flower show' which has little value from any point of view," Brewster asserted.[101]

Brewster's horticultural committee included two other ladies (Mrs. Tiffany Blake and Mrs. Clay Judson), as well as nurseryman Alvin Nelson and Paul Battey. She worked closely with Ralph E. Griswold, a fellow of the American Society of Landscape Architects, from Pittsburgh. Her goal to make the horticultural exhibition noncommercial seems to have been thwarted. Unable to compromise, at their October 6 meeting, the Century of Progress Executive Committee accepted the resignation of members of the horticultural committee.[102] Still, women played an important role in the 1933 World's Fair. Mrs. William Karchner, of Freeport, Illinois, was appointed by the governor of Illinois to represent horticultural interests for the State of Illinois.[103] Annette Hoyt Flanders, a landscape architect from New York City who designed a number of estates on Chicago's North Shore, created the landscape plan for the fair's Good Housekeeping gardens. Many garden clubs participated in the fair, and the Advisory Committee, while still predominantly male, included Dorothy Ebel Hansell, secretary of the National Association of Gardeners; Maud F. Robertson, Chicago's managing editor of the *Home and Garden Review*; and landscape architect Ruth S. May.

The proliferation of women's garden clubs in the early twentieth century actually eclipsed men's gardening groups in Chicago. In the program for the 1929 garden and flower show, an essay by Abbie S. Kendall noted, "If you have a garden club of women, then organize a club of men and women. Men are enthusiastic gardeners and are eager for a better understanding of garden planning and a better knowledge of the plants which should be grown therein."[104] It wasn't until a man won the *Chicago Tribune's* 1927 gardening contest that the first Men's Gardening Club was established. Contest winner Leo W. Nack's backyard garden was chosen over some 2,500 other entries, and observing that many visitors to his 40-by-60-foot garden were men, Nack formed a Men's Garden Club on March 15, 1928. The group expanded to the Men's Garden Club of America in 1932, and survives today as the Gardeners of America/Men's Garden Clubs of America, with nearly five thousand members.[105] Horticultural history had come full circle now that the Men's Garden Club had been founded in Chicago, where horticulturally thwarted women were at one time "antimolifiable and outrageous."

In 1933, the year of Chicago's second great world's fair, a member of the Garden Club of Illinois expressed her views on women gardeners during a WGN radio broadcast. Mrs. R. R. Hammond of Barrington had just read John Beverley Nichols's *Down the Garden Path*, and vociferously objected to the Englishman's characterization of women. Mrs. Hammond said, "Perhaps, if you have not read the book, you would like to know his accusations against us in this our favorite pursuit. He says we are too 'dainty'—calls it a besetting sin; that we cannot plant as a man can . . . that we are envious, suspicious, and even malicious when we compare our two herbaceous borders." While acknowledging that Nichols may have exercised some literary license, Hammond pointed to the garden club flower shows as examples of women's aptitude for industry and horticulture. "I think an invitation to our Flower Show should be sent to Mr. Nichols," Hammond proposed. "The Flower Show is an exhibition which takes more than an ordinary amount of cooperation, of business understanding and a very careful handling of all those characteristics which Mr. Nichols has found so objectionable in women gardeners."[106] As seedswomen, florists, landscape architects, and organizers of some of the largest flower shows in the nation, Chicago women gardeners had successfully plowed unfamiliar terrain.

WEALTHY AMATEURS

Whereas scientists, nurserymen, florists, and market gardeners contributed the brawn and brains behind Chicago's horticultural growth, and assumed the greatest risk to their livelihoods, their efforts would not have been nearly as successful if not supported by Chicago's so-called wealthy amateurs. This group, so named by horticultural professionals, were members of the city's affluent society whose estates were planned and planted by those in the green trade, but who also took personal interest in gardening themselves. Without paying clients, Chicago's horticultural professions would not have grown.

The 1840s marked the beginning of a long-standing, symbiotic relationship between Chicago's elite and the professional gardeners who worked for them. Both groups shared an interest in horticulture: one for their livelihood and the other for their personal enjoyment and the trappings of culture. Here on the prairie, class distinctions were overlooked in the pursuit and study of new flora. Gardeners of the wealthy amateurs often achieved prominence and then struck out on their own. For example, George Wittbold was gardener to Charles E. Peck about 1859, before forming Williams and Wittbold a few years later. Frank Calvert earned his reputation through his gardening efforts with realtor S. H. Kerfoot, and ultimately opened his own successful nursery in Lake Forest.

In addition to providing jobs for Chicago's gardeners, the wealthy amateurs subsidized flower shows, helped various horticultural societies, served on city and public park beautification boards, and opened their own estates to qualified garden lovers. Frequently, they were able to offer discounts through their businesses; many railroad barons, for example, offered discounts to horticulturists traveling to professional meetings, and often corporate land was leased to

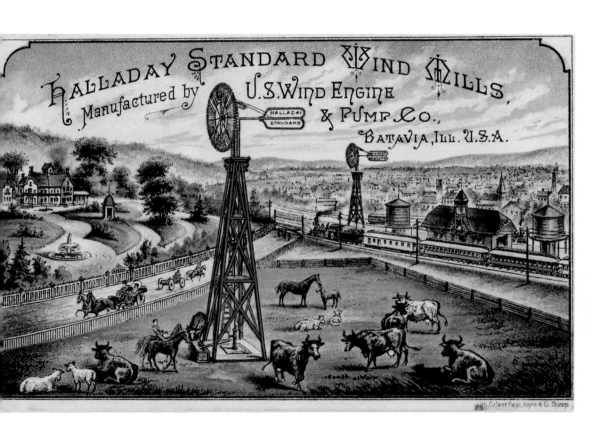

PLATE 1. Chicago, the Windy City, turned wind power to good use in the late 1800s. West suburban Batavia earned the sobriquet Windmill City to recognize its many windmill manufacturers, including U.S. Wind Engine and Pump Company. The Halladay windmill shown here was self-regulating and improved current technology by maintaining equal wind pressure on all sides of the windmill. Ca. 1890. From the author's collection.

MA PERKIN'S
OLD FASHIONED GARDEN

PACKED BY

Vaughan's Seed Store
CHICAGO ❦ NEW YORK

PLATE 3. The D. Hill nursery specialized in evergreens and other trees and shrubs. The Hill nursery thrived under several generations of management. From the author's collection.

« PLATE 2. This package of seeds showing old-fashioned perennials was enclosed in a "personal" letter written by the mythical Ma Perkin—a wonderful early marketing approach. From the author's collection.

PLATE 4. With the advent of the automobile, consumers drove to nearby roadside markets offered by local farmers and gardeners. Here, Chicago-based Armour Company's *Farmers' Almanac* featured a road-stand market on its 1924 cover. From the author's collection.

PLATE 5. Vaughan's catalog of 1890 showed the ambivalence toward women. Here, a female model is literally put on a pedestal, surrounded by flowers. Yet Vaughan was one of many nurserymen who hired women. Courtesy of Sterling Morton Library and Archives, Morton Arboretum, Lisle, Illinois.

PLATE 6. Seed packets of popular annuals such as pansies and nasturtiums were marketed by Hiram Sibley and others with the eye-catching new chromolithographs on the cover. From the author's collection.

PLATE 7. Catherine Mitchell's postcards featured native wildflowers in full color. Mitchell, a Riverside resident, was a member of the Prairie Club and the Wildflower Preservation Society. From the author's collection.

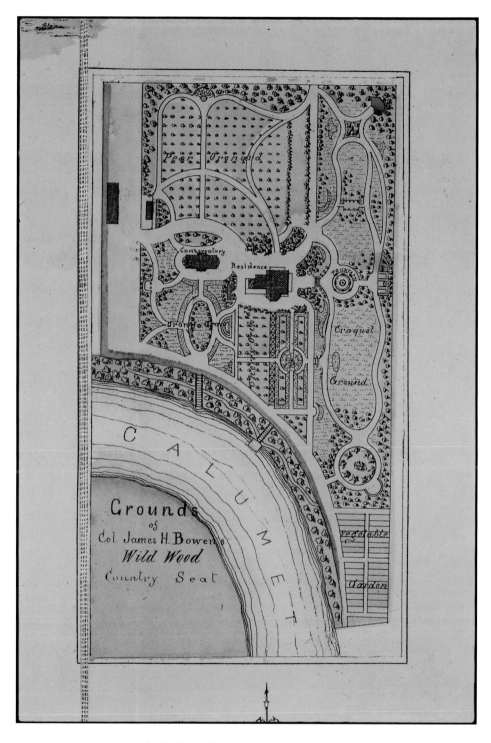

PLATE 8. In H. W. S. Cleveland's plan for James H. Bowen's home, the conservatory is almost as dominant as the residence in this drawing. A "Grand Grove" is planted between the conservatory and residence, and extensive pear and cherry orchards surround both. Extensive space is dedicated to the croquet ground, and a wooded walkway borders the Calumet River. Reproduced by permission of Chicago History Museum ICHi-29590.

PLATE 9. As indicated by its name, this La Grange nursery catered to suburban clients. By 1916, its vegetable seeds were available for kitchen gardens and local market gardens alike. From the author's collection.

PLATE 10. This 1914 image from the Leonard Seed Company catalog shows how nurserymen tried to appeal to suburban homeowners with gracious lawns and flowers and also to rural residents who might grow mostly vegetables. From the author's collection.

PLATE 11. The Chicago Horticultural Society frequently sponsored flower shows. Here, at their sixteenth annual show, the chrysanthemum predominated. Ca. 1908. From the author's collection.

PLATE 12. Pomological displays were important at the 1893 World's Fair. The Illinois fruit exhibit received critical acclaim. From the author's collection.

support city gardens. Professional nurserymen were not shy about asking for help, as noted in the 1859 minutes of the Chicago Gardeners' Society: "There are vacant rooms in some of the large blocks owned by the wealthy men of the city that can doubtless be obtained at a nominal rent and fitted up to meet the growing wants of the society. Employers of gardeners will consult their own interests if they see that rooms are provided ... Every owner of a greenhouse or hot house, and proprietor of grounds requiring the services of a professional gardener can afford to pay five or ten dollars ... We trust this hint will suffice."[107] Peer pressure was also effective among Chicago's wealthy to accomplish horticultural goals. In 1912, for example, the CHS challenged wealthy estate owners to sign pledges to grow particular plants on their own estates to boost interest in gardening. The title of "Chicago's champion gardener" would then be bestowed on the winner. (Although this was couched as a civic education effort, one wonders, since nurseryman J. C. Vaughan chaired the society's committee, if the gardeners again saw the mutual benefit of this arrangement—contestants had to pledge to spend at least $200 on plants recommended by the committee.) The reconstituted CHS of the 1890s included more wealthy amateurs than professional gardeners. Rosters of patrons of turn-of-the-nineteenth-century flower shows mirrored Chicago's list of financial success.

Ezra McCagg was one of the first directors of the reconstituted CHS in 1890. McCagg not only aided the city's horticultural prowess with his personal plant collections, sponsorship of flower shows, and tours, but also leveraged his connections with luminaries in the emerging landscape-design world. McCagg, who is often credited with bringing Frederick Law Olmsted to Chicago, met the

Ezra McCagg's conservatory, which survived the Chicago Fire, is shown here in its prefire glory. From the author's collection.

landscape architect through the Civil War Sanitary Commission. Olmsted described McCagg as a Shakespearean scholar, philanthropist, and city booster. Writing about McCagg and his wife, Olmsted noted, "They are rather sensitive about the West & Chicago lest anyone should think that people are not likely to be as well informed & cultivated there as anywhere."[108] The McCaggs lived on Chicago's north side with well-landscaped grounds and extensive hothouses. Olmsted noted, "They are passionately fond of flowers & the house is always rich in them."[109] Olmsted's observations captured the sensibilities of many of Chicago's elite; defensive about accusations of poor taste, they used horticulture as a symbol of culture.

Some of the wealthy amateurs were dedicated plant hunters. Early Chicagoans, such as William B. Ogden, roamed the nearby countryside seeking new plants. Barbour Lathrop, a world traveler and Chicago resident, developed a bamboo grove near Savannah, Georgia (now the Bamboo Farm and Coastal Gardens of the University of Georgia). Lathrop, who financed and joined explorations by famed botanist David Fairchild, believed in the potential of bamboo as a timber, food, and landscape resource.[110] Meatpacker Allison Armour, another patron of Fairchild's work and a plant hunter himself, offered his yacht to Field Museum and University of Chicago botanist Charles Millspaugh, for a plant-hunting expedition to the Yucatan. Armour and some of his colleagues joined the professor for the months-long expedition in 1898.

Other affluent garden afficianados became connoiseurs of particular flowers. As more varieties became available, collecting and growing new plants became the pastime of the rich. Take the case of John Lane, a "wealthy inventor, who of late has turned his attention for health and amusement to amateur floriculture. His pets are chrysanthemums," according to the *Chicago Tribune* in 1888.[111] Lane, a Kenwood resident, hosted flower shows on his own grounds, inviting not only local florists such as John Goode and Company, George W. Miller, and the Oakwoods hothouse proprietors, but also exhibitors from Philadelphia and New York. By some accounts, Lane's flowers bested those of the professionals.

In 1908, a *Chicago Tribune* full-page article was headlined "How Chicago Millionaires Spend Fortunes Cultivating Rare Plants." Able to hire talented gardeners and afford well-equipped conservatories, "the Chicago millionaire has taken the privilege of scientific floriculture peculiarly to himself until in a botanical sense he has become a philanthropist. New blooms have been brought out every year until even in the east the success of the Chicago flower shows have been regarded enviously." The article described the rare ferns in banker John J. Mitchell's conservatory on Woodlawn Avenue, under the care of gardener George Woodward. Attorney Alfred S. Trude reportedly spent over $12,000 annually in his greenhouse on Drexel Boulevard. Here, under Robert Miller's supervision, hundreds of roses, palms, and carnations were prepared for show.

Martin A. Ryerson, also on Drexel Boulevard, spent a reported $15,000 for an-
nual greenhouse expenses. His prized red chrysanthemums, under the care of
John Reardon, were blue-ribbon contenders. According to the *Tribune*, "the ri-
valry between Mr. Mitchell and Mr. Ryerson has always been keen, and this year
it will be a great struggle as both have produced varieties of chrysanthemums
that are excellent."[112]

Palms and orchids were exotics favored by the social elite. Florist George
Wittbold is attributed with importing some of the first palms to Chicago in 1857
from Prussia. By the 1890s, extensive private conservatories around Chicago
nurtured these exotics. George Pullman maintained a conservatory connected to
his house—complete with a piano, Turkish rugs, and comfortable divans. This
greenhouse, as well as the separate conservatory he owned across the street, featured palms, some towering to twenty feet. In the 1880s, the Lake View conservatory of William H. Chadwick was renowned for its hundreds of rare orchids.
Chadwick's nontraditional conservatory did not house plants in rows of tubs
and buckets; rather plants were said to sprout from moss-covered logs and naturalistic vignettes on the roof and walls.[113] In the early 1900s, with the aid of gardener Charles Gebhardt, Mr. and Mrs. Harry G. Selfridge of retail fame indulged
their passion for orchids at their Lake Geneva home. Selfridge, an amateur plant
hunter, brought back many of the two thousand or so plants from trips to Cuba
and other tropical countries, and others were crossed and created in situ.[114]

Many wealthy amateurs opened their estates to the public. This photo of H. N. Higinbotham's farm in
Joliet was taken during a Prairie Club visit. Courtesy of Westchester Township History Museum, Chesterton,
Indiana.

Many of Chicago's elite opened their estates and grounds for public viewing. As their homes became more removed from the public eye, no longer in the center of the city, special field trips were arranged to share the grounds and landscapes. H. N. Higinbotham's grounds in the city were said to have the only French-styled garden in Chicago in the 1890s.[115] But it was his summer home Harlowarden, near Joliet, where visitors enjoyed field trips to explore the wildflowers around his grounds and the conservatories where he raised prized carnations. Typical excursions included joint picnics by the Long, Scott, and Short societies, and a special excursion train carrying 450 Sunday school students in 1899 to enjoy the grounds. The Prairie Club, Chicago's early hiking and conservation group, was also invited to Higinbotham's estate around 1908, and enjoyed the natural beauty of the four-hundred-acre tract. Prairie Club members were also invited to explore the Cyrus H. McCormick, J. Ogden Armour, and Samuel Insull estates in the northern suburbs.[116] Continuing the tradition even into 1930, Mrs. Edith Rockefeller McCormick, and other garden club members, opened her estate, Villa Turicum, to visitors and members of the North Shore Horticultural Society for tea and a flower show.[117]

During the Progressive Era, many corporations established gardens around their headquarters or manufacturing plants as a means of civic beautification and/or employee enjoyment. Often the horticultural proclivity of the company's executives dictated the extent of their investment in beautiful grounds. Sears, Roebuck and Company, which featured extensive grounds including flower beds and pergolas, included noted horticulture patron Julius Rosenwald among its executives in the early 1900s. Both the Swift and Armour plants included gardens, which influenced their environs. The *Chicago Tribune* noted, "In various parts of the city big corporations have laid out and maintain large and handsome gardens. Wherever such a garden is started an immediate improvement in conditions for blocks around is noticed."[118]

Perhaps two of the best-known horticulture patrons, who also maintained a connection with the Olmsted firm, were Charles L. Hutchinson and his wife, Frances Kinsley Hutchinson. Banker and president of the Board of Trade, Hutchinson literally married culture and horticulture in Chicago. One of the founders of the Art Institute in 1882, Hutchinson served as its president until his death in 1924. He also served as one of the early directors of the CHS. With his interests in both art and horticulture, it is not surprising that many of the CHS early flower shows were held at the Art Institute, and in fact, between 1908 and 1918, the two institutions shared the same business manager.[119] To jump-start the CHS first fall flower show in 1891, Hutchinson offered a $50 premium and a medal for the best begonia. This Hutchinson medal has been awarded by the CHS each year to an exemplary horticulturist.

No. 46 THE FOUNTAINS AND GARDEN,
SEARS, ROEBUCK & CO., Chicago, Ill.

TOP Companies such as Swift and Armour created gardens for their employees' enjoyment and to harvest produce for their kitchen tables. Sears Roebuck and Company purchased forty acres of land for their administration building and surroundings. Sited in a residential section of the city, Sears ultimately planned to build more buildings but, according to the description on this stereoview, "in the meantime, we have devoted large sums to landscape gardening, in an effort to beautify the space not as yet occupied for business purposes." Sunken gardens with a classic Grecian pergola in the background extended an entire block. The company's thousands of employees were encouraged to stroll amid the fountains, pools, and planted urns. As the stereoview descriptions noted, "We believe it is good business to beautify our grounds . . . because a change of environment and attractive surroundings send them [the employees] back to work again greatly refreshed and forgetful of the little annoyances of the morning." Ca. 1910. From the author's collection.

BOTTOM Here, in 1932, an African American woman tends to her vegetable plot in the McCormick Works garden. The garden was intended to help underemployed and unemployed community residents. Employees with at least five years' service were eligible for garden plots. The McCormick Works community gardens were near 95th Street and Crawford Avenue, and consisted of approximately 440 acres divided into 50-by-150-foot plots. The gardens were used by over 1,800 employees. Reproduced by permission of Wisconsin Historical Society.

WILLIAM B. AND WILLIAM C. EGAN: FATHER AND SON "WEALTHY AMATEURS"

Born in Ireland, William B. Egan came to Chicago in 1833 after a stint as a teacher in Quebec and medical school at the University of Virginia. A practicing physician and avid real estate investor, the congenial Egan was a CHS charter member. He held several civic positions and gave a dedication address at the opening of the Illinois and Michigan Canal in 1836.

In 1846, Egan and his family moved from their Clark Street home to a fifty-acre homestead at Van Buren and Sangamon streets, then considered far outside the city limits. In 1847, a *Prairie Farmer* writer visited Egan's grounds and found them "improved with considerable taste."[1] There were several hundred evergreens, including firs, junipers, pines, and arborvitae. "Dr. Egan suffers his evergreens to grow their limbs to the earth. 'Pruning them up' is a piece of barbarism to which he has not yielded," declared the *Prairie Farmer*, noting that this practice was suitable only for deciduous trees.[2] Hedges of arborvitae and locust (*Robinia pseudoacacia*) were also planted, but had not yet attained maturity.

Egan had large beds of strawberries, including a 'Grove End Scarlet' variety that he brought to market. He had beds of raspberries and defied the winter with figs grown outdoors. He grew cranberries, a daring experiment in Chicago's alkaline soil, "hilled up like corn upon a sandy prairie soil." Like gardeners everywhere, Egan was plagued with pests and complained, "I lost all by the crickets and grasshoppers, whose taste was so good as to steal every berry even before my face." Still, he predicted, "like our other indigenous fruits, time, cultivation, and enlightened attention to their peculiarities, will develop them, and they will occupy a very prominent place amongst the luxuries of the garden."[3]

Egan's grounds were ornamental as well as fruitful. In the late 1840s, before bedding out became a widespread practice, his flower gardens were a cheerful mix of rare and favorite plants. One critic offered a mild rebuke: "Of flowers we saw no lack—arrayed in large circular beds, cut up into various devices. The plan here adopted is to grow them open and separately, instead of in masses, a mode—the latter—which we prefer."[4]

Egan's interest in gardens evolved from cultural practicalities to a heightened sense of design. In 1853, he traveled to England to tour English gardens, and brought back new horticultural ideas and techniques. He became a great admirer of Andrew J. Downing, spearheading a movement among local horticulturists to erect a commemorative monument upon Downing's death. In the early 1850s, he began improvements to a large tract of land south of the city, which he planned as a public park to be ultimately converted to his personal retreat. John Kennicott described the grounds in 1853: "Dr. Egan's place is the show place of Chicago, and I doubt [not] that his new 'Downing Place,' of 400 acres will prove a credit to the city, and the man. The doctor has planted, or intends planting, 20,000 evergreens there—and all for ornament."[5]

Downing Place, later renamed Egandale, was located in present-day Hyde Park and extended roughly from 47th Street south to 55th Street, between Cottage Grove and Woodlawn avenues. A "porter's lodge" was sited at the northeast entrance, and from there, a winding drive meandered through a pleasure garden. According to one account, "When its construction was first undertaken, the site was nothing but prairie land and Dr. Egan hauled carloads of evergreens and deciduous trees to the place, and there had them planted, according to the most approved method of landscape gardening."[6] The southwest corner, near 55th and Cottage Grove, contained a horse racetrack, where the celebrated Flora Temple once ran. There were two ornamental mounds: one crowned with a rustic arbor and another serving as an observatory. Nurseries were planted on the southern end of the grounds. According to A. T. Andreas, "the prominent plan of Egandale, was that of the domiciliary estates of large landed proprietors of Great Britain."[7]

Egan never achieved his manorial estate but, as a real estate speculator, might have appreciated the controversy his land caused. Following Egan's death in 1860, Egandale was opened to the public about 1863. Four years later, Ezra McCagg promoted a bill to create a public park that would subsume Egandale, but the Egan estate, now controlled by outside interests, fought the bill.[8] "The Egandale interests wanted a Park, but wanted no part of Egandale taken. They wanted Egandale to front on the Park, all around or on as many sides as possible." Ultimately, two strips of Egandale from Cottage Grove and along 55th Street were ceded to the park. By 1869, the *Chicago Tribune* was advertising "30 Elegant Lots in Egandale—50 x 160 each" to be sold at auction."[9]

William C. Egan, son of W. B., inherited his father's love of gardening, but was a late bloomer. He explained, "I am garden-bred, for in the early fifties my father's garden was one of the show-places in Chicago; but I have no recollection of a fondness for gardening during my youth."[10] A self-avowed amateur gardener, W. C. picked up the love of gardening for recreation and health after thirty-five years in business. He bought his country home, also named Egandale, in Highland Park, circa 1885, and as he confessed in his 1901 essay, "How I Built My Country Home," "I had the place, but no knowledge of how to develop it. Flowers, shrubs and trees did not grow among my business affairs. Nevertheless, I was determined that the place should be of my own creation, and so I resolved to go ahead and make my own mistakes in my own way."

With charming self-deprecation, Egan proceeds to describe the key design errors he made—"uncouth" curves in roads, a "scarecrow" of a rockery, a lawn littered with exotic plants such that "my man got dizzy dodging them with the lawnmower," and "stork-like" trees placed unattractively throughout. Through reading and experimentation, Egan soon rectified his mistakes and made his own Egandale a showplace of the North Shore. His advice was widely sought, and he published articles, served as a garden judge, and patronized a number of horticultural society functions.

This rare view of W. B. Egan's family and garden shows the European influence of manicured shrubs and garden statuary. Courtesy of Chicago Public Library, Special Collections.

1. "The Garden and Grounds of Dr. W. B. Egan," *Prairie Farmer*, October 1847, 308.

2. "Chicago Gardening," *Prairie Farmer*, August 1849, 236.

3. W. B. Egan, "Cranberries," *Prairie Farmer*, May 1850, 144.

4. "Garden and Grounds of Dr. W. B. Egan," 308.

5. John A. Kennicott, "Notes of a May Tour down South," *Prairie Farmer*, July 1853.

6. A. T. Andreas, "Egandale," in *History of Cook County Illinois, from the Earliest Period to the Present Time* (Chicago: Lakeside Press, 1884), 536–37.

7. Ibid., 537.

8. According to Andreas Simon in *Chicago: The Garden City* (Chicago: Franz Gindele Printing Co., 1894), 40, those interests included the Drexels of Philadelphia and the Smiths of Chicago.

9. *Chicago Daily Tribune*, May 9, 1869.

10. William C. Egan, "How I Built My Country Home: A Concrete Example of Landscape Gardening," in *How to Make a Flower Garden*, ed. L. H. Bailey (New York: Doubleday, Page and Co., 1914), 323.

LEGACY LANDSCAPES: HERITAGE HOMESTEADS

Regrettably, there are few homes or businesses remaining from Chicago's earliest gardening pioneers. With Chicago's transient population, the push ever westward, and omnipresent land speculation, very few homesteads remain. Sometimes a home might be preserved, but often the grounds have been parceled out to subdivisions, and the plantings are mere shadows of what they once were. Through perseverance of preservationists, and in some cases, serendipity, a few pioneer landscapes remain. While time has changed many aspects of these gardens, a visitor can still envision the footprints of our heritage horticulturists.

The Grove: Home of an Early Nurseryman

The Grove
1421 Milwaukee Ave
Glenview, Illinois
http://www.glenviewparkdist.org/fa-grove-info.htm

Groves of trees were extremely valuable assets on Midwestern homesteads in the 1800s. The trees not only provided fuel for the wood stove, but also lumber for the home and outbuildings. On flat prairies surrounding Chicago, groves also served as windbreaks, protecting field and fireside. One of Chicago's foremost horticulturists, Dr. John A. Kennicott, staked his claim at the Grove in Glenview. Kennicott, a medical doctor and nurseryman, spearheaded many of the early city and state horticultural associations. Kennicott's son, Robert, was one of America's most famous naturalists. The Kennicott family home has been rehabilitated and interpretive gardens installed. Another home on the campus once belonged to noted naturalist Donald Culross Peattie. Now a national historic landmark, the Grove has historic homes, a nature center, interpretive center, and wooded trails.

This vintage view of the Grove homestead shows the majesty of the surrounding trees and siting of the home on a rise. Today the home is planted with a kitchen garden and less lawn, more likely reflective of the original. Courtesy of the Glenview Park District.

When John Kennicott moved to Chicago in 1836, the ink was barely dry on the Great Indian Treaty. The Chippewa, Ottawa, and Potawatomi tribes had given up their land, but conditions were still primitive. Kennicott, a physician and horticulturist, settled here to join his parents and five of his fourteen brothers and sisters who preceded him from the

East. Known locally as the Old Doctor, Kennicott established his home and nursery at the Grove in what is now Glenview.

Kennicott's Grove was on high land and bordered the Milwaukee Stage Road. Here on 40 of his total 230 acres, Kennicott established one of the first nurseries in northern Illinois. The Grove nursery garden was on the northwest corner of the Kennicott acreage. An ad for the Grove in an 1854 issue of the *Prairie Farmer* declared:

> We have a fine lot of Budded and Grafted APPLE TREES, at 12<1/2> cents each, or $10 per 100, for 2 and 3 years old and double these rates for 5 and 6 years from bud and graft . . . A very large stock of healthy PLUMS, 50 cents each, (McLaughlin $1) or $37.50 per 100 to dealers . . . A fair, but not large supply of other fruit trees, and abundance of Evergreen and deciduous Ornamental trees. Hardy Roses and other Shrubbery, Herbaceous Perennials, choice Bulbs . . . We have Weiglia [weigela] roses . . . and fifty new sorts of Remontants, Hybrid China . . . now offered for the first time. Catalogues will soon be ready.[120]

The nursery was a commercial venture, but Kennicott was passionate about sharing his knowledge with his countrymen. He was known to give substantial price breaks to neighboring nurserymen who were just starting their businesses. He wrote prolifically about proper horticultural care and selection of plants for the Chicago region in the *Prairie Farmer* and other journals. Kennicott urged his fellow farmers to try native species, and enlightened them about the difference between prairie soils and the acidic soils of the East and South.

In addition to writing for many agricultural papers, including editing the *Prairie Farmer* for awhile, Kennicott was an officer or founding member of many early horticultural organizations. The CHS, Northwestern Fruit Growers Association, and Illinois State Horticultural Society are just some of the groups he led. He was endorsed by the Illinois state legislature and leading horticulturists to lead the newly established U.S. Department of Agriculture. While he did not receive this appointment, due to ill health, he was most influential in the department's origin.

Like many early horticulturists, Kennicott was a leader in civic and cultural affairs. He was an outspoken proponent of the Morrill Land Act, which resulted in the establishment of the University of Illinois. During the Civil War, Kennicott mustered up a regiment, and offered encampment space and facilities at the Grove. An 1861 issue of the *Prairie Farmer* noted, "The Old Doctor writes us . . . about twenty *Prairie Farmer* boys recruits . . . and says, 'please tell the boys up this way that the "Major's" company is not yet full, and that the Old Doctor's latch string is out, and beef, bread and good coffee in abundance.'"[121]

As one of the few practicing physicians in the northern Chicago area, he traveled a wide circuit, making house calls as far north as Waukegan and Lincolnshire and as far west as Elgin. He used his home as a laboratory for new plants. One visitor noted that many kinds of unusual shrubbery were planted outside his dooryard, so that Kennicott could observe at close hand their growth patterns. The Kennicott farmhouse was situated high atop a hill. Built in a simple cottage Gothic style, it had outbuildings nearby for chickens and other farm animals. Probably to accommodate the roving habits of the farm animals, the Kennicott personal flower garden was located farther from the house. A description of this garden was written in 1846:

> The flower garden [at the Grove] lies in front of the house, from which it is separated
> by a green lawn, beautifully shaded by some noble oaks which did their early grow-
> ing when the land was the abode of Indians, wolves and pinnated grouse and which,
> in a most un-Yankee-like manner, the proprietor has suffered to remain. These are
> joined by numbers of the black locust of astonishing thriftiness—to which are
> added a few of the ailanthus, catalpa, and mountain ash. Syringas, tall growing
> roses, and a few other flowering shrubs, give agreeable variety to this beautiful plat
> of ground. The garden slopes from the house toward the street, so that its blossoms
> are seen at a great distance. The list of roses comprises upwards of 70 varieties, a
> majority of which are now in bloom.[122]

The garden did not owe its beauty to roses alone. Herbaceous plants of great variety were
in blossom, including a large rose-scented peony, Canterbury bells, larkspurs, a variety of
foxgloves, phlox, penstemon, spirea, tradescantia, sweet william, and more.

Landscape Integrity

John Kennicott's house has been restored and sits behind a broad expanse of front lawn
as it did in the 1850s. Outbuildings, including a wood shed and chicken coop, are in back
of the house, surrounded by gardens interpretive of the 1850s. Miles of footpaths wind
around forest and ponds to give a flavor of the untouched wilderness. While the plantings are interpretations, the spatial relationships of the home to the land are accurate.

Thanks to the efforts of the Save the Grove Committee, in 1976 the Grove was declared a national historic landmark. Now with more than 120 acres of prairie, wetlands,
oak woodlands, and many varieties of flora and fauna, the Grove includes an interpretive
center and greenhouse. The Redfield Center, a 1929 home built for naturalist Donald
Culross Peattie and his wife, Louise Redfield Peattie (great-granddaughter of John Kennicott), is also part of the complex.

In 1995, forty-two additional acres of the original Kennicott property were purchased.
This land, north of the homestead, adjacent to the Tri-State tollway, was once part of the
Kennicott nursery. After crews spent months clearing out an undergrowth of buckthorn
and other invasive plants, hundreds of daffodils, narcissus, peonies, and roses thought to
date from the original Kennicott nursery began to bloom. Visitors can stroll along the old
nursery on walking paths. The Kennicott family horticultural tradition continues with
Chicago-based wholesale florists, Kennicott Brothers, operated by a descendent of John
Kennicott. The company was recently instrumental in obtaining thousands of peonies
from the original Kennicott stock. These flowers have been planted amid the lawn areas
as Kennicott might have done.

Restoration is under way on the gardens around the Redfield estate, home to the
Peatties in the 1930s. There were once over twenty acres of formal gardens around the
Redfield estate, including a large reflecting pool, which has now been uncovered. Thus,
the Grove's prairies, woods, and cultivated gardens shaped the lives not only of a famous
physician and horticulturist, but also two famed naturalists, living decades apart. With
flowers growing there today descended from original species, the Grove is a wonderful
time capsule of Chicago's horticultural history.

Wilder Park: A Woman's First Garden

Wilder Park
175 Prospect Avenue
Elmhurst, Illinois
http://www.epd.org/parkpage.asp?park=22

LEFT At the Henry King home, shown here looking at the west façade, Louisa King learned gardening from her mother-in-law. Ca. 1891. Reproduced by permission of Elmhurst Historical Museum.

RIGHT The Wilder Park conservatories, one circa 1890 and one circa 1920, are adjoined and hint at the gardening exploits of Louisa King. Ca. 1920. Reproduced by permission of the Elmhurst Historical Museum.

Once the suburban villa of an early Chicago industrialist, Seth Wadhams, White Birch, as it was then known, typified the Chicago country home of the late 1800s. Of particular importance, this home became the training ground for Louisa King, a pioneer of women gardeners and garden writers. King apprenticed under her mother-in-law, who then owned the property and who cultivated an extensive flower and herb garden. She later attained prominence for her writings and leadership in national horticulture organizations.

It is not surprising that Louisa King found fertile ground in Chicago to become a renowned gardener. Chicago women have a long and influential history in Chicago's landscaping legacy. Unfettered by the class distinctions of their East Coast or southern counterparts, and pressed into jobs usually taken by men through sheer necessity, pioneer Illinois women felt they earned a voice in many horticultural endeavors. They were frequent contributors to local papers like the *Prairie Farmer*, and, through the powerful Chicago Women's Club, launched many early environmental activities.

The typical leisure-class Victorian lady was little more than a hothouse flower herself, more apt to speak the lyrical Language of Flowers than soil her hands in down-and-dirty gardening. In Illinois, still considered the Wild West by some Victorians, women gardeners were more likely by necessity to put shoulder to plow. Still, their parlor gardens often outshone their prairie plots.

Louisa King helped American women become more than mere garden ornaments. Founding member of the Garden Club of America (1913), first president and founder of the Women's National Farm and Garden Association (1914), and prolific writer, King began her gardening career right outside of Chicago. Often associated with the English gardening luminary Gertrude Jekyll, King frequently corresponded with Jekyll, and the two women had great respect for each other. King's books stress the concepts of planned garden designs and artistic use of color in plantings—ideas that are synonymous with Gertrude Jekyll. Both Jekyll and King, on opposite sides of the Atlantic, had a "down to earth" approach to gardening that was useful both for the cottage garden and lavish estate.

The daughter of a New England clergyman, King moved to Elmhurst, Illinois, as a new bride in 1890. For the next ten years, she honed her gardening skills on the estate now known as Wilder Park. King credited her early interest and talent in gardening to her mother-in-law, Aurelia Case King, and dedicated her last book, *From a New Garden*, to her.

Aurelia King had an extensive formal garden at their Elmhurst estate, including an herb garden with over two hundred varieties. This home of Henry and Aurelia King, Louisa King's in-laws, was where Louisa served her gardening apprenticeship. She drew inspiration from this garden and from her mother-in-law's practical advice and extensive library.

When the senior Kings inherited the estate in 1888 from wealthy businessman Seth Wadhams, it was known as White Birch. Many imported species of trees had been planted to beautify the grounds. Like other estates of the era, White Birch, originally farmland, was enclosed by tall hedges of arborvitae. Aurelia established her legendary flower and herb gardens on this beautiful property. Louisa lived nearby and later, with her husband, inherited the estate.

A gardener's cottage was on the southwest corner of the estate. Cow and sheep pastures were established on the northern part of the estate. Although these pastures and the cottage are long gone, the original greenhouse used by Louisa and Aurelia King still stands at the south end of Wilder Park. Built even before Aurelia King inherited the house, this greenhouse is thought to be the oldest in DuPage County. An additional conservatory and a second greenhouse was added by the Elmhurst Park District in the 1920s.

From these modest beginnings, Louisa King became one of the foremost authorities on gardening. King's writing career spanned over fifteen years. Her first book, *The Well Considered Garden*, debuted in 1915. Her books are written for the layperson, and focus not so much on horticultural dos and don'ts, as on pleasing arrangements of garden design.

King left a legacy that included numerous books, articles, and speeches. Her impact has not gone unnoticed. In Washington, DC, the National Arboretum features extensive plantings in a memorial garden, the Mrs. Francis King Dogwood Garden. In 1937, a beautiful white rose was patented and introduced bearing her name. And in Chicago, vestiges of King's early Elmhurst years help us picture the era when women came out of the hothouse and into the garden.

Landscape Integrity

The present-day Wilder Park, now owned by the Elmhurst Park District, looks vastly different from the old White Birch estate. The conversion from private home to public park has necessitated changes. Although Wilder Park today has lovely landscaping, parking lots have replaced the large trees and shrubs near the house itself. None of the original birches that gave the estate its name remain.

The most intact feature is the original greenhouse circa Louisa King's time, which is delightful in its simplicity and utilitarianism. An adjoining conservatory was added in the

1920s. The old mansion, undergoing renovation, remains in its original site, so visitors can appreciate the relationship between it and the greenhouse, and imagine the garden walks between. The function of Wilder Park as a complex of cultural buildings naturally makes circulation and traffic patterns different than that of a private residence.

The garden to the south of the greenhouse, Elizabeth's Friendship Walk, is a very recent addition that pays tribute to Seth Wadhams's wife, Elizabeth, herself a garden lover. It is not a garden restoration, but is meant to be indicative of the era. A large garden ornament to the east of the greenhouse is of further historical note. This is one of the original urns that was outside of prefire Chicago's courthouse. The ornament was salvaged and moved to White Birch after the fire by persons unknown.

University of Chicago: Pioneering Ecology Lab

University of Chicago
The historic main quad is bounded by Ellis and University avenues, between 57th and 59th streets.
http://www.uchicago.edu/docs/gardbroNS/

In the early years, individual scientists combed Chicago's countryside to gather botanical information and share it through informal gatherings of academic societies or journals. Chicago's universities formalized and expanded the exchange of horticultural knowledge, and drew leading scientists of the day. Men such as John Coulter and H. C. Cowles, pioneers in the nation's study of ecology, established living laboratories here. This venerable campus has also benefited from the artistry of such stellar landscapers as Frederick Law Olmsted, O. C. Simonds, Olmsted Brothers, and Beatrix Jones Farrand.

The lush landscape in this early view of Botany Pond has been recently restored. From the author's collection.

When the University of Chicago held its first faculty meeting in 1892, Frederick Law Olmsted, in conjunction with the Columbian Exposition of 1893, was in the midst of designing the Midway Plaisance, a greensward that bordered the university's south side. That the university's beginnings coincided with preparations for the most influential world's fair on American soil hints at the many influences that shaped the campus buildings and landscapes. With a board of trustees that included most of Chicago's leading citizens, the university was never at a loss to find good landscape artists. From its beginnings at the turn of the century, through the 1930s, and beyond, many of America's leading landscape professionals left their impressions on the campus quads.

The first individual associated with landscape improvements was trustee Judge Daniel Shorey. Coincident with the 1893 Columbian Exposition, Shorey personally funded and had installed a double row of acacia and catalpa trees along the driveway south of the central quadrangle. He was also conservation minded, making an astute decision to preserve several existing black oaks on campus. May Thielgaard Watts, University of Chicago graduate in botany and famed naturalist, praised this decision even though the oaks, which outlasted many other trees, may have interrupted the overall landscape design.[123]

O. C. Simonds prepared and partially completed installation of a landscape plan (approved by the board) for the fledgling university at the turn of the century. By 1902, however, Olmsted Brothers was commissioned by the board of trustees to review Simonds's work and recommend changes. A variety of circumstances seems to have conspired against Simonds. An engineer working on underground steam tunnels rerouted some of the sidewalks Simonds had specified, causing inefficient and inelegant traffic patterns. Board members were also unhappy with the immature plantings and naturalistic style. The Olmsted firm spent two days examining the campus, and met with Simonds, university president William Rainey Harper, and the architects. They wrote a nine-page letter of recommendations. Although the Olmsted firm described this task of critiquing a colleague's work as "a very disagreeable duty," they did not mince words. While acknowledging Simonds's expertise for "certain branches of the art of landscape gardening," the Olmsteds commented, "We do not know that Mr. Simonds' education or experience has brought him into a thorough knowledge or appreciation of the motives of design controlling leading architects of the country . . . he has failed to fully appreciate the artistic requirements of the problem."[124] The Olmsteds' rebuke perhaps echoes the prevailing East Coast bias against Midwest design sensibilities.

Neither did the Olmsted report spare architect Henry Cobb's efforts. "Mr. Cobb's plan, which was primarily a plan for the disposition of buildings rather than for meeting all the requirements for drives and walks on the grounds . . . fell decidedly short of meeting all reasonable requirements of convenience in this regard."[125] The Olmsteds recommended a simple landscaping scheme, with straight walkways that would harmonize better with Cobb's buildings than Simonds's curvilinear walkways. They emphasized art and harmony—sure to please the report's recipient, board trustee Charles L. Hutchinson, who was also president of the Art Institute and founding member of the CHS.

Perhaps the most enchanting landscape element amid the formal quadrangles' subdued simplicity was Botany Pond, conceived by John Coulter, the university's first director of the Botany Department. Coulter, and his successor H. C. Cowles, were among the foremost names in botany. Cowles's seminal work on plant succession in the Indiana Dunes ignited the field of modern ecology and environmentalism. Both Coulter and Cowles were strong proponents of field-based learning experiences. They organized hundreds of nature-study excursions in the Chicago environs and around the nation.

Coulter's early efforts to build first-class botanical laboratories were not without struggle. In what contemporary accounts called a "war of roses" in 1898, Coulter argued with university executives for upgraded staff and facilities in the greenhouse on the roof of the Botany Building. To date, the conservatory was empty of any and all plants. Before acquiring collections representing polar, equatorial, and desert plants, Coulter insisted that a trained gardener—preferably from Europe—be hired to care for the greenhouse, whereas trustees argued that a janitor could do the job.[126] Coulter's next endeavor was the creation of an outdoor laboratory, Botany Pond, in the stately confines of Hull Court. Working with the Olmsted firm in developing the construction plans for the pond, Coulter was intimately involved in the planting specifications.

In February 1904, Coulter wrote a mildly critical note to university president William Rainey Harper, noting that steps had not yet been taken to fund Hull Court improvements. He proposed a $500 planting plan for the pond, with $250 of that paying five months' laborer's salary to help with the project. Plants would be donated from botanical gardens with whom Coulter had ties, with the balance sourced through the U.S. Department of Agriculture, nurseries, or collections of live specimens. He proposed a plant palette including 25–40 species of ferns, 60 species of shrubs, 40 species of aquatics, and 250 other plant species.[127] Even though the pond is not large, the great diversity of plant material shows in period pictures of the lush plantings around the pond. The pond was, and is, a naturalistic oasis amidst the structured formality of the quad.[128]

O. C. Simonds, Chicago's highly regarded naturalistic landscape gardener, was not out of the picture at the university. According to Richard Bumstead, the university's current university planner, Simonds was commissioned to prepare a landscape plan for Ida Noyes Hall, built in 1916.[129] In 1924, he also prepared a planting plan for Hull Court.[130] Whether this plan was implemented, or whether he collaborated with Olmsted Brothers (who continued their association with the university for years), is unclear. But in 1926, zinc labels were prepared for the Hull Court plantings—furthering education and interpretation of the outdoor laboratory. The plants included many Midwest favorites such as crab apple, common ninebark, cockspur hawthorn, fragrant sumac, Rosa setigera, and flowering plum. Favorite plants of the period such as spirea, lilac, and peonies added an old-fashioned touch.

During the 1920s, the university's property continued to develop, even as original landscaping matured. Since 1903, the board of trustees had delegated day-to-day responsibilities to a superintendent of building and grounds. From 1919 through 1953, Superintendent Lyman Flook coordinated activities related to landscaping, among others. His correspondence reflects an interesting period where rural technologies for landscape maintenance struggled to catch up with building expansion. It wasn't until 1924, for example, with the erection of some of the university's medical buildings, that the university barnyard had to be removed. This caused Flook to write a detailed proposal on relocating some large compost piles to a spot between Cottage Grove and Maryland Avenue on 59th Street, "planting it up with a screen of shrubs to hide it from the Midway and Cottage Grove."[131] Two years later, the end of an era resulted when Flook wrote, "I recommend that the Campus horse be disposed of. This horse is about 26 years old and of very little usefulness."[132] Flook compassionately proposed that the horse be sent to pasture or the Humane Society, and noted that renting a horse for winter and summer plowing and mowing would result in cost savings.

Even as post–World War I urbanism began to surround the university, it still retained undeveloped real estate, pending future expansion. The vacant areas were often leased to interested individuals or groups for crops or gardens. Flook appears to have recognized the value of these mutually beneficial arrangements, and the low-cost landscaping effect of borrowed views. One Miss Slye, for example, leased land near the new Lying-In Hospital that was built in the late 1920s. Flook interceded for Slye when it appeared that new construction would demolish her garden on university property. "In a conversation with Miss Slye last week, she did not yet understand that the Lying-In Hospital lease extended to include practically all of the garden to which she is so firmly attached . . . I think it would be very nice . . . that she be left undisturbed in this area. She has, in past years, taken such good care of vacant University property allotted to her that I think she deserves whatever consideration we can give her."[133]

The Botany Department continued to collaborate with landscaping efforts, particularly around new greenhouses built in the late 1920s. E. J. Kraus of the Botany Department wrote to Flook with specific preferences as to proposed plant material, often preferring native plants. "First, I consider Matrimony vine an absolute 'abomination unto the Lord,' and hope that not one sprig of it will be planted around the greenhouse,"[134] Kraus wrote, recommending instead *Rhus aromatica*. He also proposed *Rosa setigera* as an alternative for the Japanese barberry and pink flowering almond suggested in the plan forwarded by Flook.

For many of the smaller projects, such as the work around the greenhouses, Flook let contracts to a variety of local landscape architects. As major buildings were constructed before World War II, and as increased automobile traffic warranted better traffic patterns, the university sought the input of Beatrix Farrand, who had worked on other university campuses such as Yale and Princeton. While her advice for the University of Chicago's central quadrangle was not implemented, her plans for the Oriental Institute, International House, Eckhart Hall, and Burton-Judson Courts were completed.[135]

With just over two hundred acres today, the university campus has grown significantly since Marshall Field's first donation of ten acres of land. The campus landscape has brought together an interesting assembly of local and out-of-town landscape architects, each of whom imparted a different vision for the campus. Perhaps in landscaping, too, the university's motto unfolds: "Let knowledge grow from more to more; and so be human life enriched."

Landscape Integrity

The university is presently engaged in an ambitious campus master plan that addresses virtually all aspects of the buildings and grounds. Retaining open space and refurbishing many of the historic landscapes are stated goals of the master plan. Already under way, in partnership with the Chicago Park District and other groups, is the development of the Midway Plaisance, a historic landscape in Frederick Law Olmsted's overall design for Chicago's great South Parks (now Jackson Park, Washington Park, and the Midway Plaisance). Although this space never materialized as part of the interconnecting canal system envisioned by Olmsted, its sunken, green lawn is a local landmark. Its broad green space, heretofore dividing "town from gown," will include both passive and active recreation elements to join community residents and the university population. A skating rink, winter garden, and reader's garden are already installed. Although the Midway is owned by the Chicago Park District, the university has been among the key stakeholders in the design process. Noted Olmsted scholar Charles Beveridge consulted on the project and identified three key aspects that should be preserved in the new design: retaining the open vistas, formal plantings of trees along the streets, and sunken landforms.

The Olmsted firm's influence on the landscape is particularly apparent in the university's main quadrangle. The sunken Hutchinson Court, with its center fountain and four sets of stairs leading from the corners to the fountain, are part of the Olmsted design. (The Snell-Hitchcock Court has a mirror design, but is missing the fountain and two sets of stairways.) Within the graceful geometry of these courtyards, the Olmsted design provided multipurpose space: amphitheaters that were often used for small theater productions or ceremonies, and functional traffic patterns. The overall simplicity of the quadrangle landscape design—with carefully sited mature trees, the large garden circle at the center of the main axes, and subtle foundation plantings—retain part of the Olmsted vision.

Farrand is noted for her mastery of small spaces, and she does not disappoint in the various courtyards designed for the university. The most accessible courtyard, in the International House, features low hedges with terraced stone walls surrounding an octagonal pool and fountain. Small trees frame the perimeter. In the winter, the landscape is somewhat underwhelming, but when the trees leaf out, the height of the surrounding four walls disappears, and the garden is a sheltered cloister of simplicity. More paving has been added to Farrand's original design, including some rather unfortunate lava rocks replacing her specified colored gravel, but the essence remains.

Surely the prized landscaping element is Botany Pond, which remains in its original location in Hull Court. Calling the university "one of modern botany's most influential centers of research and teaching,"[136] America's Library of Congress Web site holds many of the university's environmental photographs. It is not, perhaps, a stretch to view this little pond as the birthplace of modern ecology. The pond has been recently rehabilitated with a mix of native and cultivated plantings meant to emulate the lush landscape of Coulter's era.

In 1995, a project was begun to establish a formal botanical garden on the campus of the University of Chicago. Two years later, the campus was admitted into the American Association of Botanic Gardens and Arboreta, recognizing the campus' landscape legacy and significant plant specimens.[137] Efforts are under way to label significant trees and plants. Still, there are many venerable trees that can be identified even without labels. Look for the old hawthorns in front of Rockefeller Chapel—planted in the 1930s from a Root and Hollister design. Many old oaks predating campus construction still remain in the Classics Quadrangle (59th Street and Ellis Avenue) and the Social Science Quadrangle (59th Street and University Avenue). The so-called World's Columbian Exposition oak—planted during that world's fair to add verisimilitude to a Native American exhibit—lies north of North Plaisance Street in front of the University of Chicago Hospital. The aralia trees near Hull Gate fence are a rare variety thought to be a legacy from John Coulter. Perhaps he would be pleased that the university is carrying on his tradition of outdoor laboratories.

THREE

PLANTING THE PRAIRIE
Defining the New Plant Palette

Settlers arriving in Chicago by prairie schooner or steamship carried precious cargo: carefully nurtured jars of seed or plant cuttings from their old homesteads. These sentimental mementos of heirloom gardens should probably have been left behind, because Chicago's climate and soil conditions were inhospitable to many nonnative plants. By necessity, early gardeners not only discovered new plant species, but new landscape uses for plants to accommodate the Midwest's often hostile climate. Through individual experimentation, and collaborative efforts in horticultural societies, Chicagoans not only created their own plant palette, but also greatly contributed to the nation's understanding of plant culture and varieties.

PLANT TREES!

Nineteenth-century author A. T. Andreas prominently featured a full-page picture of "Chicago's most historic tree" in his tome, *History of Cook County, Illinois*. Whether this is indeed Chicago's most historic tree is debatable, but there is no doubt about the critical importance of trees to Chicagoans. On the prairie where trees were scarce, they were vital landmarks and often provided essential fuel and shelter. After several decades of lumbering and plowing prairie and oak savanna for farms, Chicagoans became leaders among national reforestation advocacy groups. This environmental mindset developed gradually, after years of experimentation with trees on the prairie.

First and foremost, trees meant survival. For 1830s settlers, trees were valued for timber that provided raw material for log homes, wood-burning stoves, fencing, and tools. Later, farmers recognized the commercial value of orchards, and so the importance of fruit-bearing trees grew. Shade and windbreak trees were also needed on the windswept prairie. Lastly, as the population grew wealthier and living conditions more favorable, trees became prized for their ornamental qualities. Prospective settlers in the Midwest had grave concerns about the ability of prairie soil to support timber culture. An 1840s writer for Illinois's Union Agricultural Society observed, "The chief objections [among potential emigrants] to locating in Northern Illinois appear to be its alleged destitution of timber."[1] Thanks to discoveries by Chicago-area horticulturists, settlers learned that trees could indeed be grown on the prairie. Extensive work by Chicagoans in proper tree selection and culture helped future settlers.

This cottonwood tree, on the South Side at 18th Street between Prairie Avenue and the lake, was the alleged site of the massacre of 1812. Andreas, in his book on the history of Cook County, printed a testimonial from Aurelia King, mother-in-law of future gardener Louisa King, attesting to the historic value of the tree. From Andreas, *History of Cook County.*

At first, choosing the right tree for the right site was a matter of trial and error. M. L. Knapp, a physician from Bloomington and frequent Chicago visitor, recommended trees for timber and fruit in the 1843 *Prairie Farmer*. He praised the black walnut (*Juglans nigra*) as "the most useful timber tree in Illinois," noting its suitability for furniture and fences. He also cited white walnut, or butternut (*Juglans cathartica*²), for its overall usefulness; the bark was a good dye, and the root, as an extract, was a good physic. Native honey locust, black locust, catalpa, and Kentucky coffee trees were cited as good for timber and ornament. Native red mulberry (*Morus rubra*) was also identified as a good timber and fruit tree. For those seeking evergreens, Knapp named many pine species as good timber sources: white pine (*Pinus strobus*), Scotch fir (*P. sylvestris*), silver fir (*P. picea*), hemlock (*P. canadensis*), Norway spruce (*P. abies*), Oregon fir (*P. Douglassii*), white larch (*P. larix*), black larch (*P. pendula*), red larch (*P. microcarpa*), and cedar of Lebanon (*P. cedrus*).³ Knapp predicted that all but the Scotch fir would grow well on our prairies, although some might do well in river-bluff habitats.

By the late 1840s, homesteaders became more familiar with trees and their potential for lumber, prompting the *Prairie Farmer* to raise the question of forest conservation. As early as 1849, the *Prairie Farmer* noted, "To talk of the culture of forests in this region of prairies may seem singular; but we know of no subject more apropos nevertheless. To provide for the future wants of the country, in respect to firewood, fencing and building . . . should engage more attention than it receives."⁴ The article persuaded readers that they would indeed reap benefits from tree planting in their own lifetime—despite popular opinion that growing trees took too long. Readers were encouraged to scout their properties for "little patches of shrubs" that, without prairie fire or animal grazings, could be cultivated into trees. Crops of cottonwood and poplar, while admittedly poor trees, were cited as good stopgaps until better timber trees, such as locust (*Robininia pseudoacacia*)⁵ could be grown. Yet, even in the 1850s, proper tree culture was still far from mature. Jonathan Periam recalled, "Then the country was new indeed, with homesteads scattered only here and there . . . The roads were mere tracks across almost boundless prairies, with no timber, except there and there groves that had gotten a foothold upon rough and broken ground or along the margins of streams. Hardly an orchard could be seen in a day's journey; our great cities were mere villages or hamlets; and only the beginnings of one railway."⁶

Choosing suitable trees for our climate was a major challenge. John Kennicott observed in an 1860 lecture before the state horticultural society, "It is a great mistake to suppose that a sort [of fruit] good in New England, or even New York and Ohio, must be good here. It is often good for nothing." Homeowners were prey to false claims of traveling salesmen. These men, derisively called "tree peddlers" by western horticulturists, typically hailed from eastern nurseries and sold stock door-to-door. Fledgling western nurseries, which undeniably fostered a self-interest in disabling the competition, lobbied to discredit and disparage eastern products. Except for representatives from well-established eastern nurseries, there probably was merit to claims that tree peddlers sold "worthless sticks." Their stock was often unsuited to prairie conditions and was not properly packaged for widespread distribution. Angered by the poor public relations caused by tree peddlers, in 1867 professional Chicago-area horticulturists passed this resolution:

> WHEREAS, The progress of horticulture has been very much impeded by tree peddlers, through their ignorance and dishonesty in selling trees untrue to name, at enormous and unusual prices, sometimes delivered in cold, freezing weather, by which farmers and amateur fruit-growers have been discouraged and disgusted, therefore
>
> RESOLVED, That this Society, as a body of fruit-growers, farmers, legitimate nurserymen and others, do emphatically denounce the business as empirical and as injurious to the best interests of horticulture and the prosperity of the state.[7]

Despite this pronouncement and other public campaigns, upstart western nurseries fought a long and hard battle to establish their own credibility vis-à-vis established eastern firms.

Timber Trees and Conservation

Northern Illinois horticulturists were among the first to formally recognize the potential devastation of our nation's forests. As early as the 1860s, Chicago horticulturists were sounding the alarm about tree conservation. In their horticultural society meetings, Jonathan Periam noted that forward-thinking European countries were already engaged in widespread forest replanting. The U.S. eastern seaboard, he noted, had observed unsettling climate effects due to denuded forests, including dry streams and unhealthy orchards. He warned, "The time has come when the farmers of the great West must begin to plant forests for their own salvation from great evil and inconveniences." He urged railroad companies, themselves major timber consumers, to plant trees along railroad lines and depots, and exhorted individual farmers to plant trees in their fields.[8]

In 1868, several essays on conservation were presented at the Illinois State Horticultural Society's meeting in Freeport, and extracts of these essays were published in leading agricultural papers in the Northwest and the East. In 1870–71, the society appointed a committee to present to Congress a plan for greater timber planting and cultivation. "Pursuant to these instructions, bills were introduced; and the result has been that a stupendous system of timber planting has been adopted by Government, through the agency of the pioneer settlers of the widespread plains and prairies of the West,"[9] Elgin nurseryman D. C. Scofield wrote. Subsequent efforts sought to repeal tariffs on foreign lumber imports, but according to Scofield, "through the agency of the lumbermen of Chicago, one of their number being a representative there, the movement was defeated. Then and there the great interests of the nation were brought to bow to the lumber princes of Chicago."

Chicago thus became the locus of heated debate pitting lumber barons against vocal nurserymen. William B. Ogden, for example, a great advocate for Chicago horticulture, was also invested heavily in lumber.[10] Ogden's lumber holdings included interests in Chicago as well as northern Wisconsin. His railroad interests, which included the Chicago and Northwestern, the Illinois and Wisconsin, the Buffalo and Mississippi, and the Union Pacific, depended on a prodigious lumber supply. Even though Ogden and many others like him were strong advocates of horticulture, their business investments in timber often ran counter to nascent conservation efforts.

Scofield disputed lumbermen's counterclaims that forests were safe. He noted that the Chicago market processed thirty-five times as much lumber in 1873 (1.1 billion feet of lumber and over 500,000 wood shingles) versus that of 1847. "Nay fellow citizens, the timber famine is at our doors," Scofield warned, "and nothing but a timely relief from the Government placing lumber on the 'free list,'

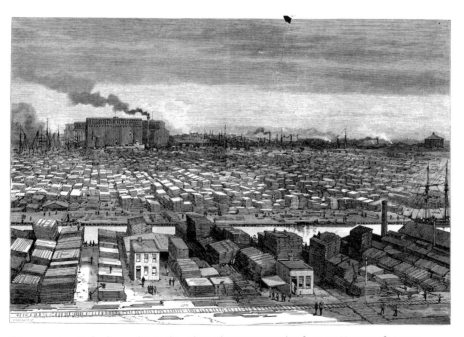

Chicago was a major lumber market, as shown here. This engraving, taken from an 1883 issue of *Harper's Weekly*, shows the extent of the lumber district on the southwest side. Horticulturists, as harbingers of the environmental movement, sounded early warnings about the destruction of Midwest forests. From the author's collection.

and encouraging and urging the immediate planting of extensive plantations of pine, will save us from the impending ruin to which we are hastening."[11]

The terrible fires of 1871 destroyed an estimated ten-year supply of timber from Illinois, Wisconsin, and upper Michigan. Jonathan Periam argued that unrestricted lumbering had actually contributed to the fires: "There can be no doubt that these terrific conflagrations, with all the attendant loss of human life, were caused partly by . . . man himself, in clearing up the country." He believed that the only way to convince people to plant trees was to demonstrate the economic benefits of reforestation. With pine lumber selling in Chicago's market at $55.00 per thousand feet, Periam asserted, "Let the money value of this enterprise be assured and its future is safe." He believed horticulturists should be leaders in the conservation movement. "The old West is passing away. Neither the farmer nor the horticulturist is responsible for the conditions precedent, not altogether responsible for the conditions present. They found the country lacking timber but clothed with rank perennial verdure . . . As horticulturists, practicing the higher art, we were the first to see the evil, and the first to take the necessary steps to remedy the evil, by propagating and planting timber trees. What we want now is cooperation of the government if possible, but especially of corporations."[12]

On September 10, 1875, a group of concerned citizens met at Chicago's Grand Pacific Hotel. Originally called the American Forestry Association, this group evolved into American Forests, one of America's oldest and still extant conservation groups. Robert Douglas, Waukegan tree nurseryman, was elected the first president, although he declined the honor in deference to a preeminent horticulturist from Ohio, John Warder. Warder had called the meeting in Chicago to capitalize on the opening of the city's exposition of 1875 and concurrent meeting of the American Pomological Society. On the American Forests Web site, historian Henry Clepper called the birth of this organization "a turning point in history. It inaugurated the conservation movement."

In 1876, the ever-vigilant D. C. Scofield proposed that all nurserymen belonging to the Horticultural Society of Northern Illinois give each customer one tree, to be known as the centennial tree. At the same meeting, one Mr. Williams of western suburban Batavia suggested that the society adopt a day as Arbor Day, when every family in the district would plant a tree. This was more than a decade before the State of Illinois passed legislation to adopt Arbor Day as a state holiday in 1887, joining other progressive states. Scofield's advice and his colleagues' legislation were important, but it was Arbor Day, the brainchild of J. Sterling Morton, that put the spotlight on tree planting. By organizing schoolchildren and community groups to plant trees, Morton captured the public's heart. He said in 1887, "Other anniversaries stand with their backs to the future, peering into and worshipping the past; but Arbor Day faces the future with an affectionate solicitude . . . and etches upon our prairies and plains gigantic groves and towering forests of waving trees."[13] The U.S. government agreed. An 1896 U.S. Department of Agriculture missive noted, "In our own country, [Arbor Day] promises to do more than anything else to convert us from a nation of wanton destroyers of our unparalleled heritage of trees to one of tree planters and protectors."[14] J. Sterling Morton's son, Joy, would carry on the family tradition and establish a world-class arboretum near Chicago in the 1920s, further championing the Morton family motto, "Plant trees!"

Shade and Windbreak Trees

Soon after the most urgent needs of shelter had been provided, Chicago homesteaders turned to trees for comfort and protection. Trees screened the open prairie from bitter wintry winds and crop loss due to summer storms. Trees offered much-needed shade in the hot summers around prairie homes and city streets. In 1847, *Prairie Farmer* editors sought the best shade tree for the Chicago area. They recommended virtually any maple tree (except the sugar maple, whose growth was deemed too slow). Elms and ailanthus (now known as the invasive tree of heaven) were suggested, with caveats to site them properly for adequate room to grow, and balsam fir, white pine, and arborvitae were recommended evergreens.

"Set out shade trees!" urged the *Chicago Magazine* editors in 1857. "No man has surely done his duty until he has planted one tree to cast a pleasant shadow after him. Set out trees."[15] Shade was often the first luxury sought, after trees were used for shelter and fuel. The next issue addressed was wind protection, and for windbreak, Jonathan Periam's top picks of deciduous trees in the 1860s were fast-growing yellow and white willows, white or ash-leaf maple,[16] and Lombardy poplar. Deciduous trees were cheaper than evergreens, and Periam prescribed three rows for an effective screen. Homeowners could profit from these plantings as much as "the same amount loaned at twenty-five per cent per annum," according to Periam. By-products of windscreen trees, such as syrup from the maples or timber from other trees, produced some profit. Additionally, windscreens would protect a homeowner's investment in orchard crops, since "the scathing effect of the prairie winds upon our fruit trees are matters of common observation."[17]

Horticulturists were aware that trees and forests positively influence the climate—reducing the extremes of temperature and attendant drought and storms. One interesting theory held by Periam and his colleagues proposed that trees actually *caused* rain showers. Periam observed, "The clouds are attracted by many millions of leaf points, and will follow this attraction unless it is overpowered by strong air currents."[18] Thus, trees became a settler's version of a rain dance. If rain dances were unsuccessful, at least shade trees made life on the plains infinitely more comfortable and imbued a sense of permanency. Jonathan Periam wrote, "To enjoy the shade of trees in the West, we must plant them. If we would seek the shelter of the woods at noon-day, we must make it. Would we leave the noblest heritage to our children that the Western farmer can,—a growing grove of timber,—all that is necessary is, each spring, plant trees! PLANT TREES!"[19]

Evergreens

Early efforts at transplanting evergreens in the Midwest were only moderately successful. An 1849 issue of the *Prairie Farmer* emphasized proper timing for transplanting. It also debunked a popular myth that evergreens shouldn't be pruned upon removal, proposing instead a trimming of one-half to two-thirds of the branches' length. Still, people were willing to experiment because thrifty evergreens were highly prized.

Evergreen trees were valued as windbreaks in Chicago, and for providing color during gray, long winters. Once considered challenging to grow, evergreens had achieved greater popularity by 1859. According to John Kennicott in an 1859 *Prairie Farmer* article, "Our people of the prairies, are beginning to show their good taste, by planting evergreen trees, and their good sense by selecting liv-

---

NORWAY SPRUCE: A MULTIPURPOSE TREE

The nonnative Norway spruce (*Picea abies*[1]) was recommended by many leading Illinois horticulturists for its multifunctionality as windbreak, timber source, and ornamental. Native to northern and central Europe, the Norway spruce was brought to America by early European settlers. Widely planted across America even today, the tree was appreciated by Chicagoans because it was fast growing, adaptable to clay soils even though it prefers acidic, readily planted from seed, and generally easy to transplant, according to Arthur Bryant. As a windbreak, the Norway spruce was unsurpassed—Bryant estimated its potential height as 120–150 feet (although current estimates such as on the University of Illinois's extension Web site suggest a smaller tree of 40–60 feet). It is easy to reconstruct the vastness of the prairie and correspondingly large tracts of farmland bordered by such majestic trees.

Bryant described the wood from Norway spruce as "strong, light, and elastic." In the 1870s, milled planks were sold under the name of White Deal, and used interchangeably with white pine. An acre of timber plantations of Norway spruce could support 640 trees.

For ornamental purposes, homeowners often used Norway spruce as a specimen tree, although its great size and formal appearance was off-putting to some. Some gardeners overlooked the ultimate size of the tree—Bryant often noted "three or four Norway Spruces growing on a cemetery lot twenty feet square."[2]

1. Arthur Bryant Sr., in his book *Forest Trees for Shelter, Ornament, and Profit* (New York: Henry T. Williams, 1871), 179, used an earlier synonym, *Abies excelsa*.
2. Ibid., 181.

ing specimens, from the nursery, instead of dead ones, on the street, or from peddlers' wagons." Kennicott praised the native white pine and hemlock, but also noted that some foreign trees, like Norway spruce, Austrian pine, and Scots pine, were quite successful. He decried the balsam fir, which "shows its old age at twenty, and is a dingy decaying eyesore, at forty." Arborvitae was the cheapest of evergreens, he noted, and most suitable in windbreak belts. Jonathan Periam endorsed Norway spruce as the best evergreen windbreak tree. In 1863, spruce cost twenty dollars per hundred for two-and-a-half-foot-high specimens. The trees were to be planted in rows, ten to fifteen feet apart, with trees in the second row opposite the spaces in the first. Ultimately, Periam noted, these rows would interlock, forming "an almost impenetrable and beautiful wall of foliage, which will effectually check the fury of our most violent storms."[20]

Robert Douglas of Waukegan was the acknowledged evergreen expert of the time. An 1867 *Prairie Farmer* reported, "It will, no doubt, surprise a great many well informed people, West and East, to learn that the most extensive planta-tion of evergreen seedlings in the United States, lies within thirty miles of Chi-cago."[21] Douglas reportedly had upward of five million seedlings growing on his property, comprising over thirty varieties. Having spent more than twelve years propagating native and foreign evergreens from seed, Douglas focused on Nor-way spruce, white pine, and Scots pine. A few years later, Douglas's expertise was acknowledged by Arthur Bryant Sr., nurseryman from Princeton, in his 1871 book, *Forest Trees for Shelter, Ornament, and Profit*. Bryant based this seminal com-pendium of tree culture on his own significant work and on shared knowledge from horticultural societies. Evergreens in particular, he felt, were valuable to westerners for ornament and function. "For nearly half the year, few dwellings of civilized men appear more desolate and uninviting than many on the open prai-ries of the West, that have for years been occupied without an attempt to plant an evergreen, or ornamental tree of any kind," he wrote. Like Douglas, Bryant highly recommended Norway spruce for windbreaks.

White pines were D. C. Scofield's particular favorite, prompting him to wax poetic about pines as "singing trees, better than the music of a thousand harps." Encouraging tree planting, Scofield noted, "I have many trees now on my forest grounds near this city, that were set there twenty years ago, that are now thirty to forty feet in height and a foot in diameter, from which plantation I cut logs in 1873, had them sawed into lumber, which was used in the finishing of my dwelling-house in this city."[22] Despite his own lumbering business, Scofield, the avid proponent of reforestation, finished his essay with the impassioned plea,

> *Plant* the *evergreens*; plant them along your borders; plant them around your homes; plant them for the sons and plant them for the sires; plant them for or-nament and plant them for *gain*; plant them for protection and plant them for

shade; plant them for health and plant them for wealth; plant them for beauty and plant them for pleasure; plant them for the boys and plant them for the girls; plant them for the wives and plant them for the little ones; plant them for the singing-birds and plant them for the doves; plant them for your dwellings and plant them for your coffins; plant them for the present and plant them for the future; plant them for the living and plant them for the dead; plant them for forest and plant them for park; plant them on the hill-top and plant them in the valley. Plant! Plant! Plant the evergreens.

Evergreen hedges became fashionable around country villas in postfire Chicago. Major estates in wealthier suburbs such as Elmhurst were enclosed by arborvitae hedges, which provided affluent homeowners with a sense of privacy, as well as some green color during the winter. D. Hill nursery of Dundee was one of the region's leading providers of evergreens. Its 1927 evergreens catalog stated that seventeen greenhouses were entirely devoted to propagating evergreen cuttings and grafts. In full color, the catalog showed mixed evergreens as foundation plantings along suburban houses, and even promoted "baby" evergreen trees in window boxes. Following design trends of the day, new uses for evergreens were

The D. Hill nursery of Dundee, Illinois, specialized in evergreens. Smaller varieties such as these from the 1920s became popular as foundation plantings. From the author's collection.

suggested, including rock garden plants such as savin juniper, Pfitzer juniper, mugho pine, and yews. Despite such modern innovations, tried-and-true ever-green uses persisted. Hill's catalog noted that "on farms and around homes all over the country, Norway spruce windbreaks are giving substantial protection to stocks and buildings." It also promoted the ever-dependable arborvitae for hedges around suburban homes.

Fruit Trees

Given today's plentiful markets, it is hard to imagine that fruit was once a luxury in early Chicago. As early as 1843, *Prairie Farmer* writers urged their fellow farm-ers to grow fruit. Among the "Six reasons for Planting an Orchard" given in an 1843 *Prairie Farmer*, essayist Edson Harkness opined that a proper orchard would leave a fine inheritance, make a home more pleasant, encourage children to love their home, and cultivate thankfulness toward God. Scarce as it was, fruit was a strong temptation to Chicago youth and immoderate adults alike. An 1846 *Prairie Farmer* reported, "One of the worst of the obstructions in the way of fruit cultivation is the thieving propensity of grown and half grown boys, who are sure to infest every neighborhood where it is attempted."[23]

But to the first settlers, shelter was more important than fruit. John Kenn-icott recalled the early days:

> For six or eight years after the first settlement of this region, very little was thought on the subject of fruit culture, and next to nothing accomplished. We had our "claim" rights to defend; and many of us felt insecure, and therefore de-layed testing the question, then considered doubtful, of fruit growing in the open prairie . . . Immediately after the land sales, the farmers . . . commenced "sticking in fruit trees" . . . Unfortunately, most of these trees were seedlings, or doubtful grafts, sent from friends east, or purchased of southern "tree pedlars" . . . and of course, disappointment, and often disgust, have been the result.[24]

Contrary to the prevailing planting practice, Kennicott sited his own orchard di-rectly on the prairie in the 1840s, instead of near woodlands. Most of his plants, including sixty varieties of apple, were obtained from the Benjamin Hodge nurs-ery in Buffalo, New York. For prairie planting, Kennicott recommended stocky plants with very low heads, and preferred trees from the borders of Lake Erie. He thought peaches would grow here, given poor, dry soil and minimal prun-ing, and was also hopeful about the plum, although his stock had succumbed to curculio, the highly damaging fruit pest. Pears were bearing fruit, although Kennicott feared "frozen sap blight" and had not had major success with cher-ries. Kennicott identified four local nurseries, besides his own, as good sources for fruit stock: the St. Charles Nursery, Mr. Gooding's nursery near Lockport, the Carpenters of Chicago, and Hastings' Chicago Garden and Nursery.

Leading the bustling pace of horticultural innovation, Kennicott began with his sixty varieties of apples and an assortment of other fruit in 1842. In five short years, he nearly doubled his apple varieties. Reporting on the family business in 1847, he said, "We (the boys) have now fit for planting out from 6,000 to 8,000 engrafted apples of 4 to 6 years growth and of near 100 varieties . . . Of other fruits fit for planting, we have, say 2000 peaches, 500 cherries (mostly Mazzards), 400 plums, 300 pears, 200 orange quince, &c, &c." His 1847 report modestly noted, "We have the reputation of one of the best nurserymen in New York, to back our own slight pretensions to practical knowledge in this department of science." Yet, as a self-assured westerner, Kennicott's 1852 ad for the Grove noted, "Our Eastern connection ends with the next season . . . The undivided half of all our young stock of trees is now for sale, at a bargain, and on terms open for negotiation. Apply to B. Hodge, Buffalo, N.Y." Presumably, Kennicott's confidence in his own grafting and growing techniques allowed him to sever professional (but not personal) ties with established nurseryman Benjamin Hodge, formerly his supplier.

In the 1860s, the Chicago Historical Society was asked by the U.S. Bureau of Agriculture to conduct a survey and assess the progress of fruit culture in Cook County and surrounding towns. The results were discouraging as compared to other cities, causing one writer to ask, "Is the fault in our soil and climate disadvantages, or is it in ourselves, not yet comprehending the conditions on which success may be assured?" According to the survey, the Barrington, Jefferson, and Elk Grove areas led in acreage and production of fruit. A Mr. Arnold of the town of Rich voiced a major concern of fruit nurserymen about tree peddlers: "If Eastern nurserymen had not swindled us, there would have been twice as many trees in the township."[25] Noting how the tide was turning against eastern-grown nursery stock, an 1860s *Prairie Farmer* writer wrote this epitaph to unsuitable fruits:

Here Stands
The last vestige of
An Eastern Orchard;
Most Worthy Fruit Tree, born, 18—,
Diseased by Strong Drinks, and Stimulating Food,
In its native home, Western N.Y.
It was sent West
Where hopes were entertained of its recovery;
But the *Heart Disease* was so strongly
Imbibed through its entire system,

That it became dangerously ill in its passage West.
With good Earth protection it withstood
The rigors of the first season,
But the following Spring
The scorching sun and piercing winds,
Drew the Life Blood from its vitals.
It Died in March, 18—.
Every needful care was granted for its relief;
Its Body was embalmed in lye and soap suds,
And was wrapped in straw,
But the Disease was previously so firmly seated,
The remedies were insufficient,
A *change* of *climate*, and *exposures*,
Proved inexpedient.[26]

After decades of trial and error, Chicago orchardists had achieved demonstrated success. The display of Illinois fruit at Chicago's 1893 World's Fair was the talk of the nation. Pyramids of neatly stacked apples, from multiple seasons' growth, testified to Illinois's ability both to grow and preserve orchard products. Forty-six varieties of pears, 75 of peaches, 59 of apples (of the 1893 crop

Illinois horticulturists were proud of their fruit display at the Columbian Exposition. 1893. Courtesy of Chicago Public Library, Special Collections.

alone), 128 of grapes, 90 fruits and nuts of native plants, and many other exhibits of small fruits such as strawberries and raspberries were displayed by 48 of Illinois's 102 counties in over 1,600 square feet of exhibit space.[27] The Illinois State Horticultural Society secretary said, "I think it can be truthfully said of the Illinois fruit exhibit, that during the first four months of the Exposition it was unequaled by any other state exhibit in the building, with possibly the exception of California."[28]

Chicago's agrarian achievements were often overshadowed by its industrial prowess. By 1917, despite the fact that Illinois was the third largest apple producer in the United States, the average consumer was unaware of the state's fruit-growing prowess. To counteract this image, the Illinois State Horticultural Society staged the Illinois First Great Apple Show, in Chicago in November 1918. The president of the society urged members to launch a big publicity campaign showcasing Illinois's apples. He noted, "When your committee began work for this show they found hotel and newspaper managers in Chicago who were surprised to learn that Illinois grew fruit for market in any appreciable quantity. They asked, 'Can Jonathan apples be grown in Illinois?' 'Are they as good as Western apples?' and other similar questions."[29] Ironically, Chicago's fruits were now compared not with eastern produce, but with that of the Pacific Northwest—Chicago itself had become the "eastern" market.

Publicity campaigns for Illinois apples and other fruits were still needed to raise public awareness in the 1920s. The Illinois State Horticultural Society

To reestablish Illinois as a great apple-growing region, the first Illinois apple show was hosted in Chicago in 1918. From *Transactions of the Illinois Horticultural Society*, 1918.

secured space at the 1924 Illinois Products Exposition in Chicago and displayed a ten-foot-high pyramid of apples. Free Jonathan and Grimes apples were handed out with wrappers boasting "This Apple Grown in Illinois." Visitors were still amazed that apples could be grown locally; one overheard comment was that "all Illinois fruit must be shipped out of the State because we never get it here in Chicago."[30] By 1933, with input from the University of Illinois extension

SALOME: STORY OF AN ILLINOIS APPLE

The story of the Salome apple contains all the luck, persistence, and hard work embodied in horticultural discovery in the Chicago region. The apple, considered heirloom and scarce today, is a firm, crisp, and juicy eating apple that ripens in November. It is particularly well adapted to our latitude. The story begins with an overgrown thicket of a former nursery in Ottawa, Illinois, about eighty miles west of Chicago.

Elias C. Hathaway, who came to Illinois as a young man from Massachusetts, bought this abandoned nursery in the 1850s. Hathaway began to clear much of the land for his own vegetable and fruit nursery. In 1869, he started grubbing out a heretofore neglected portion of the property, identifying and saving fruit trees that he thought bore fine fruit. Unfortunately, one of the prized trees (ultimately the Salome parent) was mislabeled and mistakenly chopped down. To Hathaway's delight, the stump sprouted the following spring, and he carefully nurtured the largest sprout.

The new apple tree bore consistently and abundantly. He began to show it at various horticultural societies, particularly the Horticultural Society of Northern Illinois, of which he was a long-time member. According to fellow nurseryman Guy Bryant, "It attracted a great deal of attention wherever shown for the western fruit growers were looking for an apple that would be hardy as well as a heavy cropper."[1] The parentage of the apple was not known, although it was thought to have descended from another great winter apple, Seek-No-Further. Hathaway named the apple after his mother, Salome Cushman Hathaway.

In the 1890s, Hathaway sold his nursery to developers who subdivided the land. Earlier, in the 1880s, he had sold virtually the entire stock of Salome trees to the esteemed Princeton nurseries of Arthur Bryant Sr. The Bryants introduced the Salome apple to the general public. Neither Hathaway nor the Bryants achieved overnight success in marketing the apple. Hathaway did not introduce it on a large scale because he wanted to continue observing and testing its hardiness and bearing qualities. The Bryants encountered the typical issues in introducing any new product—lack of name recognition, and a quiet period while the young whips came into fruit-bearing age.

The apple itself, like many heirloom varieties, was not as physically attractive as many commercially viable fruits. The conical yellowish fruits are medium to small, mottled with pink and red striping. Some early critics doubted its hardiness in latitudes such as northern Iowa.[2] But by 1912, Guy Bryant of Princeton stated that sales of the Salome apple were doing quite well, and that "we of the middle west are certainly greatly indebted to Mr. Hathaway for his care in preserving the Salome to us and for the thorough testing which he gave the variety."[3]

1. Guy A. Bryant, "The Salome Apple: Its Origin and History," *Transactions of the Horticultural Society of Northern Illinois*, 1912, 261.

2. J. Sexton, "The Salome Apple," *Prairie Farmer*, June 9, 1888, 374.

3. Bryant, "Salome Apple," 264.

stations, a very specific list of apples and other fruit was recommended for northern Illinois. Only thirteen apples made the list—vastly trimmed down from thirty-one possibilities listed in 1846 (see appendix 1). This pruned list reflected not only decades of pomology experience, but also Chicago's shift from fruit grower to fruit marketplace.

Early Ornamental and Street Trees

With utilitarian needs satisfied, Chicago homeowners were ready to decorate their landscapes. In the 1850s, the *Prairie Farmer* offered suggestions for ornamental trees, highlighting maple trees as clear favorites. Sugar maples, despite their slow growth, were recommended deciduous trees, followed by more rapid-growing silver maples. Red maples were suggested for wet soil. American elms came next, followed by slippery elms—with the caveat that the latter were hard to transplant. Bur oaks and chestnut oaks also were desirable. Magnolias were dubbed "glorious trees, though not common in the West."[31] Buttonwoods,[32] despite earlier popularity, were deemed susceptible to blight. Butternuts and black walnuts were thought handsome, although prone to insects. Catalpas were declared too tender for the Chicago region. Native buckeyes were recommended, as were horse chestnut and the Osage orange. The favorite understory tree was the dogwood, with American mountain ash a close second.

Nurserymen and growers responded as their customers gradually shifted from gardens of sustenance to ornament. Reflecting the shift from fruit to ornamental trees, Lewis Ellsworth's 1876 spring catalog showed only 5 types of apples, 2 types of pears, and 1 type each of cherries and peaches, instead highlighting 15 ornamental trees in different calipers, and nearly 30 ornamental and climbing shrubs. This was a dramatic shift from food-producing plants highlighted in his 1853 catalog with its 123 types of apples, 40 pears, and 16 cherries.

Many real estate speculators planted street trees as one of the first steps in beautifying new suburban developments. But in the suburbs, as well as the city of Chicago, street trees had a short life expectancy. J. H. Prost, the first forester for the city of Chicago, reported that until parkway trees came under the city's jurisdiction, street trees were subject to all sorts of "mutilation," including usage as hitching posts, wholesale cutting of roots for new sidewalks and curbs, and gas leaks, which could cause a tree "to be defoliated in two weeks."[33] Accordingly, until maintenance responsibilities were budgeted, street trees often were short-lived—beautiful only if homeowners took a personal interest in their parkways.

One such homeowner was Mrs. H. S. Newton, the so-called tree mother. Newton, who lived for more than thirty years at 1706 Warren Avenue in Chicago, took it upon herself to compile an index of property owners who lacked trees on their parkway lots. Chairman of the committee on streets of the Outdoor Municipal Art League in 1913, Newton made daily calls to treeless lot owners on Warren Avenue and oversaw the planting of nearly three hundred trees. She claimed the city forester had given her permission to plant trees wherever she wanted on the parkways. Her recommended tree was the Carolina poplar, which she quoted the forester as saying was the best for the soot and smoke of Chicago. Although the poplar was indeed fast-growing and cheap, Newton came under fire from at least one tree planter, J. F. DeForest, who said the poplar was a misnomer for the common cottonwood, a weak and straggly tree. Instead, this critic suggested planting the box elder—by today's standards also a weed tree. Knowledge of worthy trees, and a willingness to invest in their growth, was still ongoing.

Trees and Shrubs of the Twentieth Century

By the turn of the twentieth century, as a second generation of settlers grew up in the newly developed suburbs of postfire 1870s, trees became an accepted part of the landscape. Twenty to thirty years of growth on mature trees helped suburbs age gracefully, and imparted a sense of permanence to the frontier land. Great strides had been made in determining which trees grew well here, and which did not.

In 1901, landscape gardener O. C. Simonds and "wealthy amateur" William C. Egan wrote the chapter "Trees for the Home Grounds" in Doubleday's book *How to Make a Flower Garden*. While Egan wrote about weeping trees, Simonds emphasized native trees such as swamp white oak, wild plum, wild "thorn-apple" (or hawthorn), serviceberry, crab apple, and tulip trees. He also allowed for some nonnative species with particular seasonal interest, such as *Magnolia stellata* for spring, and Norway maple or sycamore maple for fall. By the 1920s, professionals such as Highland Park retail grower Fritz Bahr felt sufficiently confident in recommending "Sixteen trees for Lawn and Street," ostensibly foolproof trees that would work in any climate. Lacking a crystal ball, Bahr said, "Of the eight or ten varieties [of elms] none is more desirable than *Ulmus Americana*, the American Elm."[34] He, and the many others who recommended American elms, would live to see the trees' demise in the 1930s at the hand of Dutch elm disease. Research for hardy, disease-resistant trees continued, and with it was established such organizations as the Morton Arboretum in Lisle, Illinois, in the 1920s.

Of woody plants, shrubs and vines became important landscape elements in twentieth-century Chicago. Increasing populations led to smaller property sizes, which in turn necessitated smaller plant-material options. Shrubs became popular alternatives or additions to trees, providing greenery and privacy without needing a lot of space. Shrubs were largely ornamental—used as foundation plants or privacy hedges. These were the landscaping requirements of a population much less concerned with growing trees for shelter, food, and windbreak. Japanese barberry, forsythia, flowering quince, and hydrangea were popular choices in the first few decades of the twentieth century.

Vines were always favored by Chicagoans, particularly for their vertical height when space was at a premium for trees. As the nation became more industrialized, Chicagoans had more leisure time to spend on their porches during the summer, and vine-draped entryways became fashionable. Garden structures such as pergolas and arbors, which enjoyed popularity in the first quarter of the century, offered additional opportunities to accessorize with vines. W. C. Egan offered his "Select List of Vines" in the early 1900s, which included Japanese chocolate vine (*Akebia quinata*) interplanted with varieties of clematis for porch pillars. While written for a national audience, Egan's list contained specific tips for Chicago gardeners, including vines with winter interest such as Oriental bittersweet (*Celastrus orbiculatus*) with its winter berries. For trellises, Egan recommended either wolfberry or matrimony vine (*Lycium barbarum*) as a stronger grower for the Chicago region. Tongue-in-cheek, he suggested growing the abundant Virginia creeper (*Parthenocissus quinquefolia*) on a neighbor's fence, where "the caterpillar may feed upon its leaves and not become familiar with

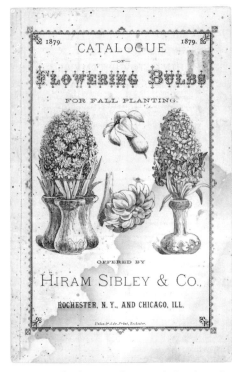

A local presence for tree and other plant sellers was important in order to avoid being labeled as a "tree peddler." New Yorker Hiram Sibley, one of the founders of Western Union, saw the importance of linking to Chicago through a branch seed company. His 1879 catalogue offered bulbs, which were more readily shipped than trees.

and attached to you."[35] In the late 1920s, vines were still recommended by the *Chicago Tribune*'s garden writers to "cover up soot covered walls." Boston and English ivy were the vines of choice, to be planted on any wall but the south, where strong winter sun could kill it.

Care of Trees and Plants

Because of their long growing cycle, trees represent a huge investment of time and money. Mistakes are expensive. In his 1861 essay "Mistakes of Tree Planters," John Kennicott shared twenty years of tree-growing wisdom. Here are some of his key lessons learned.[36]

> Wrong soil type and preparation: Kennicott debunked the prevailing myth that "any good corn soil will do for an orchard." He opined that the lighter the color of the soil, the better for orcharding. Lime and clay were critical. Dry soil, obtained either through proper site selection or through artificial drainage, was deemed a key to success.
>
> Planting time and technique: Autumn could be as good a time to plant as spring. Kennicott opined that most people didn't dig deep enough in preparing the soil—just made little "post holes." Never put manure in the bottom of the hole, and always heel in your plants, he advised. He cautioned farmers about being too hasty to plant trees—he assured them that trees could live even if the roots froze or dried out a bit.
>
> Pruning: "Nearly our whole system of pruning is a mistake, a barbarism," Kennicott warned. He advocated only light pruning of dead wood and interfering branches, or light cutting back.

By the 1850s, unsuccessful experiments with shade trees had become apparent. The *Prairie Farmer* noted, "No person can walk our street without being astonished at the multitude of ill-shapen, marred, half dead and dying shade trees which he sees along the streets."[37] Many of these trees had been dug from nearby forests and hadn't adapted to roadside conditions. The newspaper recommended new cultural practices such as planting nursery-grown or open-field-grown trees instead of digging shade trees from the forest. This would more closely emulate a transplant's natural environment, it was thought. If a tree was taken from the woods, its branches should be made to start low, unlike forest trees whose crowns are leafy, but whose trunks are bare. How low should the limbs be? The *Prairie Farmer* advised, "For the street, they must be out of the reach of cattle, or if in a lawn, out of the way of whatever animal runs in it."[38] Lewis Ellsworth prepared some "hints for transplanting" in his 1853 catalog. "Many persons plant a tree as they would a post," he began. "Many an orchard of trees, rudely thrust into the ground, struggles half a dozen or more years against the adverse condition, before it recovers, and is never equal to what it would have been with proper treatment." Ellsworth gave detailed directions, including planting-hole sizes, root pruning, and trimming limbs after planting, "remembering in this Western Country low heads for trees are preferable."[39]

Even into the 1870s, homesteaders were still inexperienced at tree growing, and perpetuated practices from their land of birth. Robert Douglas recommended that every prairie farm have at least one-tenth of its acres in timber. With proper spacing, tree growth would improve, Douglas explained. Douglas urged his colleagues to depart from old European notions of "planting thick and thinning quick." This practice was needed only on English moorlands, which weren't plowed, and where saplings were cut down quickly for use as rustic fences and sheep hurdles. Prairie-state farmers, with ample space, could instead work horses between forest plantings, and "would be loth [sic] to cut down their fine, thrifty saplings before they were of much value."[40]

Tree health improved through the evolution of insecticides and fungicides from homegrown brews to a sophisticated chemical industry. In 1889, Princeton nurseryman L. R. Bryant described his methods of spraying for codling moth and cankerworm. Using a homemade concoction of arsenical poisons, Bryant sprayed two or three times in the growing season. He cautioned that too much spraying would cause severe leaf burn, and proposed subsequently weaker applications as the season progressed. His standard mixture was one pound of London purple to 160 gallons of water, and he was experimenting with kerosene mixes with London purple. Bryant described his process: "My apparatus for spraying is simple. Two wagons are fitted, each with a 160 gallon cask mounted on its side, and a Field's Force Pump, which can be changed from one cask to the other. In operation, the poison is put in the casks before filling with water, and by the time the wagon gets to the orchard, the contents are well mixed."[41]

Poison dilutions and timing of spraying continued to be topics of discussion and experimentation throughout the last quarter of the 1800s and the early 1900s. Gardeners were still learning about different insects and their effect on plants. In 1899, for example, gardeners on Chicago's north and west side were mystified by widespread decimation of trees on public and private grounds. Elms, maples, and basswoods were particularly affected, to the extent that profound leaf drop caused residents to think a particularly early autumn had arrived in early August. The South Side parks and gardens seemed to escape the tree devastation; however, caretakers had problems with withered lawns. With the help of the growing entomology department at the Field Museum, a red spider and boll worm were identified that were thought to have caused the problem. The spider (*Tetranychus telarius*) and worm (*Heliothus armigera*) were accused of the tree and lawn damage, respectively. Experts were split on proper methods of eradication. Charles Fuhrmann, head gardener at the Potter Palmer mansion, recommended Paris green spray. John Algota, head gardener for the Pullman grounds, thought the leaf drop was not due to the spider, but to an earlier hard frost. John Sell, head gardener for Garfield Park, felt that although the trees were badly hurt, the damage was not fatal.[42] By sharing techniques, gardeners and scientists identified best practices in lawn and tree care.

World War I brought about the first use of chemical weapons and, as a by-product, new uses of agricultural chemicals. L. A. Day of Chicago, in a 1917 presentation on chemical manufacture before the Illinois State Horticultural Society, credited University of Illinois professors with advancing research in this area. Day noted that the young chemical industry was fraught with difficulties from supply shortages to labor woes. Dry powdered arsenate of lead "has almost entirely replaced the old paste goods with the up-to-date consumer," according to Day, and dry lime sulfur was another modern choice. Factory conditions were dangerous and unacceptable to even the most industrious of workers, and Day noted, "I have known cases where men have been hired in the morning and supplied with a clean outfit at the front door and walk out the back door without doing a stroke of work."[43] In the 1920s, completing a circle of invention, Vigoro, a commercial lawn fertilizer, was introduced by Swift and Company, the meat-packing plant of Chicago. Whereas the Swift stockyards had once provided raw manure as fertilizer in the late 1800s, the same company advanced to produce a more sanitary, chemical fertilizer for the "modern" gardener.

GARDEN FLOWERS MAKE A HOME

While trees, because of their utilitarian value, were among the first vegetation planted by Chicagoans, flowers helped settlers feel more at home. Early *Prairie Farmer* magazines soon featured a column called "Flowers for Farmers' Wives and Daughters to Cultivate." Even though most writers were then men, flower gardens were women's domain in pioneer times. Flowers listed in this column were the best known to date for prairie conditions. An 1846 column, for example, highlighted the benefits of nasturtium (*Tropaeolum majus*). Nasturtium, which only came in yellow and orange, was to be planted in mounds. Its seeds were said to be good for pickling. Poppies, hollyhocks, and China asters (*Aster chinensis*) were also recommended. Editors also praised the common dandelion (*Leontodon taraxacum*) hoping not to "shock the nerves of any sentimental florist."[44] Were it not so readily available, the editors thought it would be more highly esteemed. They endorsed its "beautiful and modest blossoms," and noted its medicinal properties for "dropsical complaints and bilious derangements."

Flowers that were readily available and easy to cultivate made the recommended list. Peonies, iris, and dahlias met this test, as did certain spring and summer bulbs like tulip, hyacinth, narcissus, gladiolus, and daffodil. Common double scarlet peonies were highly recommended. Pansy (*Viola tricolor*) and stone crop (*Sedum anacampseros*) were good candidates for mound displays. In 1846, the *Prairie Farmer* noted, "The Dahlia has perhaps held its sway as a fashionable blossom longer and more universally than any other, and is now every where known and cultivated."[45] In Chicago, the only concern was overwintering the bulbs, and to solve that problem, a good cellar came in handy. Annual flowers were a mainstay of Chicago gardens, with seeds purchased locally or by mail. In 1853, a dollar bill or "fifty cents in postage stamps" would buy Chicago garden-

OLD-FASHIONED GARDEN FLOWERS

As with garden fashions across the United States, the popularity of individual flowers waxed and waned in Chicago. Often the introduction of newer varieties caused original species to fall out of favor. Global and regional competition induced local nurserymen and seedsmen to discontinue unprofitable stock. Transportation and technology influenced plant choices—improvements in greenhouses, shipping, aquatic gardens, and other advancements shaped a homeowner's garden. In Chicago suburbs, change occurred as homeowners' trees matured. Larger trees meant shadier gardens—and a proclivity for shade-loving plants rather than sunny choices that may have predominated earlier.

Many old-fashioned flowers grown in Chicago are not easily found in today's seed and plant catalogues. Marguerites (*Chrysanthemum frutescens*), once called Paris daisy, for example, fell out of favor

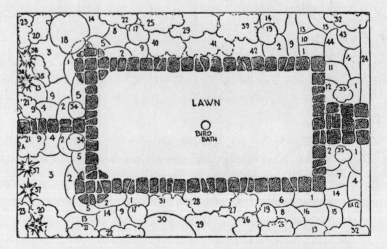

This 1920s plan from the *Chicago Tribune* for a 25-by-40-foot garden shows the enduring love for tried and true garden flowers. Around a rectangle of lawn, the plan calls for a flush of pink, yellow, lavender, and white tulips (4–7 on plot) mixed with white, pink, and blue hyacinth (8–10), and masses of forget-me-nots, peonies, iris, bleeding hearts, and pansies (2, 13–16). Following the spring show, the bulbs were replaced with cosmos, pink and yellow zinnias, larkspur, white and lavender petunias, pink snapdragons, bachelor's buttons, *Phlox divaricata*, and lily of the valley. Shrubs such as lilac, spirea, forsythia, and fragrant honeysuckle (21–23, 25) rounded out the border. Copyright 1929 by Chicago Tribune Company. All rights reserved. Used with permission.

ers eight to twelve varieties of flowers from such notable firms as Joseph Breck of Boston, J. M. Thorburn of New York, Ellwanger and Barry in Rochester, or Starkweather and Hooker in Chicago. Locally, any seed store could also provide popular varieties such as morning glories, balsam, rocket larkspur, and portulaca (plate 6).

In the early 1850s, garden flowers were typically grouped in conspicuous displays. Small trailing plants such as phlox and portulaca were sown in 3-to-8-foot round beds. To display petunias, a raised mound was often created by stacking a tower of successively smaller boxes from four feet wide to fifteen inches for

when newer varieties sported larger flowers. The intensely fragrant mignonette (*Reseda*) was sown under glass and transplanted by home gardeners in the 1880s, according to Jonathan Periam. By the 1930s, however, Fritz Bahr noted that no one grew it for its flowers, and few florists grew it because it was unprofitable. Prettyface (*Calliopsis*), the two-to-three-foot yellow daisylike wildflower, is a rarity in today's garden. The evocatively named love-in-a-mist (*Nigella damascena*), snow-on-the-mountain (*Euphorbia marginata*), and wallflower (*Erysimum asperum*) are all scarce today. At the turn of the century, with the carpet-bedding fad all but over and with perennial borders in fashion, many annuals were considered passé. By 1911, author Lena McCauley listed sixty-six annuals as old-fashioned, but worthy of reintroducing in the garden, including ageratum, alyssum, snapdragons, columbines, asters, marigolds, morning glories, phlox, larkspurs, cosmos, hollyhocks, stocks, and violets.

This was not the first time old-fashioned flowers were mourned. In 1889, Edgar Sanders remarked on the prolific introductions by the city's growing number of florists. Noting that while they turned out millions of plants, florist introductions were "gay, but ephemeral," and florists "have driven out the old-fashioned flowers. Sweet william, larkspur, London pride, southernwood, primroses, hollyhocks, Canterbury bells, hardy roses, and sweet briars are getting lost in the shuffle," according to Sanders.[1] It seemed every generation wistfully looked to the flowers of youth.

Perennials and shrubs became popular at the beginning of the twentieth century, and many old-fashioned favorites have survived today, in an improved form. Spirea, lilacs, mock orange, flowering almond, and barberry were favorite shrubs at the turn of the twentieth century. Delphinium, foxglove, hibiscus, peony, and phlox were then popular perennials. In the late 1920s, the *Chicago Tribune* announced "Grandmother's Garden Is Again in Style." Based on input from H. B. Dorner, a University of Illinois professor who "made a study of the old-fashioned garden," grandmother's garden was described as a hodgepodge, created from gardeners swapping with each other to get a great variety of the old standbys. Zinnias were a staple in the garden, particularly the colossal and dahlia-flower kind. Small and large marigolds, delphiniums, bachelor's buttons, lupines, painted-tongue (*Salpiglossis sinuata*), scabiosa, coxcombs, amaranths, four-o'clocks, petunias, and verbenas were essential, per Dorner. Japanese morning glory, sweet pea, climbing nasturtium, baby's breath, candytuft, and annual phlox could be added for more variety.[2] Nostalgia for yesteryear's garden persisted even into the modernistic 1933 World's Fair, where several "old-fashioned gardens" were displayed.

1. Edgar Sanders, "Thirty Years in Western Horticulture," *Transactions of the Horticultural Society of Northern Illinois*, 1889, 235.

2. Frank Ridgeway, *Flower Gardening by Farm and Garden Bureau* (Chicago: Chicago Tribune, 1929), 55.

a total height of four feet. This mound, situated in the middle of a grass plat, would then be covered with plants, and offered three to four months of bloom. Toward the end of this decade, Edgar Sanders began to bemoan the fate of some old-fashioned perennials such as phlox, which had been eclipsed by the current fads for verbena and other bedding plants. He allowed that some of the older

CHRYSANTHEMUMS

In 1966, Chicago's city council and Mayor Richard J. Daley designated the chrysanthemum as Chicago's official flower. The plant has enjoyed long popularity in Chicago, bringing color into the midst of autumn and the onset of winter. Chrysanthemums have ancient origins in China and Japan, and were taken to England in the mid-1700s. In the early 1800s, it is said that there were less than one hundred varieties in England, whereas by 1900, there were more than three thousand varieties in Europe and America.

Simple chrysanthemum species were prevalent in the early Chicago area—the common white daisy (*Chrysanthemum leucanthemum*) grew freely on the prairie. The popular old-fashioned marguerite in dooryard gardens was *Chrysanthemum frutescens*. The eye-catching autumn-blooming species, *C. sinensis*, was the focus of much interest in the late 1800s in Chicago. Chicago florists were greatly involved in promoting new varieties, from the work of Samuel Brooks, the Father of the Chrysanthemum, to the early chrysanthemum shows of the newly formed Chicago Florists' Club. Florists hosted their first exhibition of chrysanthemums in 1884, at Chicago's Inter-State Exposition Building.

Mum mania was sweeping late Victorian America as travel to Japan showcased its national flower. As with many things Japanesque, the chrysanthemum became a new fad in the United States. At the Columbian Exposition, the Japanese Garden on the Wooded Island drew more attention to the flower. The newly reconstituted CHS began hosting annual chrysanthemum shows. Their November 1894 show listed sixteen exhibition categories under cut chrysanthemum flowers, and twelve display categories of chrysanthemum plants. Various introductions were revealed at these shows, with homage paid to the Chicago-based patrons of the exhibit; 'Mrs. George Pullman', a deep yellow flower, received accolades in the June 1894 *Garden and Forest* magazine. 'Mrs. W. C. Egan', a light pink and yellow mum, won a silver medal, and 'Mrs. M. A. Ryerson', a large, pure-white flower, was well reviewed by *Garden and Forest* magazine in 1896. *Garden and Forest* carried generally favorable reviews of these early shows, although in 1895 it noted, "The exhibit, as a whole, showed a tendency to breed and cultivate the more compact or even ball-like forms instead of the wider or more open heads, which give the Chrysanthemum one of its most distinctive characteristics. As one expressed it, they are too Dahlia-like. It is the commercial or transporting qualities which are in the ascendant."[1]

At the turn of the century, after several years of mum shows, it seemed as if the chrysanthemum craze might have worn out. A 1901 *Chicago Tribune* headline declared, "Chrysanthemums Are Going out of Fashion," and the new styles in flowers were carnations and roses. A representative of the Peter Reinberg greenhouses called the chrysanthemum coarse and lacking fragrance—a mere fad. The prediction was short-lived: at the November 1902 flower show, chrysanthemums were again declared the queen flower, with even more named after Chicago's society women. A pink Japanese variety was named 'Mrs. H. N. Higinbotham', a large white variety named 'Mrs. C. L. Hutchinson', and a yellow Japanese type was christened 'Mrs. Potter Palmer', while other varieties took the names 'Mrs. Arthur J. Caton', 'Miss Florence Pullman', 'Mrs. Moses J. Wentworth', and 'Mrs. John J. Glessner'. The chrysanthemum held its own for the next several years.

varieties of the tall growing phlox such as *Phlox decussata* might be considered coarse, but the newer dwarf varieties of *P. suffruticosa* were most promising.

Florists of the 1860s were so prolific that Sanders found it necessary to explain the "Definition of Gardeners' Language." According to Sanders, *florist flowers* were those "which by cultivation and extraordinary care and skill in raising,

Even as roses and carnations were added to the CHS's nineteenth annual show in 1910, the chrysanthemum still captured an entire day on the program and had more display categories than the other flower types. Chrysanthemums became available in so many colors and forms that entire decorating schemes—from households to weddings—were built around the flower. Mums were even coordinated with fashion and clothing styles. In 1911, a "fuzzy chrysanthemum" ('L'Enfant des Deux Mondes') was developed to accessorize the latest craze for the so-called fuzzy coats, hats, and vests made of textured fabrics and feathers.[2] More institutions held chrysanthemum shows—the University of Chicago hosted its first such event in 1914. Cementing Chicago's relationship with the mum, the Chicago Park District hosted chrysanthemum shows beginning in 1912 at the Garfield Park Conservatory, and in Lincoln Park. The fall chrysanthemum shows continue today, popular oases of warmth for the city battening down for winter.[3]

Crysantheum Show
Lincoln Park Conservatory
Chicago, Ill.

"Take your choice, aren't they fine?" wrote the author of this postcard to a Wisconsin flower lover. This Lincoln Park Conservatory chrysanthemum show was part of an autumn tradition that continues today. Ca. 1908. From the author's collection.

1. *Garden and Forest*, November 20, 1895, 469.
2. "Fuzzy Chrysanthemum Newest Fad," *Chicago Daily Tribune*, November 6, 1911, 3.
3. "University of Chicago Has Its First Chrysanthemum Show," *Chicago Daily Tribune*, October 30, 1914, 11.

have very much altered their character as shown in the original state."[46] Examples were carnations, dahlias, verbenas, and pansies. Sanders described *bedding out plants* as an improvement over old-fashioned flower beds with mixtures of permanent plants. Bedding plants, such as verbenas, cupheas, geraniums, lantanas, heliotropes, petunias, and salvias, were grouped as masses of similar color and were typically annuals. *Foliaged plants*, an even newer innovation, referred primarily to houseplants with variegated leaves, and *double flowers* resulted from breeding for larger, but sterile, flowers. Sanders recommended annuals with trailing habits for hanging baskets, and mignonette, sweet alyssum, and white candytuft for winter flowering in the greenhouse. *Phlox drummondii* he dubbed the "cream of all the Annuals," second only to the verbenas. Even though a few years earlier he'd decried the flower, he then called verbena the "prince of bedders, a flower that everybody who has a garden should grow."[47]

After the Chicago Fire of 1871, many relocated Chicagoans were anxious to settle down in their new homes in the suburbs or in renovated city homes. Once again, quick-growing plants helped make the transition easier. Annuals were popular for their instant effect. For fragrance, mignonette, sweet peas, stocks, tea roses, tuberoses, and *Verbena montana* were popular. For continuous color in beds, dianthus, portulaca, *Phlox drummondii*, verbena, pansies, stocks, and zinnias were recommended. Tall orange marigold, *Celosia huttonii*, alyssum, *Ageratum mexicanum*, and white candytuft were popular choices for ribbon-bed gardening. Bulbs also offered multiseason possibilities, with traditional favorites such as crocus, tulips, gladiolus, and hyacinth.

In the late 1870s and early 1880s, ornamental grasses became a popular fad for the garden border or for the parlor. A Waukegan gardener recommended a wide variety of grasses, noting, "I think every lady who has a garden might, if she chose, grow her own grasses for drying for ornamenting of the parlor."[48] Japan maize, which "created such a furor a few years ago," could be grown with low-growing plants or among shrubbery. Pampas grass was not deemed hardy in Chicago, but could be a good tub plant. *Eulalia japonica*[49] was said to grow six or seven feet tall and was recommended as a lawn specimen plant. "Quite a number of our common Summer grasses which grow so quickly and spread all over the garden, if not repressed, are very pretty when dried," the gardener observed.

During the 1890s, Chicagoans' interest in flowers was piqued by the displays at the 1893 World's Fair, and by the profusion of elegant chromolithographed seed catalogs. Pansies enjoyed renewed popularity, thanks to many of the Columbian displays by J. C. Vaughan and others. In favor were large flowered varieties such as 'Giant Trimardeau Purples' and Cassier plants from France. Instructions for planting flowers were tailored for homeowners (often of city lots) who did not have experience in gardening. Whereas pioneer Chicagoans, as a rule, had had an agricultural background, the new generation of born-and-raised city dwellers represented a market for garden writers and suppliers. Easy-to-grow annuals became favorites among these people, including sweet peas, poppies, mignonette, candytuft, and snapdragons.

Detailed instructions for planting biennials and perennials by seed (e.g. coreopsis, columbines, larkspur, foxgloves, and hollyhocks) were offered in popular newspapers along with the suggestion that immediate results could better be obtained by purchasing plant roots from reliable nurserymen.[50] Helping the novice gardener design his own landscape, the *Chicago Tribune* printed an article in 1895 called "Artistic Gardens: A Paper for Amateurs." Here the writer offered possible combinations of plants—poppies with an undercover of forget-me-nots, or white lilies framed by dark evergreens. Presaging the trend for smaller gardens, the author opined that amateurs should not try to create carpet bedding, "vegetable butterflies, elephants, and rugs" as these did not complement any current house style or smaller garden, and required more care than the novice gardener had.[51]

The early twentieth century saw two distinct trends in garden flowers in Chicago. Two- and three-flat rental properties became more prevalent in the ever-burgeoning city, and thus quick-growing plants again became popular for the renter. Annuals such as nasturtium, portulaca, four-o'clocks, marigolds, and petunias were promoted not only for ease of cultivation, but also for their ability to stand up to poor soil, smoke, and dust of the city. Sunflowers and caster oil plants were suggested where height was needed to cover a fence or other unsightly object.[52]

A second trend was the move to specialty gardens and consequent display of newly introduced flowers. As more varieties proliferated, gardeners became connoisseurs of rare flowers. Paralleling the movement of national and regional flower societies dedicated to particular plants (e.g. the American Rose Society [1892] and American Iris Society [1920]), some avid Chicago-area gardeners started their own specialty gardens. Gardens dedicated to peonies, lilacs, or roses, for example, became popular. Trends in novelty landscapes such as rock gardens or Japanese gardens brought nonnative plants to Chicago, but a concurrent trend toward native plants helped preserve our native species.

NATIVE PLANTS: CELEBRATING CHICAGO'S ROOTS

Even as some gardeners sought to grow the latest cultivated flower, many became enamored of native prairie and woodland plants. As early as 1847, the *Prairie Farmer* noted, "Nothing is more common than to hear from strangers exclamations of surprise on seeing for the first time a prairie landscape, before settlement has marred its beauty; and to hear them wish that some of the beautiful flowers they see might be domesticated."[53] Attempts to transplant prairie flowers were largely unsuccessful, possibly because of the plants' long roots or inadequate understanding of culture requirements. "Many of the blossoms which show well on the prairie prove quite inferior, when first transplanted to the parterre," the *Prairie Farmer* reported. Some natives were deemed suitable for garden culture, however, including blue spiderwort (*Tradescantia virginiana*), false cowslip (*Dodecatheon meadia*), partridge pea (*Cassia chamoecrista*[54]), button snakeroot (*Eryngium yuccifolium*), dragonhead (*Dracocephalum virginianum*[55]), fringed gentian (*Gentiana crinita*[56]), and, despite its odor, meadow garlic (*Allium canadense*).

While amateur botanists struggled with identifying the bewildering array of new prairie flora, it wasn't until the 1850s that more scientific investigations began. Field trips sponsored by the newly formed Chicago Academy of Sciences and the Illinois Natural History Society formalized the study of native plants. Pioneering naturalist George Vasey wrote a series on botany in the *Prairie Farmer*. Noting that "a full development of the vegetable productions of the State has not yet been accomplished," Vasey began writing about different plant families found on the prairie. His articles in the 1859 *Prairie Farmer* proposed to include the most prominent plant families, including the buttercup family (*Ranunculaceae*), mustard family (*Cruciferae*), and others.

Mrs. Chatwitt, an 1860s *Prairie Farmer* correspondent, sounded an early cry for wildflowers. "This morning I transplanted a lot of wildflowers," she reported. "In looking around a year ago for some old favorites, I found they had disappeared—a fine large kind of wild pansey, a beautiful wild white flower, which grew spirally upon a single stem, and some others." Chatwitt rescued valerians, native larkspur, wild geraniums, columbine, and bluebells, among others. "There is one beautiful thing vulgarly called 'Dutchman's Breeches,' which I have never yet seen successfully transplanted, but I presume it could be," Chatwitt encouraged. The Illinois State Horticultural Society also recognized the value of native plants, and urged their preservation. In promoting the 1862 Horticultural Society Fair in Chicago, Chicagoan C. Thurston Chase suggested contest entries of natives, noting, "The prairies are gemmed with wild flowers, which are vanishing with the advance of the plow. Many of them are possessed of intrinsic worth, superior to many cultivated exotics. The day is not far distant when, if rescued, they would rank among the most highly prized sorts, and possess a historic interest to those who may come after us. The Society desires to encourage their preservation."[57]

In 1867, Mathias Dunlap argued for native plants in his report on the status of horticulture. "We should ask our groves and river forests to give up the gems in their keeping before we send to distant states for novelties of less value. But even now we buy back our own, and thus cheat ourselves into the belief that distance lends enchantment to the view."[58] Dunlap wryly noted that unsuspecting gardeners bought plants with fancy names from faraway dealers, unaware that they were natives from their own backyard. He pointed out the irony where Illinois plants

> are taken from beneath their native shelter at a year old, at a cost of six dollars a thousand plants, and returned to us after three or four years at a dollar each, but under the imposing name of Tulip tree, or botanically, *Liriodendron tulipifera*. Our beautiful foliaged Chickasaw plum comes to us under the name of *Cerasus chicasa*, or in plain English, Cherry Plum. We receive our own beautiful Red Bud under the name of Judas Tree, and generally at a cost of a dollar, and please ourselves with the idea that it is a distinguished stranger (*Cercis Canadensis*). The nurseryman's *Amelanchier botryapium*, the "waahoo of our woodlands," useful in warding off fever, came back from the nurseryman as strawberry tree or burning bush.

Dunlap recommended

> a few drives into our forests, a little more study of the names of plants and trees will save us many a foolish outlay. There is more in a name than we are willing to admit, as we too often order plants with high sounding names without knowing their real value. Who wants a Waahoo, so common in our groves? No one, for we seldom see it in its winter garb of scarlet berries; but when we see a picture of the Strawberry Tree, with its delicate foliage, green wood and beautiful berries, we are ready to subscribe for a tree at a dollar. How many times do we meet our own beautiful Spiderwort that has come from the distant nurseries at the cost of a dollar a plant, when a ride out on the prairie would give us a thousand for the digging.

Chicago nurserymen's improved ability to grow plants under glass in the 1860s and 1870s made exotic varieties accessible and affordable to the masses. Indeed, when Jens Jensen installed native plants in his American Garden in Union Park in 1888, park visitors "exclaimed excitedly when they saw flowers they recognized; they welcomed them as they would a friend from home."[59] Native plants, ironically, had become as rare as exotics once were.

Europeans were aware of this disconnect. Landscape gardener O. C. Simonds recounted his 1892 visit to the country estate of esteemed UK landscaper William Robinson. Simonds remarked on native American asters at the estate, prompting Robinson to remark, "You Americans do not appreciate your wild flowers. We have to bring them over to England and cultivate them awhile before you will notice them."[60] The times were ripe for Jensen and O. C. Simonds to showcase the benefits of native plants. Even as nurserymen like J. C. Vaughan flaunted major displays of cannas and pansies at the 1893 Columbian Exposition in Chicago, momentum was building across America to save native plants.

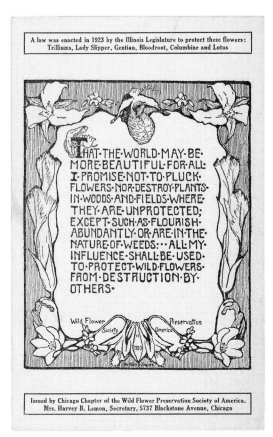

A law was enacted in 1923 by the Illinois Legislature to protect these flowers: Trilliums, Lady Slipper, Gentian, Bloodroot, Columbine and Lotus

THAT·THE·WORLD·MAY·BE· MORE·BEAUTIFUL·FOR·ALL· I·PROMISE·NOT·TO·PLUCK· FLOWERS·NOR·DESTROY·PLANTS· IN·WOODS·AND·FIELDS·WHERE· THEY·ARE·UNPROTECTED; EXCEPT·SUCH·AS·FLOURISH· ABUNDANTLY·OR·ARE·IN·THE· NATURE·OF·WEEDS;···ALL·MY· INFLUENCE·SHALL·BE·USED· TO·PROTECT·WILD·FLOWERS· FROM·DESTRUCTION·BY· OTHERS·

Wild Flower Society Preservation America

Issued by Chicago Chapter of the Wild Flower Preservation Society of America. Mrs. Harvey B. Lemon, Secretary, 5737 Blackstone Avenue, Chicago

Local chapters of the Wildflower Preservation Society of America took a pledge to protect native plants. From the author's collection.

Highlighting the work of the Wildflower Preservation Society of America, popular newspapers such as the *Chicago Tribune* began running articles with headlines such as "There may be no more wild flowers." The Wildflower Preservation Society, founded in 1902, included Chicago members in its early chapters (ca. 1914). Like many other outdoor groups popular in the Progressive Era, the Chicago Wildflower Preservation Society led hikes around the city to introduce the general public to the value of wildflowers (plate 7). The Prairie Club of Chicago, founded in 1908 with Jens Jensen as a charter member, urged its members to "take only photographs and leave only footprints" on their walks around Chicago. O. C. Simonds, whose daughter Gertrude and business partner Roy West were members of the Prairie Club, wrote an essay on "Wild Flowers" in 1908. "Let us take for granted the fact that our native flowers are beautiful, and that they may be used to advantage in home grounds, parks and other areas,"[61] he wrote. In a telling comment of the times, he observed that many natives could now only be found along railroad lines or roadsides, "unimproved" areas where nature was left undisturbed.

After almost a decade of publicity and effort toward saving wildflowers, the battle was still not won. In fact, it's possible that the increased interest in botany

The Prairie Club, Chicago's oldest regional environmental group, sponsored trips around Chicago's countryside and into Indiana and Michigan. Here, members explore fields of daisies in suburban Chicago. Ca. 1910. Courtesy of Westchester Township History Museum, Chesterton, Indiana.

and nature during those years may actually have inspired more people to explore the woods and pick wildflowers. In 1912, professors H. C. Cowles and E. J. Hill agreed that it was increasingly harder to find wildflowers near Chicago. Hill noted that whereas he used to be able to botanize on the South Side of Chicago on his way to school, now he had to go as far as Gary to find wildflowers.[62] The only remaining places to "botanize" near Chicago were some tracts of land southwest of Western Avenue, the Skokie slough—although it was soon destined to be drained—along railways, and a few remnants near the Calumet. Particularly scarce were trailing arbutus, hepatica, pink lady's slipper, fringed gentian, yellow lotus, pitcher plants, and ferns. The more conspicuous the flower, the more likely it was to be harvested.

The Forestry Class of the Chicago Women's Club held an exhibit of wildflowers in 1917 whose purpose was "to acquaint people with our rich and varied flora and to arouse public sentiment to save our wild flowers which are fast disappearing in the so-called march of civilization."[63] Displays of over four hundred native plants arranged in order of bloom sequence delighted local crowds. The women's club urged that "the demands on the public purse because of the war should not lead to a neglect of the immediate demand of saving for generations to come the rich and varied flora in and about Chicago." Local railroads got in the act, and offered wildflower excursion tours, and train routes dedicated to botanical "rambles." Chicago's elite became avid proponents of wildflower preservation, and grew native plants on their estates. By the 1933 World's Fair, naturalized gardens and exhibits on the Indiana Dunes were mainstream attractions.

VEGETABLES: THE MEATPACKING CITY IMPROVES ON "HOG AND HOMINY"

Vegetables provided food for the family dinner table and later a source of income as Chicago's truck and market gardens became big business. Common root vegetables such as turnips and salsify were readily planted, but other vegetables were not as immediately popular. Tomatoes, for example, while cultivated by the French in Kaskaskia, Illinois, for years, were regarded skeptically in northern Illinois. There were then generally available two varieties of large reds and two of large yellows. Recipes were publicized to familiarize readers with tomato tarts, pickled tomatoes, and tomato figs (flattened and dried). The preferred cooking method recommended stewing peeled tomatoes in a slow fire for three or four hours, with brown sugar to reduce the acid along with a little salt.[64] Eating tomatoes raw was considered dangerous. By 1852, one writer revealed that he had dared to eat it raw, and gave detailed instructions for tomato culture (plant seed in deeply trenched soil in the fall, near a fence for training), and food preparation (red are for cooking, yellow should be sliced like cucumbers).[65] The same writer noted that a western medical professor prescribed a daily tomato in the diet for dyspepsia, indigestion, and various bilious disorders. Slowly, the tomato and other vegetables earned their place on the table.

Myths abounded regarding the amount of work required to sustain vegetable gardens. John Kennicott proposed sheltering hedges of hemlock or American arborvitae to protect vegetable gardens from winds. In the early 1850s, he recommended London's Charnwood and Cummins wholesale price list with over four hundred varieties of vegetables. Kennicott emphasized quality over quantity in vegetables and opined that the secret was to maintain good growing conditions and to harvest vegetables in their prime. He suggested early peas, lettuce, spinach, asparagus, radishes, turnip-rooted beets, and Early Horn carrots.

"Farmers, stand up for your rights," an 1860s front-page *Prairie Farmer* article urged. "Would you preserve health and study economy, mind the vegetable garden."[66] Vegetables (cooked, as there were still fears of the link between uncooked vegetables and various illnesses) were associated with health, moral character, and balanced minds, according to the editors, who proclaimed, "We are not vegetarians, but we are fond of well cooked vegetables and fruits of all kinds."

For years, Jonathan Periam was the acknowledged local vegetable expert, and often delivered lectures or articles on kitchen garden topics. In the late 1860s, he wrote, "Did it ever occur to you that you are perhaps not doing your duty to yourself and family in not cultivating a kitchen garden properly? That a well kept and properly planted acre in vegetables would make one-half the living of your family, which, with the product of your fruit garden adjoining, would render you happy and contented, and ward off dullness and disease, produced, perhaps, by a too liberal use of hog and hominy?"[67] The healthful benefits of vegetables for personal consumption were sometimes overlooked by their market potential. Periam's garden calendar for the year 1868 proposed the use of hotbeds and cold frames for sowing early lettuce, eggplant, tomato, cabbage, cauliflower, and radish in February. Succession planting of these crops and more continued throughout the season. In November, root crops were to be gathered and preserved in the cellar or pits. A big proponent of natural fertilizer, he asserted, "If you do not believe in heavy and annual manuring, double the dose of well composted manure upon your best land, and be converted; one hundred tons yearly will not hurt *good* land."[68]

In the mid-1860s, *Prairie Farmer* editors, still on the bully pulpit, recommended preserving vegetables for the winter. Most western farmers, the editors thought, subsisted on meat and but one vegetable, the potato. Simple preserving methods, without extensive sugar, would produce more satisfactory results, they advised. Preserved autumn and summer fruits could then be had liberally in winter months. Surely, the editors argued, a vegetable and fruit diet would be better than the "hog and hominy that habit and indolence have made the diet of too many in the West."[69]

James King featured vegetable seeds for both home and market gardeners. His 1899 catalog noted that he grew many of his seeds himself, but even more were procured from "specialists in various parts of this country and Europe." From the author's collection.

Vegetable gardens were important elements as suburban properties developed in the last quarter of the nineteenth century. Edgar Sanders devised a plan for the typical "sixty by forty" suburban garden. In this he proposed rows of melons, cucumbers, cauliflower, cabbage, carrots, beets, spinach, lettuce, and radishes. But more importantly in the later 1800s, vegetables were grown as cash crops. Certain vegetables were particularly good for market gardens. Onions were popular market garden crops. Chicago's name, after all, was said to derive from the Native American word for a local onion. In the 1860s, Danvers Yellow, Red Dutch, Early Red, and Silver Skinned onions were the preferred varieties.

Celery became a major export, grown by truck farmers in the 1880s on Chicago's north side. Edgar Sanders observed, "The Lake View Ill. soil … is of a black boggy material, plentiful mixed with sand, and but a few years ago … was every Spring a slough … There is no land which can compete with such as this for celery culture."[70] Sanders started celery from seed in greenhouses and sowed in the garden in May and June. Celery, tied in bunches of twelve stalks, was shipped from early July through December to all parts of the country. Once a rarity, celery sold for fifteen cents a bunch in 1888.

Buckbee's celery was featured in its 1906 catalog. Celery, once a delicacy, was grown widely in Chicago, particularly in North Side market gardens, and in nearby areas such as Rockford. Buckbee claimed it as the "world-beating" celery with a rich nutty color. From the author's collection.

In 1902, kitchen gardens were still popular in the city, with the *Chicago Tribune* reporting twenty thousand amateur kitchen gardens within the city limits, ranging from ten square feet to twenty-five thousand square feet.[71] The *Tribune* identified many of the kitchen gardeners as recent immigrants who raised vegetables as a matter of economic necessity. Home gardeners were urged to plant those vegetables that were most perishable—radishes, onions, new beets, lettuce, and parsley. Other staples such as sweet corn, potatoes, squashes, and beans took too much space and effort and could be better found at the grocer. The grocer would not miss the business, according to the article, and could devote the shelf space to more profitable items with longer shelf life.

In the first decade of the twentieth century, as new methods and products for canning became available, homeowners and "domestic science" schools emphasized vegetables that could be canned. New recipes were developed that did not rely as much on the pickling or salts needed in previous methods. The victory gardens of World War I revived interest in homegrown vegetables, which had become victims of Chicago's own success as a market center. Local nurseries, such as Vaughan's, promoted vegetable growing by offering recipes using "modern" methods of cooking and canning.

WINDOW GARDENS

With Chicago's long winters, indoor gardening was always popular. Wealthy amateur horticulturists constructed elegant conservatories to nurture plants through the cold months. But even modest homes could afford window gardening. Many middle-class homes in the city and suburbs featured large bay windows on the south side to capture the lingering rays of the sun. Displayed in

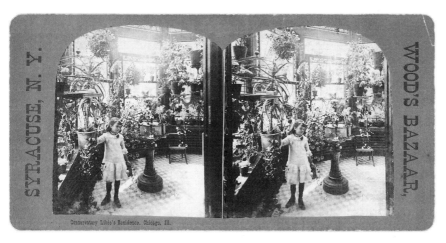

This view of a residential conservatory shows the profusion of hanging plants, windowsill plants, and even a terrarium. Ca. 1910. From the author's collection.

pots and Wardian cases (an early type of terrarium), or draped around windows,
houseplants became great entertainment for the lady of leisure.

Edgar Sanders acknowledged the unique difficulties of window gardening in
Chicago. He observed that typical household heat generated in Chicago homes
would force plants to grow tender shoots, leaving them as vulnerable as exotic
plants. "Window gardening, with the rigors of our winters, we confess to be attended with difficulties . . . and we opine the case will be frequently given up as
hopeless."[72] He suggested that town and city residents, "owing to the construction of the dwellings" would have an easier time than those living in country
homes, and suggested well-lighted basement windows. His favorite, most reli-

J. C. Vaughan chose to feature houseplants on the cover of his 1899 autumn catalog. Plants are displayed in
window gardens, more affordable than conservatories. From the author's collection.

able houseplants included Red Vein Indian mallow (*Abuliton striatum*) and nasturtium (*Tropaeolum lobbianum*).

Sanders also provided cultural information for specimen plants, which more often than not were exotics. Earlier greenhouse conditions were far too hot in winter, said Sanders, causing plants to create an indifferent show of flowers over a long period of time. To create a "brilliant display of blossom," Sanders recommended hard-wooded plants, small pots, and a greenhouse temperature of 40 to 45 degrees.[73] He also offered advice on training greenhouse plants, naming three training shapes: bush, pyramidal, and standard. Each plant was suited to a particular type; for example, flowers with drooping character should be trained with a single stem topped by a bushy head. He deemed this standard shape the ideal, and noted it was particularly hard for amateurs to achieve.[74] Rustic hanging baskets were considered high fashion for greenhouses or windows in the 1860s. Gardeners were urged to make baskets by covering wooden butter or soap bowls with acorns. If well-varnished, the baskets, strung with wire, were said to last two or three years. "Besides these," Edgar Sanders wrote, "there are also highly ornamental hanging baskets, sold at the fancy or variety stores, and some seed stores."[75]

The window gardens of Mr. Hochbaum of Chicago, circa 1884, were especially admired. The Hochbaum family lived in a "modest homestead erected immediately after the great fire," on North Wells Street near Division. In summer, the home was surrounded by masses of bedding plants. In winter, the "large windows, evidently built to accommodate the flowers as well as to give light to the room, are excellent examples of window garden embellishment."[76] On the main, or "parlor," floor, there were geraniums, callas, begonias, and palms. The dining room and kitchen also had windows filled with plants. Most impressive was the screen of heliotropes, which Hochbaum had propagated by himself in the fall and kept in a frame until the cold weather. Hochbaum built a homemade plant shelf to give seedlings better light and air. Special construction techniques were needed to adapt the typical window design from England for Chicago's colder climate (e.g. insulated sides and bases, and rolling blinds or outside shutters to protect against westerly winds).[77]

By the late 1880s, old favorites such as the waxplant (*Hoya carnosa*) were still popular, but garden writers were experimenting with new ideas in house gardens. Water plants, situated in a window aquarium, were suggested for interest and to block unsightly views in towns. Arrowheads (*Sagittaria*), bog bean (*Menyanthes trifoliata*), Cape pond-flowers (*Aponogeton distachyon*), eelgrass (*Vallisneria spiralis*), variegated arums, marsh marigolds (*Calthus palustris*), and "any of our small-growing native ditch plants" were recommended for the water garden. Growing mushrooms in a greenhouse at home was also recommended in the later 1880s. One simply needed to collect horse droppings and spread under the greenhouse stage—about a foot deep. The top was to be spread with dirt, and upon that, mushroom spawn, sold in bricks from seedsmen.

One of the key advantages to indoor gardening for Chicago gardeners was getting a head start on the growing season. Sanders recommended indoor seed starting as an economical way to obtain the very popular verbena. "If sown early in the spring in a pot kept in the window…the young plants will spring up, grow in the pot, while rude winds still sweep without doors."[78] Budget-conscious gardeners could also increase their supply of stock and geraniums, long-time favorites, by taking cuttings during the winter and spring months.

In addition to interior window gardens, Chicagoans, particularly in the city, were enamored of exterior window boxes. Window boxes offered a nice alternative for those with little yard space or small incomes. Wealthier individuals ordered preplanted gardens directly from the florist—window box and all. Others planted their own with either a purchased wrought-iron or wooden box, or a homemade planter using an old wooden soap or starch box. Trailing plants such as petunias or nasturtium were used to hide the rough-hewn boxes. Dwarf annuals, geraniums, snapdragons, and other readily obtainable plants were the usual fare. As more tenement districts sprung up during the early 1900s, civic organizations recommended window boxes for the poor as a means to beautify the city. These efforts were so successful that other cities around the nation emulated the window-box fad in Chicago.[79]

Indoors and outdoors, Chicago-area horticulturists explored plant possibilities and adaptations to the regional climate. Many new plants were discovered, many new methods created to cultivate old favorites, and perhaps the greatest treasure found was in Chicagoans' own backyard—with the beauty and hardiness of Chicago's native plants.

LEGACY LANDSCAPES: SHOWCASE PLANTS

Chicagoans' experimentation and love of plants was often seen in the extent of their private and public greenhouses. With improved greenhouse technology, it became possible for individual homeowners to extend the growing season and showcase exotic plants. The Garfield Park Conservatory is a unique juxtaposition of exotic plants showcased in a naturalistic environment designed by Chicago's famous prairie landscape architect, Jens Jensen. The Morton Arboretum, originally designed by an equally noted prairie landscape gardener, O. C. Simonds, also shows plants from around the world, in an outdoor museum. Lilacia Park, once the private home of Colonel William Plum, is an example of a twentieth-century trend toward specialty gardens featuring many varieties of a single plant.

Garfield Park Conservatory

300 North Central Park Avenue
Chicago, Illinois
http://www.garfield-conservatory.org/

This exquisite example of Jens Jensen's naturalistic interiorscaping is experiencing its second Renaissance. In the early 1900s, Jensen designed the new glasshouse to resemble a prairie haystack, and the interior planting scheme to mimic planted prairie river bluffs. Under the early leadership of Chief Florist August Koch, the conservatory established one of the greatest collections of plants in the world. After several decades of neglect, the conservatory has recently benefited from renewed public attention and funding and now hosts educational programs and world-class exhibits.

TOP A pre-Jensen view of Garfield Park Conservatory's old Floral Hall exhibits typical Victorian architecture and exterior plantings. From the author's collection.

BOTTOM Jensen's Garfield Park Conservatory was styled after a prairie haystack. The interior shown here included a naturalistically designed fern room, framed by sculptures by contemporary Lorado Taft. From the author's collection.

In eighteenth- and nineteenth-century England and America, greenhouses were little more than ugly museums for plants. Plants were artlessly displayed in plain pots. Heating pipes snaked around utilitarian tables used to hold the plants. Nondescript tubs of water provided plants humidity. There was little to distinguish the public greenhouse from commercial greenhouses.

Enter Jens Jensen. In 1911, *Country Life in America* magazine heralded Jensen's Garfield Park Conservatory as revolutionary: "The best new idea about greenhouses is landscape gardening under glass. For this inspiration we are largely indebted to Mr. Jens Jensen, a landscape gardener of Chicago." The idea was deceptively simple. Instead of stockpiling plants in an artificial environment, Jensen made the greenhouse as naturalistic as possible. He created a lawn of moss, and green hills of plants where none existed before. He planted directly in the ground, and hid unsightly pipes and heating equipment behind outcroppings of stone and concrete that blended seamlessly into the overall picture. Jensen also applied his trademark use of water elements. *Country Life in America* explained, "The ordinary greenhouse has an ugly tank under the benches in order to give moisture enough to the air. Jensen thought that a nature-like lake would make a prettier centerpiece for a greenhouse than a lot of benches covered with pots. This is so obvious that the world seems incredibly stupid not to have discovered and acted on the suggestion long ago."[80]

The pond in the conservatory's Fern Room was not only picturesque but also provided the humidity plants need. Today, the lovely pool and waterfall in the Fern Room hardly seem radical, but Jensen did not have a particularly easy time selling the concept of a new conservatory to his bosses. In the early 1900s, each of Chicago's three west parks had a conservatory. Yet all three in Douglas, Humboldt, and Garfield parks were in dire need of upgrades and maintenance. Jensen spoke to this effort in the March 1930 *Saturday Evening Post*: "I conceived the notion of putting all three conservatories under one roof, and since this was Chicago, and Chicago is accustomed to doing things on a big scale, I thought while we were about it, we might as well build the largest publicly owned conservatory under one roof in the world ... Many persons called it Jensen's Folley [*sic*]: they predicted that the glass wouldn't hold up in a building of such size, that a windstorm would smash it, that it couldn't be heated."[81]

Jensen also made a compelling economic argument, citing reduced maintenance and improved plant culture. With this logic, and by horse-trading new gardens for Douglas and Humboldt parks, Jensen got approval to build the new Garfield conservatory. The exterior of the conservatory was unlike anything previously attempted. Jensen thought the typical ornamentation on European glasshouses was inefficient for lighting and inappropriate for American architecture. Collaborating with Hitchings and Company, an engineering firm, an entirely new look was achieved. Jensen explained: "We did not want our greenhouses to look like a palace, a chateau or a Renaissance villa, but like a glass house for the cultivation of plants which everyone might enjoy. In order to fit them into a prairie landscape I thought they might well take the outlines of the great haystacks which are so eloquent of the richness of prairie soil."[82] Although this haystack icon from the agrarian era is less familiar today, the building's shape is still more aligned with its surroundings than would be a typical "wedding cake" style greenhouse.

On the inside, Jensen's artistic spirit led him to create a naturalistic scene appealing to all the senses. Instructing workers on how to build the waterfall in the Fern Room, he suggested they listen to Mendelssohn's "Spring Song" for inspiration. He struggled with details such as the green hill between the waterfall and brook, taking six months to get the contour of the slope correct. He also incorporated stratified stone into the edges of the room, to hide the mechanicals and echo the native rockwork of the Illinois countryside.

Jensen's so-called folly was a hit. Not only did most other public conservatories follow suit and make their displays more naturalistic, but private homeowners also adopted the new style. A *Country Life* reporter asked how this new landscape idea could be adopted to people of moderate means. Jensen replied that "all that is necessary is for clients to get out of the old rut and use a little imagination ... The conventional idea of a [home] conservatory is to have a few palms with which to decorate the house during festivals and some permanent plant decorations that are replaced from time to time. To break away from this slavery is a very easy matter if the architect will suggest a conservatory connected with the house, and will say that this will be a splendid place in which to serve tea or entertain friends with music and song."[83]

Conservatories thus became a place for plants *and* people. Jensen's "landscape under glass" idea had a profound effect on both public and private spaces, with homeowners reconsidering their own gardens—both indoors and out. But it takes plants to make a conservatory come to life. In 1912, the West Parks Board appointed August Koch, formerly of the Missouri Botanical Gardens, in charge of the conservatory and all outdoor plantings in the West Chicago Parks. Koch was able to obtain an impressive collection of plants—of interest both to scientists and to the general public—in a short period, while still remaining true to Jensen's aesthetic ideal.

Landscape Integrity

Throughout the years, the conservatory has been updated to reflect new educational opportunities. The various rooms in the conservatory have undergone changes in both name and plants. From that standpoint, the conservatory's Fern Room and Aroid Room are most historically intact. When a pipe burst in 1994 and destroyed much of the Aroid Room, the Chicago Park District and concerned citizen groups committed to a multimillion-dollar restoration program. Through this restoration, creative programming, and outreach, Garfield Park Conservatory is again becoming one of the city's most popular destinations. Although many of the spaces have been reconfigured to appeal to different audiences, the essence of Jensen's interior landscape retains his innovative spirit.

The redesigned main entrance leads to the soaring Palm House. This structure, although rebuilt in the 1950s, has always housed palms. On the north end of the central planting area, a huge Scheelea palm (*Ahalea phalerata*) towers about forty feet overhead. This tree was grown from seed collected in Brazil in the 1920s. The double coconut palm (*Lodoicea maldivica*) is one of only two in the world growing under glass (the other is in the conservatory in Kew, England). The double coconut was grown from seed—this plant has the largest seed in the plant kingdom, about a foot across—at the conservatory.

Two lovely statues by famed Chicago sculptor Lorado Taft frame the doorway to the center garden of the conservatory, called the Fern Room since at least the 1920s (once the Aquatic Room). Of the sculptures, Taft wrote, "The little whimsies have no great burden of significance. They represent care-free maidens and faun-like youths of some remote period as they might idle and play in the forests of Arcadia."[84] The Fern Room displays some of the conservatory's oldest plants, the cycads, which are about two to three hundred years old. Although the cycads were not original plantings in Jensen's day, many of the other plants in the room are likely offspring from some of the originals. The Fern Room exquisitely portrays Jensen's idealized image of a prehistoric swamp, with touches of Illinois prairie limestone outcroppings throughout. North of the Fern Room is the Show House, a room that has retained its same purpose since the conservatory's beginnings. Here seasonal flowering plants are brought in from the working greenhouses and placed among stationary background plants. In addition to showcasing year-round flowering plants, the conservatory today maintains the tradition of hosting major plant shows: chrysanthemum, holiday, and spring. In 1926, the chrysanthemum show drew over 170,000 people, and the Christmas show drew 45,000 people. The shows continue to be popular today.

West of the Show House is the Aroid House, featuring plants of the *Araceae* family (typically with showy spathes, such as our native jack-in-the-pulpit). This room began as the Conifer House, but was replanted in the 1920s as the Aroid House with more than three hundred species.

South from the Aroid House is the Desert House. This room was originally designated the Economic House,[85] where plants were grouped according to their trade value, that is, edible fruits, medical, perfume-producing, rubber- and resin-producing plants, and more. Still farther south, in the southwest room of the conservatory, is the present Children's Garden. This area was once called the New Holland House and, in the 1920s, became the new home for the Economic House. East of the Children's Garden is the Sweet House (formerly a Stove House, or Warm House, used for exhibiting plants with colored or variegated foliage), showcasing cocoa and other enticing plants. In 2007, this room was redesigned to house a major exhibit on photosynthesis, "Sugar from the Sun."

Despite decades of change, Garfield Park Conservatory continues to be a refuge for Chicagoans weary of urban or wintry sights. Thanks to Jens Jensen, August Koch, and the many dedicated staff and supporters of the conservatory, Chicago has a landscape beautifully preserved under glass.

The Morton Arboretum

4100 Illinois Route 53
Lisle, Illinois
http://www.mortonarb.org

The Morton Arboretum includes 1,700 acres of trees, shrubs, and plants collected from around the world and artfully arranged in botanical and geographical collections and horticultural displays. Founded in 1922 by Joy Morton, son of Arbor Day founder J. Sterling Morton, the arboretum showcases mature plant specimens, examples of early-twentieth-century naturalistic landscaping, and many preserved and managed native-plant habitats. The Morton Arboretum is the second oldest of the major arboreta in the United States.[86]

As an outdoor museum, the Morton Arboretum invites visitors to immerse themselves in its collections—without velvet ropes cordoning off the exhibits. Like a traditional museum, it contains priceless accessions, gathered through years of exploration and discovery. Yet in this living museum, the accessioned objects grow and change—not just through the years, but in every season. "An Arboretum, however, should be more than a museum. It should be a work of art, showing to advantage the hills and valleys, with the former emphasized by the growth of tall trees, retaining large open areas so that the foliage, the sky lines, and the reflections in water can be seen to advantage. In short, it should be a beautiful place, affecting one like a beautiful painting, a great musical composition, a poem, a magnificent building, or the work of the greatest sculptor; only the landscape art should surpass all the others in beauty of lines, color and composition."[87] So wrote O. C. Simonds, who developed the Morton Arboretum's first overall site plan. Simonds, an established landscape architect, or landscape gardener, as he preferred to be

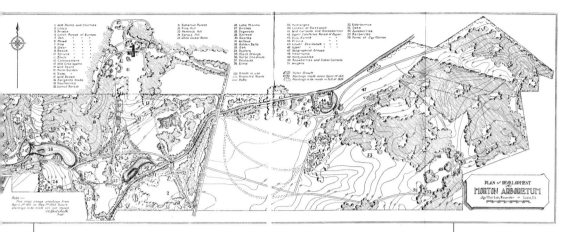

TOP This sketch of the Morton Arboretum shows the initial plant families and design elements included in Simonds's design. Ca. 1920. Courtesy of Sterling Morton Library and Archives, Morton Arboretum, Lisle, Illinois.

LEFT Today, the plantings around Lake Marmo have matured into a beautiful expression of naturalistic design. Ca. 2004. Courtesy of Morton Arboretum, Lisle, Illinois.

called, who championed naturalistic plantings and style, worked closely with Joy Morton to transform his country farm into the world-class arboretum that it is today. Prior to planning the Morton Arboretum, Simonds designed the Nichols Arboretum in Ann Arbor, Michigan, in 1906. At 123 acres, it was much smaller than Morton's property, but Simonds's extensive experience designing cemeteries, parks, and large estates; his visits to the Arnold Arboretum and famous parks and gardens abroad; and successful collaborations with botanists and former clients, convinced Joy Morton that Simonds was the man for the job.[88]

Joy Morton, a successful businessman, searched the Chicago environs before settling upon acreage near Lisle in 1909. The land included diverse exposures and soil conditions, and although ostensibly it was to be used for farming and a country retreat, the land ultimately made an ideal site for his arboretum. Morton's manor house, Thornhill, was named for the mature hawthorns standing in the hilltop meadow he selected as a home site. In 1921, Morton began planning the arboretum in consultation with Charles Sprague Sargent, the revered American botanist and director of Harvard University's Arnold Arboretum. Sargent offered formative advice, including recommendations on model European gardens and the hiring of Simonds as landscape architect. He also sent many rare plants and important books, and suggested the hiring of Henry Teuscher, the arboretum's first botanist.[89]

From the outset, Simonds's plan included botanical and geographical groupings, arranged naturalistically rather than in rows or blocks as had been done in some European arboreta. The map reproduced here, highlighting plantings from 1921 through 1925, shows a significant portion of native growth left undisturbed. Simonds arranged the growing collections around vistas wherever possible, as detailed in his early landscape plans for Pine Hill, Lake Marmo, and collections south of Thornhill. A native-plant lover, Simonds praised the remnant woodlands located throughout the grounds, which then, as now, included a rich understory of native shrubs and spring ephemerals such as hepatica and bloodroot.

The Morton Arboretum is bisected by Illinois Route 53, which divides the grounds into an east side and west side. The original Thornhill manor, located on the west side, was razed by the family after Morton's death and replaced with the Education Center in 1942; only the library wing remains. Now called the Founder's Room and filled with historical exhibits, it is a perfect setting to absorb some of Thornhill's period ambience. Another original building, a small stone outbuilding hidden in the trees just north of the Education Center (the former Thornhill site), was originally the chauffeur's cottage. On the arboretum's east side, near the intersection of Route 53 and Interstate Tollway 88, stand some of Thornhill's original barns. Originally referred to as "the south farms," they have been retrofitted by the arboretum and are used for equipment maintenance and other grounds-related purposes.

The real beauty and historical significance of the arboretum lies in its extensive plantings. As envisioned by Joy Morton and designed by Simonds and his successors, the young trees planted in the 1920s grew to become magnificent specimens, now framing the landscape and suburban skyline. Perhaps the best way to start a historic tour of the arboretum is on the west side, where, radiating out from Thornhill mansion, Simonds designed numerous landscape effects and features, in addition to preserving impressive stands of indigenous woodland. At the arboretum's original entrance, currently the southbound Route 53 exit from the west side, stands a grove of shade trees underplanted with giant yews. In 1925, Simonds wrote of this spot, "As one goes through the entrance, he faces ground descending to the east fork of the DuPage River . . . this valley, extending from the northeast to the southwest across the Arboretum and bounded by hills which rise in places to over seventy feet . . . is the leading landscape feature of this young institution. It is well to pause at the entrance and study the various outlooks."[90]

Slightly northwest of the original entrance, near Lake Jopamaca (named for Joy Morton and his brothers Paul, Mark, and Carl), Simonds designed the Ozarks Collection, which he described as "a series of low hills—the Ozarks—with valleys between. These hills will in time grow to be over 100 feet [presumably to the tops of the trees] in height but the valleys will remain low with here and there deep shade." Today, as Simonds predicted, the trees have grown, and together with the varied associated understory plantings, suggest the terrain and habitats of the Ozarks region: an innovative "immersion exhibit" created in a time when most plant-study collections were lined out in rows.

Lake Marmo, located just west of Lake Jopamaca along Lake Road, perhaps best reflects Simonds's naturalistic design approach. The lake, created by damming a tributary of the DuPage River east branch, is contoured into gentle curves around the base of the surrounding hills, and surrounded by a footpath that offers interesting vistas at every turn. Simonds's plan shows a mixture of deciduous and evergreen trees, and lush understory plantings of native elderberries, raspberries, Indian currant, pasture rose, and other woods-edge species. As Simonds noted, "There is hardly a point on this drive from which an attractive photograph might not be taken, showing the beauty of tree growth, varied topography, reflections in the water, and water margins of foliage. Spring beauties, wild violets … and blue phlox cover the ground in places."[91]

The arboretum was one of the first public gardens to include driving roads designed for the relatively recently introduced automobile. Joy Morton was a proponent of many new technologies in his day, and owned one of the first automobiles in the Chicago area. His personal library included books such as *Motoring Abroad* by Frank Presbrey, a 1908 book on taking a motor car through Europe. Morton later contributed to "the ideal mile" of the Lincoln Highway: the nation's first coast-to-coast concrete highway.[92] The sheer size of the arboretum, 735 acres at the time of Morton's death in 1934, almost necessitated the use of motorized vehicles.

Simonds created gently winding roadways that followed the natural curves of the land. In designing the arboretum, he proposed many walking trails because "one cannot study trees and shrubs successfully by riding. To learn their characteristics he must walk and examine the leaves, buds, bark and flowers carefully, and so paths have been planned. From these paths the landscape effects should be pleasing and varied. From certain points there should be distant views and from other points one should be able to study compositions of trees and shrubs showing how they can be used to advantage in parks or private grounds."[93]

Where roads were necessary, Simonds proposed attractive landscaping such as a sweep of cherry trees along the Joliet Road (now Route 53). "The highway divides the Arboretum into two parts, and cherry trees of all kinds are to be planted on both sides of this thoroughfare, so that in years to come pilgrimages may be made to it at cherry blossom time just as at present such excursions are made in Japan." In the 1940s, when Joliet Road was developed into a state highway, it was lined with drifts of flowering crab apples—many of which survive today.

Working with Simonds's job-site superintendent, Clarence Godshalk, Morton created semiformal enclosed gardens of peonies, roses, and plant oddities along Joy Path, the trail that runs south from Thornhill alongside Sargent's Glade. Although these gardens no longer exist, remnants can be seen along Joy Path: beds of peonies and other perennials at the top of the trail near Thornhill, and lower downhill, a large arborvitae hedgerow—and a tree-form trumpet vine from 1922, the last of the "plant oddities" still remaining.

On the east side near the visitor center, several horticultural displays remain from the early decades of the arboretum, including the formal Hedge Garden (planted in 1934) and the original portion of the curvilinear Ground Cover Garden (planted in 1936). Although these gardens and the nearby Pinetum (planted in 1934) have changed and expanded over the decades, all retain their original design character and intent. Throughout the arboretum, there are many historic plants and plant groupings on both the west and east sides.

On the west side, Hemlock Hill and Spruce Hill, located west of Lake Marmo, were planted in 1922 and offer a dense glen of mature evergreens, delightfully shady in the summer and beautiful when lightly dressed with winter snow. The two-and-a-half-acre Arborvitae collection east of Lake Marmo, started in 1934, contains many older varieties of this popular Midwest landscape plant. Other collections on the west side that date to the 1920s and 1930s include the ash, cashew, and olive families. There are some remarkable native-growth oak trees, including those surrounding the Morton family cemetery; the State of Illinois millennium oak, growing just northwest of Lake Marmo and whose mighty branches might have sheltered the Potawatomi tribes; and several scattered throughout nearby Daffodil Glade—some of them likely dating back three hundred years.

On the east side, some of the best collections of elms and lindens can be seen. The arboretum is especially known for its work in developing new elm trees. The elm collection, north of Meadow Lake, was started in 1926, and now has more species than any similar collection in the United States. The adjacent linden collection, begun in 1924, has many mature specimens—including one Zamoysk's linden (*Tilia* × *zamoyskiana*) with a trunk diameter of sixty inches. The juniper collection, located on the far southern edge of the Pinetum, was begun in 1934.

The plants of China and plants of Japan collections were started in the mid-1920s. Areas in both countries have climatic conditions similar to the upper Midwest, and arboretum-sponsored seed- and plant-collecting expeditions have been made there since the beginning. The five-hundred-acre East Woods, located on the far eastern part of the arboretum, originally was named King's Grove for Sherman King, a second lieutenant in the Black Hawk War of 1832, who briefly settled in the area. Most of the East Woods was divided into hundred-acre woodlots at that time, and a few parcels were logged off completely. These were later planted with experimental forestry plots, as Joy Morton wanted to see which tree species were commercially viable in the Chicago area.

Landscape Integrity

As a living museum, the arboretum is expected to grow and expand, while conserving its priceless plant treasures. Throughout the years, new landscape features have been added (e.g. ponds for improved drainage, and new horticultural displays to address contemporary themes in gardening), but in general, the Morton-Simonds landscape aesthetic of curving lines and naturalistic plant arrangement continues. New plant accessions are added each spring and fall, with planting locations determined by the curator, landscape architect, and horticulturist. With only four successive generations of directors and landscape architects since its founding, the arboretum has benefited from a continuity of vision and design philosophy.

Lilacia Park

150 South Park Avenue
Lombard, Illinois
http://www.lombardparks.com/parks/lilacia.htm

As greater success in hybridization occurred, more people became passionate about creating gardens with a single theme or collecting a single type of plant. Specialty gardens became prominent, such as this lilac garden for Colonel William Plum. Plum, an inventor, war hero, and amateur

horticulturist, had one of the most extensive collections of lilacs in the world. Today only a few out-
buildings remain of his original home; however, much of his lilac collection remains. Jens Jensen,
foremost prairie-style landscape architect, redesigned much of the grounds when it was turned
into a public park.

Colonel William Plum tends
to just a few of the lilacs in his
extensive collection. Ca. 1920.
Courtesy of Lombard Historical
Society.

Lilacia, the estate of early horticulturist William Plum, was a world-class attraction
in the 1920s. One of Lombard's leading citizens, Plum helped put the town on the map
along with its moniker as the Lilac Village. Today, Plum's homestead showcases not only
his exquisite collection of lilacs, but also the handiwork of Jens Jensen, who later transformed the grounds into a beautiful public park.

In 1867, newlyweds William Plum and his wife, Helen, bought a small house on two
and one-half acres in what was then called Babcock's Grove. A native Pennsylvanian, the
twenty-two-year-old Plum had just started his law practice in Chicago after serving with
distinction in the Union Army. He and Helen, both garden lovers, preferred a rural setting to city life. Babcock's Grove, renamed Lombard in 1869, suited the Plums perfectly: a
small farming community linked to Chicago by the Chicago and Northwestern Railroad.

The Plums began influencing the Lombard landscape in 1876, when they deeded part
of their property to the village contingent on preserving all of their black walnut trees.
This tree-saving trend continued. Many old black walnuts lining the streets' parkways
today may have once been on private property. But the Plums' most dramatic impact on
Lombard began after a European trip in 1911. Their itinerary took them to the Le Moine
Lilac Gardens in Nancy, France. Here Plum came across 'Mme. Casimir-Perier' and 'Michael Buchner', respectively white and lilac-colored varieties of lilac. There the lilac love
affair began.

Plum brought two bushes home and, over the next decade, began a collection of lilacs said to have exceeded four hundred cultivars. French, German, and Asiatic varieties
were imported. Perhaps because they reminded him of his Civil War days, Plum included
lilacs with American names like 'General Grant', 'General Sheridan', and 'Abraham Lincoln.' The Plum residence, which they renamed Lilacia, became a popular tourist spot.
Local historians equated the popularity of Lilacia with major horticultural meccas in the
United States. Thousands of garden enthusiasts from around the world visited the Plum

home each year, to see the hundreds of lilac bushes planted on two and one-half acres. Inevitably, Plum found himself drawn into an informal plant supply business. Gardeners from all over Chicago and beyond wrote to ask for tours or lilac cuttings from Lilacia. Professionals from the Morton Arboretum and Harvard's Arnold Arboretum exchanged correspondence as well as cuttings with Plum.

At one point, after his wife's death in 1924, Plum considered selling his home. He produced a circular describing the property, reproduced below. (Interestingly, although Plum describes his extensive gardens, he lists only four lilac bushes. Local historians speculate that he planned to take his beloved collection back to Ohio.) It is clear from the emphasis on horticulture in this circular that Plum felt his landscaping was integral to a successful sale.

Corner lot

350 ft. Front by 264 ft. Deep,

also large lot in rear:

Shade trees: 1 salisburia, 1 cut leaf birch, 1 extra large and imposing silver leaf poplar in front lawn; 1 yellow poplar, 2 catalpas, a splendid row of large bearing black walnuts, large wild cherry trees, a schwerdler and other choice maples, 1 mountain ash, also thorn redbud, horse chestnut and ash trees.

Fruit trees: About 40 bearing apple trees mostly very choice and all good bearers, 7 young trees, 5 choice pear trees, 5 cherry, 3 plum, 1 mulberry.

Shrubs and bushes: Snowball, acacias grafted on locust, a superb sight in bloom, bridal wreath,... holly, snowberry, elder, almond, burning bush, tamarisk, barberry, *4 kinds of lilac*, 8 bushes, 3 rose gardens...

Vines: Wild grape, clematis, Virginia creeper, scarlet coloring woodbine that clings under the east and south cornices, and also droops over the front piazza, myrtle, trumpet, cinnamon.

Flowers: 20 peonies, various colors and shades, lilies of the valley, and various other lilies, rudbeckia, phlox and many other perennials; asters, 6 week stocks, cosmos, scarlet salvia... and many other bedded plants.

Two excellent vegetable gardens in fine state of cultivation.[94]

In his will, Colonel Plum left his beautiful Lilacia estate to the village of Lombard, to be maintained as a public park and library. In 1929, two years after Plum's death, the village acquired additional acreage next to his property. Jens Jensen, Chicago's celebrated landscape architect, was retained to transform the space into a park. Jensen turned the existing lily pond into a stunning pool surrounded by his characteristic stratified stonework. Winding paths, massed groupings of spring bulbs and shrubs, and open vistas grace the eight-and-one-half-acre Lilacia Park today.

The central attraction remains the profusion of lilac bushes throughout the park. More than one thousand lilac bushes, representing about two hundred varieties, bloom in the spring. Some of the lilac bushes trace their roots back to Colonel Plum's original plants and are still in their original positions. Positive identification of many of the lilacs is a challenge. Park horticulturists have noted that, like many gardeners, Plum did not maintain very precise notes. Early plans show that Plum lined his lilacs up in rows like a nursery. Other notes show where Plum rather unhelpfully described a plant's location as "four steps from a chicken coop." Despite these "hen-scratched" notes, park horticulturists have been laboriously accumulating information about each plant's provenance. They can now identify about half of the lilacs in the park. Some of the lilacs in Lilacia Park could be extremely rare. Modern-day nurseries have discontinued many of the varieties in the collection. 'Dame Blanche', a double white cultivar, is one example. The Lombard Park District is committed to its stewardship role in preserving these unique, historic plants.

Landscape Integrity

Evaluating the preservation status of this landscape requires an awareness of two significant time periods: Plum's original plantings and Jensen's renovation. The Jensen-designed rockwork and pool are exquisite, and the amount of original or descendant lilac plantings is outstanding. With only one outbuilding remaining from the original homestead, however, the spatial relationships among garden features are drastically changed. Circulation patterns have naturally been altered to accommodate the property's use as a public park.

FOUR

CHICAGO BY DESIGN
A Blank Canvas for Cultivated Gardens

No portion of our country offers a wider or more hopeful field for the landscape gardener than the rich prairies of the West. In those parts of the country where mountains lift their green, wooded sides and high peaks, where the rocks lie piled in bold cliffs, where the rivers pour down in cataracts, or the broad sea stretches out, there is less need of man's hand . . . But here, on the prairie, nature has left a broad, gently undulating and usually treeless surface—a wonderful canvas for man to make pictures upon.

EDMUND HATHAWAY

Yet I was repeatedly told on first coming to Chicago, two years since, that I should find but little demand for my services; that the people of the West were too much occupied with the material necessities of a new country to think of ornamental work, which must be left for a later generation possessed of more leisure and means to devote to artistic luxuries.

H. W. S. CLEVELAND

The art of gardening on Chicago's blank landscape canvas was at odds with the science of making money. Real estate speculators and immigrant farmers vied for any available acreage. Whirlwind land swaps were made by absentee landlords whose goal was selling the land at a profit—not cultivating it for posterity. Farmers viewed their land more permanently, but still kept a vigilant eye on the bottom line. Cash crops were king with Chicago farmers; ornamental gardens were a luxury. Despite this short-term outlook, beautiful gardens did take root in Chicago. Horticulture was viewed as a sign of culture, and wealthy landowners nurtured a strong partnership with their hired gardeners. Through shows and experimentation, knowledge and design concepts were freely exchanged for the benefit of all Chicagoans, regardless of financial position.

Early Chicagoans were very aware of national landscaping trends. Many wealthy Chicagoans had permanent, beautifully landscaped residences in New York, and were well traveled. Chicago horticulturists such as John Kennicott, Edgar Sanders, and Jonathan Periam wrote for multiple national horticultural journals and presented or attended many national or state horticultural society meetings. The writings of Frederick Law Olmsted were reviewed in local papers as early as the 1850s, and Andrew J. Downing was a household name in many western homes. Recognizing the pioneering landscape architect's importance, both the Northwest Pomological Society and groups spearheaded by W. B. Egan sought to memorialize Downing upon his death in 1852.

Yet Chicago's first gardens were utilitarian and lacked aesthetic coherence. Pioneer conditions and the 1837 financial panic curbed enthusiasm for luxuries such as garden design. The economic downturn was not sufficient, however, to permanently dampen Chicagoans' desire to improve their lots.

PLANS AND PLOTS OF THE NINETEENTH CENTURY

In the 1840s, individual finances had sufficiently recovered from the 1837 financial panic to turn attention to landscaping. Goals for these early landscapes were modest. Successful cultivation of fruit-bearing trees and simple garden flowers were considered major accomplishments. Garden design was the province of local nurserymen, whose main interest was selling the flowers they grew. One such early nurseryman and designer in Chicago was F. H. Hastings, proprietor of the Chicago Garden and Nursery, located two and a half miles south of the city on the Chicago River. Hastings had previously worked with renowned nurseryman Benjamin Hodge of Buffalo, New York. In the 1840s, Hastings continued to order stock from Hodge's nursery, ranging from six-foot-tall horse chestnut trees to dahlias, a popular flower at the time.

In the earliest Chicago landscape designs, Hastings and his contemporaries advocated simple flower beds designed in geometric shapes such as rectangles or circles. In 1847, Hastings published a garden plan for an 80-by-40-foot plot. The plan showed a formal arrangement of concentric circles bisected by 4-foot-wide gravel walkways. The center bed was raised above the rest, and arbors marked the entrance to the garden. Grass plats interspersed with shrubbery surrounded the flower beds. Hastings recommended easy-care plants such as tulips, hyacinth, and narcissus. Crown imperials, gladiolus, iris, and lilies were also among his favorites. Spring flowering bulbs were to be displayed in beds to showcase the blending of colors. Pansies (*Viola tricolor*) were prized, and local tastemakers suggested mounded plantings in elliptical shapes. Annual vines such as morning glory were trained under windows, where "the effect is pleasing, especially when the house happens to be of rough logs."[1] Landscaping suggestions were thus provided for simple abodes as well as for grand ones.

The 1840s brought renowned American horticulturist Andrew J. Downing's seminal treatises on landscaping, and these were interpreted through Chicago's European-trained gardeners and local nurserymen. An 1849 *Prairie Farmer* article featured private gardens in Chicago, including those of Mayor William B. Ogden, W. B. Egan, and John Kinzie, then president of the fledgling CHS.

Like many eastern investors, William B. Ogden maintained a primary home on the East Coast. This New York home, newly built for Ogden in the 1870s, was dubbed Boscobel after a romantic forested landscape, and was designed by Calvert Vaux, Frederick Law Olmsted's partner. Note the sweeping, curved driveway and newly planted trees placed in a graceful arc. Ca. 1870. Courtesy of William B. Ogden Free Library, Walton, New York.

Touring Kinzie's grounds at Michigan Avenue and Cass Street, the writer commented on Kinzie's horse chestnut, Siberian crab, pear trees, quinces, and catalpa trees. Most had attained a good size in the sixteen or so years Kinzie resided there, but many lacked fruit. Dahlias and rose of Sharon were mentioned briefly as adorning the Kinzie's garden. Midwestern horticultural writers of this age were often more intent on plants' cultural requirements than on garden design.

The garden design for Justin Butterfield, a Chicago lawyer, was praised for its layout, "so as to give the visitor a chance to wander about in it for a long time and still find something new; an end first to be studied in laying out such a garden."[2] Butterfield's garden, on Michigan Avenue and Rush Street, was admired for its variety and profusion of flowers. Grafting experiments were also in evidence in his garden, with perpetual roses such as 'Prince Albert'. Butterfield's garden showcased evergreens and larches and, like Kinzie's garden, also favored balsams. Outbuildings in both gardens included a summerhouse, covered in native grape, *Vitis incisa*.

Like Mayor William B. Ogden, W. B. Egan was both a civic leader and an avid horticulturist. This view shows the extensive garden trellises and vertical accents. Egan's gardener, most likely trained in Ireland, would have planned the garden to include attractive views as seen from the second floor. Ca. 1860. Courtesy of Chicago Public Library, Special Collections.

The gardens of William B. Ogden and W. B. Egan were undisputedly hailed as Chicago's showplaces of the 1840s. Both Ogden and Egan were charter members of the CHS, and both had gardens designed and maintained by European gardeners, probably from Ireland. Ogden's four-acre estate on Chicago's north side was remarkable for its heavily forested appearance. Egan had both a private residence in Chicago and a country farm on the South Side. Their gardens, and their written contributions to horticultural publications, influenced homeowners of all income and social classes.

Cemeteries Bring Life to Gardens

The many statewide horticultural organizations launched in the early 1850s (e.g. Illinois State Horticultural Society, Northwestern Fruit Growers Association) may have diverted membership in local gardening groups. John Kennicott noted that nearly half of the dozen Chicago nurseries had closed "for want of patronage . . . Our Chicago Horticultural Society, which started off so handsomely at first, meets . . . semi-occasionally only, in fact, once in two years."[3] In fact, the perceived decline in gardening prompted the *Prairie Farmer* to opine that Chicago's Garden City nickname was no longer appropriate for "a place with 50,000 people and scarcely fifty good gardens for the whole."[4]

In the mid-1800s in America, old churchyards gave way to planned, "rural" cemeteries. Throughout the country, a spate of cholera epidemics and other diseases caused health reformers to promote relocating cemeteries outside of city limits, in more rural areas. Boston's Mount Auburn Cemetery was the first, in 1831, supported by the Massachusetts Horticultural Society. Philadelphia followed suit with Laurel Hill in 1836. In the Midwest, the Cincinnati Horticultural Society created Spring Grove in 1844, and St. Louis's Bellefontaine was established in 1849.

Mount Auburn, a national historic landmark, was designed by gifted amateurs from the Massachusetts Horticultural Society. Using Père-Lachaise Cemetery near Paris, France, as a model, the designers also borrowed from English picturesque gardens to transform this erstwhile mature woodland.[5] The forerunner of the rural cemetery movement, Mount Auburn was designed for the enjoyment and serenity of the living, and adorned with statues, fences, and lush plantings. It has been described as a horticultural collection of period rooms. The founders of Laurel Hill, also a national historic landmark, employed a competition to identify the winning designer of the new cemetery. John Notman, a Scottish architect, won the contest, and conceived the site to take advantage of Schuylkill River views and rolling terrain. Adolph Strauch, a Prussian landscape designer, is credited with much of the design at Spring Grove, Cincinnati's historic cemetery, which is said to demonstrate the birth of the cemetery "landscape lawn" plan. Open green spaces and small lakes integrate the landscape, and an attempt was made to sublimate statuary and manmade structures to natural features. Strauch was consulted by Chicagoans in designing their cemeteries.

Cemetery landscapes revitalized Chicago's interest in horticulture. Midwesterners emphasized public cemetery landscaping in the 1840s and 1850s, following both European and American trends and increased emphasis on public health. After Chicago's first steam-engine train, the *Pioneer*, made its inaugural run in 1848, burgeoning railroad systems made outlying areas, including public cemeteries, accessible. Rosehill and Calvary cemeteries were formed in 1859. Rosehill Cemetery was located along the existing Northwestern railway, and Calvary, a Catholic cemetery, was reachable by Sheridan Road. Graceland Cemetery was established in 1860, along the improved Clark Road. On the South Side, the Oak Woods Cemetery Association was formed in 1853, and became even more accessible via a new Illinois Central rail line in 1866.

Rosehill was hailed as Chicago's first public cemetery in the new landscaping style. Its dedication on July 28, 1859, drew an estimated ten thousand people.[6] On advice from John Jay Smith, president of Philadelphia's Laurel Hill cemetery, Rosehill's board of managers hired Philadelphia landscaper William Saunders, a frequent guest at Chicago's local garden and horticultural society meetings, to lay out the grounds. Unlike eastern cemeteries such as Mount Auburn or Spring Grove, which were backed by horticultural societies, in true Chicago form, Rosehill and Chicago's other major cemeteries were products of real estate speculators and industrialists. Charter members of the Rosehill Cemetery Company included William B. Ogden, John Kinzie, James H. Reese, Charles V. Dyer, Andrew T. Sherman, and thirteen other businessmen and industrialists.

Rosehill's dedication pamphlet described the picturesque advantages of the site—available to its somnolent denizens at a price. Family lots ranged from $200 to $600, depending on location. The 1859 brochure noted, "As in cities, and elsewhere, *location*, and the beauty and other advantages of particular spots, is regarded in fixing the prices." Typical of Chicago real estate, choice lots were deemed those near the artificial lakes Saunders created, and near groves of trees. Since much of Chicago was swampland, finding elevated sites for cemeteries was much harder than in East Coast cities. Rosehill was located on a gravel ridge, and its promoters touted good soil drainage and elevation. "There is *no possibility of water standing*," the promotional brochure boasted. "Graves of full depth may thus be dug at all times of the year."[7]

Besides the practical advantages of a ridgetop, Rosehill's site made good use of borrowed views—one of the first recognitions of scenic prairie landscape. In his dedication address, J. V. Z. Blaney, president of Rosehill Cemetery, noted, "At the east of us and so near at hand lies the lake, that its waters become a very important feature in the landscape. At the west the prairie spreads away to the setting sun, now dotted with the signs of a dense population where but a few years ago the Buffalo was monarch of the grassy plain."[8]

Rosehill Cemetery, although a product of real estate speculation, had a marked influence on Chicago society. A modern cultural institution and recreation outlet, Rosehill bylaws specified new visitor etiquette that placed a premium on the landscape. Dogs were not allowed, nor picnics nor refreshments. Smoking and firearms were also prohibited. Children had to be supervised and horses tethered securely, not hitched to precious trees. Visitors were to stay on the walkways and "not pluck flowers nor injure any plants, shrubs, or trees, nor remove anything from the grounds." Citing the success and ennobling influence of predecessor American cemeteries such as Mount Auburn, Laurel Hill, and Spring Grove, Blaney noted, "Rather let me point you to the cemeteries which adorn the suburbs of many of our sister cities, where horticulture and landscape gardening, and monumental sculpture have achieved triumphs . . . there can be no question but that the rural cemetery . . . satisfies every demand of a Christianized and enlightened humanity."[9] In addition to creating a much-needed civic

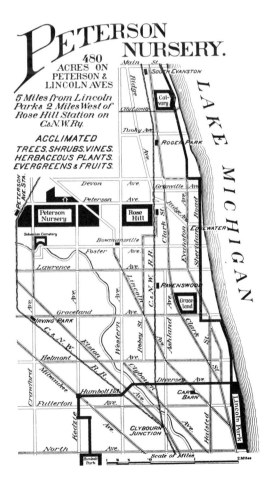

Like many early florists, Peterson Nursery advertised its location near several cemeteries, particularly Rosehill. In addition to providing flowers for the overall landscape, plants for the home could be ordered from Peterson's. From the author's collection.

improvement, Chicago cemetery founders hoped to obtain a nice return on their investment. Profit motives raised criticism from some horticulturists, and may have resulted in poor landscape maintenance in the early years. Edgar Sanders noted: "The best examples [of cemeteries] in the country, possibly, are Forest Home of Milwaukee, Laurel Hill of Philadelphia, Spring Grove of Cincinnati and Greenwood of New York. We have some fair bits of grounds in Chicago, not like these, however. Ours were hardly started right; that is, there are [sic] too much money speculation. This should not be in God's acre; they pay no taxes, and every dollar over and above costs of management should be expended in adornment of the grounds."[10]

Despite Sanders's criticism, cemeteries offered great opportunities for Chicago's nurserymen and florists. Rosehill's dedication brochure described "large and handsome greenhouses and conservatories" on the grounds for "raising plants, vines and flowers for the ornamentation of family burial lots, urns and vases."[11] Nurseryman P. S. Peterson was associated with Rosehill, and many other nurserymen and florists obtained a steady business from cemeteries.[12] Like many of his contemporaries, Peterson's own grounds were well appointed with trees and examples of his design and horticultural prowess. M. N. Angelsberg maintained an office at Calvary's greenhouses, head gardener Alexander Reed maintained Oak Woods Cemetery, and noted horticulturist Willis N. Rudd managed Mount Greenwood Cemetery.

Graceland Cemetery proved strong competition to Rosehill. Formed in 1860 by Thomas B. Bryan and colleagues, Graceland became the fashionable address for the final repose of many of Chicago's leading citizens. Despite Edgar Sanders's accusation that cemetery investors were absentee landlords, Bryan was a hands-on manager, and made daily visits to the new cemetery. In a letter to a disgruntled investor, Bryan wrote that "I rise by dawn, and travel twenty four miles to the Cemetery every morning (Sundays excepted), before other people are generally out of bed. I remain on the ground often all day without even stopping for dinner, and for my personal services, (greatly neglecting other business), I charge you nothing."[13]

Chicago's cemetery landscaping evolved from European or East Coast traditions to regional styles. Initial landscape plans for Rosehill and Graceland were drawn by an "imported" landscaper, Philadelphian William Saunders. Oak Woods was designed by Adolph Strauch, creator of Cincinnati's Spring Grove. Most early designs featured curvilinear walkways and drives, artificial lakes, and broad expanses of lawn so typical of the rural style. But nonnative plants were used, and individual lot owners persisted in personalizing their relatives' graves. Blaney's Rosehill dedication remarks celebrated typical exotic plantings: "Here cedars from Lebanon, and firs from Norway, the pine of Austria, and the arborvitae of Siberia shall mingle their shadows together."[14]

Popular taste lagged behind leading cemetery landscape ideals of a pastoral garden with smooth lawns unbroken by individual monuments. At Rosehill, the issue became heated, as reported in an 1877 *Chicago Tribune*: "There came up yesterday before Justice Meech the assault and battery case of Mr. Hanson, the Rosehill florist, vs. Mr. Anderson, the person to whom the Rosehill Cemetery Company had given the privilege of watering and fixing up all the lots there." Hanson, hired by lot owners, apparently engaged in fisticuffs with Anderson, the Rosehill Company–approved landscaper. Judge Meech upheld lot-owners rights, and according to the *Tribune*, "the Company has rescinded its obnoxious order," and unified cemetery design took a step backward. A few years later, Graceland struggled with lot owners' individual tastes. In 1885, an irate lot owner wrote to the *Chicago Tribune* about Graceland's new policy against planting flowers on Sunday. The writer accused Graceland Company of trying to "monopolize the flower business" with its own greenhouses. Bryan Lathrop, then president of the Graceland Company, argued that the new law was to prevent the disruptive practices of digging on the Sabbath.

Author Andreas Simon described Rosehill in the early 1890s, and identified the "erroneous practice of earlier days" when individual graves were fenced in. He observed, "We see plainly the difference between the old and new system [of landscaping]. On the one side we behold the irregular mass of grave stones forming an unsightly chaos with the rusty, partly broken down iron fences, the dilapidated and crumbling stone walls, the wild shoots of grass and the neglected graves, and beyond the bright beauty and symmetry of smooth and green patches of lawn, by which the graves are enclosed."[15] Of Graceland, Simon noted, "At the time when the older sections of this cemetery were first laid out . . . it was still the fashion to surround the family lots with low stone walls or fence them in with iron railings or natural hedges and then to adorn them with monuments and gravestones . . . Of course at that time this ancient system had not as yet been recognized as a mistake." Simon traced the improved, "modern" style of landscaping in Graceland to the hiring of prairie landscape architect O. C. Simonds as landscape gardener in 1878.

Simonds, in turn, credited much of the design sensibility to his patron and mentor, Bryan Lathrop, Thomas B. Bryan's nephew. Lathrop served as president of Graceland Cemetery and first met with Simonds in 1878. As Simonds described the cemetery then, "the new and undeveloped portion [of the cemetery] was then low—partly swamp, partly slough, and partly a celery field."[16] Simonds provided his much-needed engineering skill, and worked with then chief Graceland architect and landscape architect William Le Baron Jenney. Simonds listed the plants brought in for this new area, mostly natives: elms, oaks, maples, ashes, lindens, cherries, hackberries, pepperidges, and hawthorns. Shrubs included dogwood, viburnums, prickly ash, elderberries, chokeberries, sumacs, and oth-

Curved roadways, massed tree groupings, and wide lawns were celebrated as the parklike qualities of the "modern" cemetery. 1894. From Simon, *Chicago: The Garden City.*

ers brought from the local Naperville nurseries and Ellwanger and Barry. At Lathrop's suggestion, Simonds visited Adolph Strauch, the landscape architect for Cincinnati's Spring Grove cemetery, as well as other cemeteries and parks in Cleveland, Buffalo, Troy, Boston, New York, Philadelphia, and Washington. Simonds thus brought the best of American landscape design to Chicago, accented with a regional twist.

Chicago's cemeteries served to introduce the middle class to the latest style in designed gardens. Influential local garden writers started promoting aesthetic garden planning. John Kennicott's writings began to emphasize design style along with nuts-and-bolts horticultural practices. Landscape designer Andrew J. Downing's influence on Kennicott is apparent in his recommendations for a more natural style of gardening. "In laying out a flower garden, some TASTE is necessary, or much of its effect will be lost," wrote Kennicott in 1854. He advised against straight lines, and proposed curving walks and "beds of a *natural* outline ... Let your walks be broad enough for a lady to pass, without brushing ... plants with her dress."[17] He recommended siting flower gardens near the dwelling, in full view of the room most occupied.

Grass lawns made their debut in Chicago in the 1850s. Greatly admired from the European tradition, velvety lawns were signs of leisure and taste. Local horticulturists penned advice on adapting lawns for our region. Local nurseryman John Ure, not completely enamored of the lawn trend, observed that European lawns were valued for contrast with the rest of the land under cultivation. "But, here we have lawns—natural lawns enough, and large enough," Ure observed of the boundless prairies. He prioritized healthy vegetable gardens over lawns. Still,

TOP & BOTTOM The lawn of the Bowen home in Elgin included a wide-open expanse in the front for croquet or simply strolling, and a secluded flower garden on the side for cut flowers. Courtesy of Gail Borden Public Library District, Elgin Illinois Postcard Collection, Digital Past (www.digitalpast.org), initiative of the North Suburban Library System, Wheeling, Illinois.

if lawns were desired, Ure suggested grazing sheep as an inexpensive alternative to constant mowing or trimming with a scythe. Lawns were useful, he conceded, because they showcased flower beds, "like a diamond set in gold."[18] The *Prairie Farmer* praised lawns "soft as velvet and green as the Emerald Isle," but noted, "To have a satisfactory lawn in such a climate as ours, where the winters are so changeable and the summer sun at times so overpowering, it is absolutely necessary that the soil be thoroughly prepared."[19] Readers were also advised to underdrain and protect their lawns with belts of evergreen shrubbery or of native dwarf-growing trees such as redbud, crab apple, and wild plum.

Bedding plants also enjoyed a heyday in the late 1850s with Edgar Sanders recommending over thirty varieties of verbena alone. He cautioned that some "new foreign kinds" might burn during Chicago's hottest suns. Unproven newcomers included 'Charles Dickens', a rosy lilac with a dark center that was fine for pots; 'Chieftain', a dark maroon crimson with a light eye; 'Le Gondolier', a soft rosy carmine; 'Lady Palmerston', a delicate pale blue with large white eye; 'Monarch', a showy crimson purple; 'Purple Perfection', a fine maroon purple with light eye; 'Prince of Wales', bright ruby crimson with large lemon eye; 'Rosy Gem', a showy, brilliant deep rose; and 'Sir Joseph Paxton', a light rosy crimson with large lemon eye.[20]

By the end of the 1850s, garden design had sufficiently matured that writers proposed gardens with year-round interest for the Midwest's harsh winter climate. In an essay on winter gardens, one *Prairie Farmer* writer suggested an oval garden in a sheltered nook, "made to resemble a forest glade or opening." Thick belts of evergreens would surround the circle: Norway spruce on the outside and dwarf arborvitae on the inside. Specimen trees with interesting bark such as golden willow would provide contrast. Instead of grass, evergreen vinca would be used, and narrow borders of early blooming flowers (crocus, Persian iris, and Due Von Tholl tulips) were recommended. For autumn interest, the writer recommended autumn crocus (*Colchicum autumnale*), a fall-blooming bulb. A winter garden would provide for wintering birds, and even promote healthy outdoor exercise. "Though you might not succeed in growing roses there, the daily walk in winter time would plant beautiful ones in many poor pale cheeks where perhaps it would rejoice you even more to see them blooming."[21]

Horticulture was again on an upswing in the latter part of the decade. After a hard winter in 1855–56, professional gardeners regrouped and began to seriously experiment with plants that would survive Chicago's harsh climate. The Chicago Gardeners' Society was formed in 1858, with members of the working gardener trade like John Ure, Harman Khlare, John Blair, Robert Douglas, Jonathan Periam, Samuel Brooks, and Arthur Bryant Sr. This society had fewer "wealthy amateurs" than the CHS, and was dedicated to improved growing and agriculture techniques. With the 1857 financial panic over, the rise of the railroads, and Chicago's burgeoning immigration rate, landscaping and horticulture ideals could be promulgated to the middle class. "There is a great amount of innate love of the beautiful in the masses," said Kennicott in 1858. "We need to make good taste democratic. Teach the masses to replace coarser flowers with perpetual roses, beautiful tulips, pinks, and all the finer flowers."[22]

With almost two decades of experience behind them, Chicago gardeners were now learning from their mistakes. In a *Prairie Farmer* essay, "Popular Errors in Ornamental Gardening," the author reiterated the three most important words for western gardeners, "underdraining, trenching and protection." Common attempts to hasten nature's timetable resulted in overplanted, fussy gardens. The author pointed to tony Michigan Avenue gardens as examples of poor landscaping. "It is a noble street . . . and contains the homes of some of our wealthiest citizens . . . Thousands of dollars have been spent within the last ten years to fill those little yards with trees and shrubs, and some of them at this moment contain fine specimens enough to stock several acres. Now, this is not only folly, but the worst possible taste." Whether homeowners overplanted their yards due to ignorance, an untutored desire for plant variety, conspicuous consumption, or desire for an "instant landscape," the author decried the unattractive result.[23]

With fine examples of public landscaping in the cemeteries, and with more disposable income, homeowners now looked for professional advice on private landscape design. This need was met in the 1860s, when Chicago became an attractive business location for emerging landscape artists, largely through the efforts of suburban developers.

Dawn of the Developer's Garden

The 1860s brought trained landscape designers to Chicago's public and private spaces, coincident with the beginnings of suburban development and the Chicago park system. Leading Chicagoans with East Coast ties had access to professional landscape artists from Boston and New York. It is also likely that Frederick Law Olmsted, as a U.S. Sanitary Commission official, made connections with Chicagoans during the war years. Many of Chicago's horticultural leaders were involved in the Civil War, either by organizing troops or by sending produce to the Union Army.[24] Thomas B. Bryan, for example, was president of Chicago's Great Northwestern Sanitary Fair in 1865, where he engaged the services of landscape gardener John Blair.

Bryan was among the first to establish a gentrified home in the nascent suburbs of Chicago. His contributions to Chicago's horticultural scene went far beyond the limits of his property. In addition to his prominent role in creating Graceland Cemetery, he was a generous promoter of local floral shows and exhibitions and donated space in downtown Bryan Hall to the Chicago Gardeners' Society. By building his Byrd's Nest estate in then remote Cottage Hill (now Elmhurst), Bryan proved to Chicagoans that civility through horticulture could be achieved on the treeless prairie. Even though Cottage Hill was only eighteen miles west of the city, and railroads made the outskirts more accessible, *Prairie Farmer* editors marveled at Bryan's daring move: "Men do build homes in the

PRAIRIE FARMER.

NEW SERIES, VOL. 5—NO. 12.]   CHICAGO, ILL., THURSDAY, MARCH 22, 1860.   [OLD SERIES, VOL. 21—NO. 12

TOP In 1860, Thomas B. Bryan built a gracious home in the "wilderness" prairie of today's Elmhurst, meriting this front-page article in the *Prairie Farmer* with the notice, "Trees do grow on the prairie." From *Prairie Farmer*, March 22, 1860.

BOTTOM Byrd's Nest proved that culture could exist on the prairies. By 1900, the grounds had been improved to include water features, extensive gardens, and mature trees. Reproduced by permission of Elmhurst Historical Museum.

West—right out on the unprotected prairie. And it is not regarded an unfit place to make a home. If nature has not provided trees and shrubs, she has provided a natural lawn of magnificent extent, of gentle and pleasing undulation, and in the matter of ornamental trees, . . . [t]he 'Birds Nest,' [*sic*] . . . is emphatically a *prairie* home—'not a tree to begin with; but in place of it, an unobscured view for thirty miles towards the setting sun.' But the trees are growing there now—and right where the proprietor wants them."[25]

Bryan's home was essentially a stand-alone country villa. In antebellum Chicago, this form of summer retreat for wealthy citizens was typical. Early suburbs took the form of summer resorts or gentlemen's country retreats. But, in the 1860s, as Chicago's population blossomed along with improvements in transportation and town infrastructure, real estate speculators saw prime opportunities in outlying land as commuter suburbs. Gentlemen's retreats were soon carved up into suburban towns.

S. H. Kerfoot was one of Chicago's earliest real estate developers. He was also one of the first to use landscaping to lure prospective homeowners. Kerfoot's Lakeview home included about twenty acres of undeveloped property extending from Green Bay Road on the west to Lake Michigan on the east. A portion of the property was developed into a pleasure ground open to the public. One account from 1867 describes the area: "This place with its artificial ponds, rustic bridges and fine green-houses, was said to be the most artistically arranged place west of New York City. It extended from the Lake Shore to the main road [Halsted Street], and was styled 'Dawn.'"[26] Kerfoot hoped the attractive landscaping would entice wealthy investors to buy nearby acreage. He employed a series of professional gardeners to design and maintain his property. In 1859, Frank Calvert was his gardener. Calvert was noted for his elaborate rustic-work creations, and was a prominent member of the Chicago Gardeners' Society in the early 1860s. Calvert was later hired to plan the estates of Lake Forest, another early planned suburb. Lake Forest, on the North Shore, was designed in the late 1850s by a professional landscaper who recognized and accentuated the unique features of its ravines, hills, and lake views. The suburb's residents, mainly wealthy Chicago-based businessmen, recognized the value of artistic landscaping, and surrounded their country estates with the latest fashions in gardening. Calvert, who had a nursery there, designed or maintained several of the landscapes for these early gracious homes.[27]

Kerfoot subsequently hired John Blair to continue his landscape improvements. John Blair, a native of Scotland, came to Chicago by way of Rockford, Illinois, where he'd had a successful landscaping business. He established a nursery in Oak Park (later sold to Frank Lloyd Wright for his own home) and was the personal gardener to real estate developer H. W. Austin. Blair's landscaping style reflected both his European heritage and an increasing familiarity with

TOP RIGHT H. M. Thompson, a "wealthy amateur" who often entered exhibits in Chicago's horticultural fairs, built his Italianate house in Lake Forest in the 1860s. The home abutted a ravine, which was terraced and crisscrossed with bridges. The rustic work on the Thompson estate is most likely that of Frank Calvert, who was well known for such ornamentation. Courtesy of Lake Forest College Library Special Collections.

LEFT John V. Farwell, partner in a major dry goods firm, built his Lake Forest home in the late 1860s. According to the authors of *Classic Country Estates of Lake Forest*, Chicago architect George H. Edbrooke designed the Farwell greenhouse. The Victorian cutout flower beds in the foreground of Farwell's greenhouse were typical of the era. Courtesy of Lake Forest College Library Special Collections.

BOTTOM RIGHT In addition to trained horticulturists to design and plant the grounds and greenhouses, large estates such as these in Lake Forest mandated a substantial workforce for maintenance. The narrow strips of lawn bordering this pathway could only be hand-mowed with reel lawn mowers. Courtesy of Lake Forest College Library Special Collections.

prairie growing conditions. He used, for example, Osage orange for hedging, a new prairie practice. At the same time he relied heavily on evergreens, many marginal in our region, for screening. In his plan for a three-acre home, curving driveways wind through groupings of evergreen and deciduous trees. Blair urged homeowners to plant trees immediately, to allow them to grow. He thought trees were important as "shelter for both man and beast from the rude blasts that sometimes sweep over them."[28] His noted that his plan was adaptable to larger estates, but even one this size would require at least one full-time gardener dur-

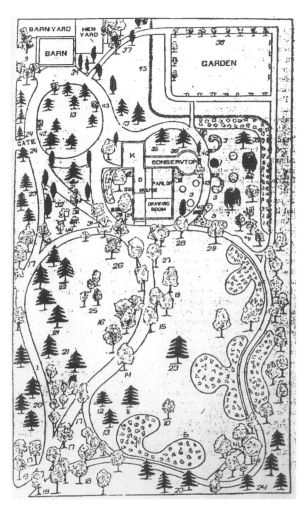

John Blair's design for a typical suburban yard shows refined restraint in that the front approach to the house is uncluttered with statuary and instead showcases open vistas with gracefully arranged groupings of trees and shrubs. On the side of the house, with views from the conservatory and parlor, Blair indulges the fanciful and locates fountains, vases, and flower-bed arabesques. Here he places specimen trees, reverting to the Victorian love of weeping trees like willows. Blair located the main entrance on the southwest corner of the property (nearest the town) and put a summerhouse on the property's southeast corner. Utilitarian barns and vegetable gardens anchor the northwest and northeast points of the yard. Ca. 1862. From *Prairie Farmer*.

ing the summer. Nonetheless this was a modest labor requirement compared to the resources needed to maintain large gardens, such as those in Lake Forest.

Blair landscaped many private residences, and, under the auspices of his patron, H. W. Austin, landscaped some of the early suburbs in Chicago. Austin was a self-made manufacturer, banker, and early real estate developer. Among his manufactured products was a pump that drained Chicago's swampy environs, making them suitable for development. Austin's home, on six acres in Oak Park, was landscaped by John Blair, as was property he'd bought in Colorado. Austin also developed a suburb bearing his name on the west side of Chicago, and profusely landscaped the streets with trees and greenery. Both for his personal residences and his real estate developments, Austin clearly saw the value in greenery and gardens. A page in his personal handwritten journal lists potential names for either a suburb or his own home. Each candidate name evokes a sense of nature and, interestingly, sometimes a reference to foreign lands. Representative

H. W. Austin's home was extensively landscaped. In the front of the home, John Blair designed curved walkways around groupings of deciduous trees. The back of the home was sectioned into areas for pastures, a vegetable garden, and lawn tennis. Ca. 1890. Courtesy of the Historical Society of Oak Park and River Forest.

names included Alhambra, Certias, Hazel Hill, Forest Hall, Mountain Meadow, Paw Paw, Home Grove, and Woody Way. Upon his death, several leading figures were called upon to speak at Austin's memorial service. Attesting to Austin's lifelong love of horticulture, the most heartfelt eulogy came from his then gardener, a German immigrant named Mr. Priess.

Railroads were clearly important in the development of Chicago suburbs—not only for improving commuter accessibility, but also in offering better shipping options for market gardeners. Railroad companies often hired landscape gardeners to plan and maintain the railroad depots, the first welcoming sign of many villages. Gardens and flowers around depots became a matter of civic pride and made a statement about the level of culture to which the village aspired. A location on the railroad line could make or break a town's growth, and, in the building boom of the 1870s (following the Civil War and the Chicago Fire), towns sprang up all along the railroad lines. Greenery was a key element of the design for these suburbs. Within the city of Chicago itself, the park movement offered greenery and new ideas in landscape design.

Parks and Professionals

As early as 1853, John Kennicott and other horticulturists lobbied for more open green space in Chicago. "Is there anything deserving the name of Park or Public

Garden in Illinois?" Kennicott asked rhetorically.[29] He opined that although the railroad had already usurped prime acreage along the lakefront, "yet the Garden City has still opportunity to secure grounds for parks and gardens somewhat worthy of her heroic appellation and prospective population." In the late 1860s, two key factors combined to further propel Chicago's horticulture and landscaping efforts. In response to growing concerns about public health, and to alleviate crowded city living conditions, the park bill was signed in 1869, authorizing Chicago's major public parks. Public parks became showcases of the latest horticultural trends, now that many wealthy residents were building their country villas farther out from the city.

It is virtually impossible to pick up a Chicago city guidebook written in the late 1800s without reading pages of superlatives describing Chicago's park and boulevard system. Chicago's expansion was timed fortuitously with the growth of parks across the United States. With the much publicized success of New York's Central Park, the improved post–Civil War economy, the general agitation for better public health, and the overall rebuilding after Chicago's 1871 fire, the city was perfectly poised to embark on a major park-building effort. Land, while never cheap, could still be obtained because the city had not been entirely built out. After predictable and ongoing political haggling, the city's three park districts in the west, north, and south created large public pleasure grounds for recreation and healthful exercise.

In his 1874 book, *Chicago and Its Suburbs*, Everett Chamberlin observed the indisputable profit potential for real estate adjacent to proposed parks.[30] At the time of his writing, the South Parks, designed by Olmsted and Vaux and successors such as H. W. S. Cleveland, were under way. In 1871, plans were completed by William Le Baron Jenney for the West Parks system, including Douglas, Garfield, and Humboldt parks. Lincoln Park, a former city cemetery, was expanded into the largest park on the North Side. Chicagoans embraced their new parks as wholeheartedly as did out-of-town visitors and city planners.

By the 1890s, when a plethora of publications were written about Chicago in anticipation of the Columbian Exposition, there were six major parks in Chicago (Lincoln, Humboldt, Garfield, Douglas, Washington, and Jackson), and six smaller parks (Union, Jefferson, Vernon, Wicker, Gage, and Midway Plaisance), comprising over 1,900 acres, connected by more than thirty-seven miles of boulevards.[31] Olmsted and Vaux famously redesigned Jackson and Washington parks in the 1890s to accommodate the Columbian Exposition and subsequent use. During the Progressive Era at the turn of the nineteenth century and beyond, smaller neighborhood parks were created to improve accessibility of green space to individuals of all income levels.

The Chicago parks, and their designers, had a direct impact on garden design for private residences. Many of the foremost designers of Chicago parks also created landscapes for Chicago's wealthier homeowners. Frederick Law Olmsted and later his successor firm, H. W. S. Cleveland, William Le Baron Jenney, Jens Jensen, Swain Nelson, J. H. Prost, and Alfred Caldwell were among the prominent park designers who also planned private gardens. Other park horticulturists such as August Koch and Edwin Kanst were highly respected authorities, frequently quoted in the city's major newspapers and on the lecture circuit.

TOP The designs in many of Chicago's large parks were emulated in home landscapes. Lincoln Park's Lily Pond was a favorite destination. From the author's collection.

BOTTOM Drexel Boulevard included islands of greenery and elaborate floral decoration. H. W. S. Cleveland planned not only this boulevard but also residences nearby. From the author's collection.

Today, the parks continue to evolve based on the needs of the city's constituents, and continue to be a source of civic pride—most recently with the unveiling of the city's Millennium Park near the site of the old Fort Dearborn.

Hearing of the prospects in Chicago, other notable landscape artists arrived in the late 1860s, many of them Olmsted colleagues or disciples. Jacob Weidenman, a onetime Olmsted partner, worked with William Le Baron Jenney and became superintendent of the ill-fated Mount Hope Cemetery in 1886.[32] H. W. S. Cleveland, a landscape architect who had worked with Olmsted, opened an office in Chicago at 115 Madison Street in 1869. His ad in an 1869 issue of the *Land Owner* offered surveys and designs for the arrangement of villages, parks, cemeteries, and private estates. Among the references he listed were David Landreth and John Jay Smith of Philadelphia. His Chicago references included the city engineer and city surveyor as well as James H. Bowen and seedsman A. H. Hovey.[33]

Cleveland was involved in many public projects, including the execution of Olmsted's designs for Drexel Boulevard, the South Parks, and his own plan for suburban Hinsdale's Robbins subdivision. Inspired by some of the unique design considerations of the Midwest, he wrote *Landscape Architecture as Applied to the Needs of the West* in 1873. He was among the first to recognize the effects of severe southwest winds upon plants. He chastised Chicago's early city planners for grid-system streets, and tongue-in-cheek described the artificial decorations of public squares: "Mountain ranges are introduced . . . lakes of corresponding size are created apparently to afford an excuse for the construction of rustic bridges . . . A lighthouse three feet high, on a rocky promontory the size of a dining room table, serves to warn the ducks and geese of hidden dangers of navigation, and this baby-house ornamentation is tolerated in a great city which aspires to an artistic reputation."[34]

Cleveland believed in maximizing a site's natural aspect, and worked in concert with civil engineers such as William M. R. French to survey and grade prospective sites. In his 1871 treatise, *Landscape Gardening in the West*, Cleveland likened the indiscriminate use of "fancy flower beds, fountains, statues, vases, grottos, arbors and rustic seats" to the poor taste shown in "a profuse display of jewelry on a plain person."[35] Like Olmsted, Cleveland obtained clients in the private sector, such as attorney J. Y. Scammon. The Scammon landscape plan was drawn in advance of the house plan, with Cleveland determining the best location for the home site. Recognizing Chicago's real estate mania, Cleveland sited Scammon's Hyde Park home off center, not only to afford sweeping vistas, but also to provide for future subdivision of the property. This provision for future development was characteristic of early Chicago designs, where land was plentiful and estates were often purchased with intent to subdivide. The more mature East Coast estates would not have sited homes this way, since much of the land had already been claimed. Scammon's design included a natural belt of woods that was left undisturbed and underplanted with shrubbery to screen the kitchen and laundry yards. The overall landscaping plan was designed for low maintenance because, according to Cleveland, Scammon did not dabble in horticulture and wanted only a "tasteful disposition of the natural features, involving the least possible need of watchful care and labor."[36] The croquet field and greenhouse hint at Scammon's wealthy status, for he had the leisure to play croquet and, if he himself was not a greenhouse hobbyist, had the funds to hire a gardener to tend to the plants.

Recognizing the importance of developing suburbs, a short-lived newspaper, the *Land Owner*, was locally published in the early 1870s. The *Land Owner* targeted land investors by highlighting key suburban developments. The Olmsted firm contributed many drawings to the publication and, using sketches of European parks, raised public sensibilities about landscape design. Cleveland also supplied plans to the *Land Owner* magazine and showed similar use of broad vistas, simple elegance, and natural adornments. His "Design for the Grounds of a Private Gentleman" on Drexel Boulevard showed the house sited near the boulevard with a large lawn to the back. A simple curved driveway led from the boulevard to a turnaround at the entrance. The *Land Owner* reported, "The shrubbery is to be grown in vistas, suited to the topography of the ground, and the entire ornamentation is natural and unostentatious, reminding one of the elegant country seats to be seen along the London, Chatham, and Dover railway in England."[37] Cleveland's circa-1870 design for James H. Bowen's home, Wildwood, showed more formality, perhaps due to his client's wishes, or perhaps due to his own evolving design aesthetic. Here substantial land was dedicated to orchards and formal gardens. The house was sited so as to face the Calumet River, then a pleasing vista (plate 8).[38]

Cleveland was perhaps the first designer to formally codify rules and guidelines addressing the Chicago region and beyond. His treatise, *Landscape Architecture as Applied to the Needs of the West*, outlined some of the unique horticultural conditions in Chicago, then considered part of the western frontier. The encroachment of trained professionals such as Cleveland caused some friction with Chicago's self-taught nurserymen. Although Cleveland acknowledged that he himself had learned much from an unnamed amateur landscape gardener, he was harshly critical of untutored landscapers, calling them "uncultivated men, who have no idea of the development of natural beauty as a means of rendering a place attractive, but rely solely upon artificial decoration."[39] Because most landscaping in Chicago was then done by florists or nurserymen who were often motivated to create designs using their own stock, Cleveland's comments could not have endeared him to the professional green trade. In 1886, following the death of his wife and loss of business in the Chicago Fire, Cleveland left for a successful career in Minneapolis. In 1891, he wrote to Edgar Sanders noting "it is so long since I met any one who knew anything of me or the events of those days that it seems like a communication from another world."[40]

The parks exposed the typical Chicagoan to new plants and planting plans. They also fueled further real estate speculation because land parcels near green space commanded higher prices. Chamberlin ascribed much of the suburban land boom to park development: "the south and west parks have been for five years the principal stimulus to land speculation and investment and the key to the situation of the Chicago real estate market."[41] In the year the park bill was signed, 1869, at least nine villages were incorporated or newly developed by so-called improvement companies (Ravenswood Land Company; Rogers Park Land Company; Wilmette, Winnetka, Glencoe, and Washington Heights by the Blue Island Land and Building Company; Maywood, Irving Park, and Norwood Park by the Norwood Park Land and Building Association; and South Chicago). Planned green space and natural features were important elements for developers in choosing these sites. Wilmette, Winnetka, and Glencoe enjoyed interesting topography, including wooded ravines and Lake Michigan. Washington Heights, built on a high ridge, included 1,500 acres where the Blue Island Land and Building Company planted 11,500 trees on the 480 acres known as Morgan Park. Norwood Park was designed with curvilinear streets like Riverside, and in fact one promotional brochure claimed that the town was laid out by "the same landscape artist who designed the [1893] World's Fair."[42]

Capitalizing on existing, albeit manmade, greenery, Ravenswood was built on what was once Wood's Nursery. Located in the middle of the block bounded by Hermitage, Wilson, Sunnyside, and Ravenswood avenues, Wood's residence included a personal garden of rare plants and two ponds. A Ravenswood resident described the area in the 1870s: "The woods and surrounding prairies were peopled with lilies, Johnny jump ups, large and small white and yellow and purple lady slippers, a dozen kinds of blue and white and yellow violets, daisies, phlox, fire flowers, puccoons, shooting-stars, spring beauties, star grass, white and purple trilliums, cranebill, anemones, Solomon's seal, mandrake, Jack-in-the Pulpits, *Collinsia verna*, blubells, etc."[43] But, like many of the first generation of settlers, Wood retired and sold his nursery to the Ravenswood Land Company.

The parks directly influenced the design of early suburbs in that the high-profile landscape professionals commissioned for the parks were also hired to lay out early suburbs. Olmsted, for example, was hired to develop a plan for the suburban village of Riverside at about the same time he worked on the South Parks. Olmsted's Riverside was a speculative development, funded by eastern investors, remarkable for the emphasis placed on landscaping and greenery.

Nearly half of the village's acreage was set aside for public green space in parks and "longcommons." Against the prevailing grid pattern of Chicago's city streets, Riverside's roads were curvilinear, winding around green spaces and parks. Gen-

erous setbacks ensured more green space in lawns. Olmsted encouraged homeowners to plant trees in the public parkway in front of their homes to add to
the parklike feeling of the village. As was the case with other suburban towns
laid out by landscape professionals, private residences often benefited from the

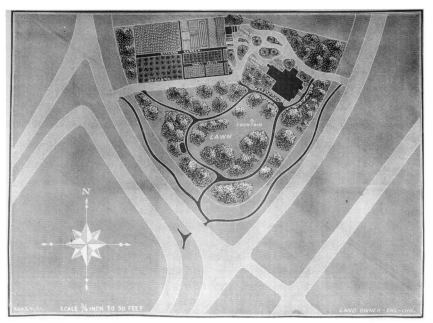

TOP Landscaping was integral to Olmsted's Riverside. Riverside's curvilinear streets wound around "triangles" of greenery and large, open parks. Along the river, broad swaths of parkland were created. Ca. 1870.
From the author's collection.

BOTTOM This plan for the Murray residence is said to have been "from the hand of Olmsted." It shows
a large kitchen garden and orchard behind the home. The home, sited off center, has a broad expanse of
lawn, and massed trees and shrubs leading from the corner of the property. A fountain serves as a focal
point—perhaps at the wish of the client. A serpentine walkway even traverses the public parkway near
the street, suggesting additional plantings there. Ca. 1870. Reproduced by permission of Chicago History
Museum qF38NA.L2.v.2.

lessons learned in community planning. In Riverside, for example, Olmsted's firm landscaped at least one home, and William Le Baron Jenney, Olmsted's colleague and a landscape artist and engineer, himself lived in Riverside.

Also emphasizing green space, the Highland Park Building Company hired Olmsted protégé H. W. S. Cleveland and his partner, engineer William French, to plat the new North Shore suburb in 1872. The Cleveland and French team designed curvilinear streets to accentuate the rolling landscape of ravines and wooded forest. As Chamberlin noted, "Much of the beauty of the place is owing to the fact that the Company, immediately after making its purchase of 1,200 acres, called in the aid of good landscape gardeners to lay it out."[44] Similarly, although much of Hinsdale was laid out in grids, Cleveland found opportunity to create visual interest in its Robbins subdivision.

This first generation of suburbs placed strong emphasis on green space, and catered largely to upper-middle-class residents. As populations burgeoned and commuter-rail transportation improved, more suburbs developed for the working class.

Fire and the Growth of the Suburbs

The Chicago Fire of 1871 destroyed more than two thousand acres of prime Chicago property. The fire spread from the southwest side northward, destroying most of the downtown business district and residences of the poor and rich alike. Edgar Sanders noted that, among the casualties of the fire, many prominent florists lost their businesses.[45] In true Chicago fashion, one of the first businesses to reopen was a real estate company. The fire, while devastating, paved the way for explosive suburban growth. New city regulations aimed at reducing fires prohibited less-expensive wooden construction within the city limits, so developers sought cheaper suburban land parcels. About fifteen outlying communities were founded or incorporated in the 1870s and more than ten were founded or incorporated between 1880 and 1890.[46]

The 1880s and 1890s also brought some interesting experiments in community planning by individuals, rather than by investment groups. The town of Pullman, for example, a self-contained model workingman's town created by railroad-car magnate George Pullman, employed Nathan Barrett as landscape architect. In the 1880s, Barrett created magnificent public gardens near the town's social and market centers. Despite these showcase gardens, Pullman landscaping was homogenized. Much like today's condominiums, the 20-to-30-foot-wide front yards of individual workingmen's homes were planted and maintained by the Pullman Company. The land and home itself were rented—surely a disappointment to immigrant workers, many of whom had escaped feudal land systems in Europe. Backyards were the only areas where workers could indulge their personal gardening tastes. These yards, enclosed by tall wooden fences, were screened from all but the most inquisitive neighbor—further limiting exchanges of garden ideas.

The northern suburbs of Zion and Kenilworth, built in the 1890s, were also unique experiments in community development by individuals. Zion was founded as a religious community under John Alexander Dowie. All roads led to a central square of greenery, like rays around the central sun. Kenilworth, built by Joseph Sears, was based on exclusivity, and used wide-open lawns and large lot sizes to draw affluent homeowners. Unlike Pullman, homeowners owned their properties, but Kenilworth and Zion were similarly bound by design standards that reflected the personal philosophies of their founder.

Pullman, Kenilworth, and Zion were experimental developments that fit the personal agendas of Pullman, Sears, and Dowie. More widespread were communities created by developers such as Samuel Eberly Gross, whose name became a household word and who built homes for the working classes. By the 1890s, Gross owned more than 150 subdivisions, an empire built by subdividing tracts along railroads both within the city and in newly established suburbs, and offering lots on installment to working-class families. Green space figured prominently in Gross's brochures, with line drawings of plantings or water fountains adorning the most modest of cottages. A promotional novelette called *The House That Lucy Built* featured a wily young bride, Lucy, as she induced her husband to buy a Gross home. A key selling point Lucy pitched to her husband was the ability to grow flowers and have greenery around her home.

Yet reality often fell short of promotional brochures. In his much-ballyhooed Hollywood subdivision, which adjoined Olmsted's picturesque Riverside, Gross attempted to create curvilinear streets like those in Riverside. Whereas Riverside's streets curved naturally among green parks and open vistas, Hollywood's seemed to tortuously bend around triangles and hairpin turns. A local history suggests that, when presented with his engineer's estimate for making curved roads, Gross insisted on cost-cutting and curve-cutting measures. Carl Mendius, Gross's engineer, was quoted: "I drew my first plan for Mr. Gross's subdivision using Olmsted's 1869 design but it did not please Mr. Gross. I then decided to 'draw the craziest sub-division plan I could cutting the area into 25 foot lots.' This attempt was much more to his liking."[47]

Ironically, Gross sought advice from the Olmsted firm in his 1910 development of a North Shore property he called Northlake. In an exchange of letters, personalities appeared to clash, with Gross ostensibly looking for free advice, and Frederick Law Olmsted Jr. (the son) pointedly listing his fees and expenses. In his memo to the file, Olmsted noted, "He [Gross] is a garrulous old man who has recently married a young and rather pretty, but uncultivated wife ... He nearly talked me to death ... He would be a hard man to work for I guess."[48]

According to this ad from S. E. Gross, by 1894, he had already sold 43,500 lots and created seventeen towns and cities. The "New Queen Suburb of Chicago: Picturesque Hollywood" was a subdivision of Brookfield (then called Grossdale). The ad featured the subdivision's proximity to Riverside and an etching of a heavily wooded, curving street. From the author's collection.

In the 1870s and 1880s, suburban villas became popular, a precursor to the country-estate era of the early 1900s. Responding to this trend, local nurserymen and florists expanded their businesses to include design consultation. Among some local nurserymen, an undercurrent of resentment continued against professional landscape gardeners—whether as a xenophobic diatribe against eastern landscapers or because of perceived threats to business.[49] This comment by Jonathan Periam, in his *Home and Farm Manual*, didn't name names, but sheds light on prevailing opinions: "Some landscape gardeners are so puffed up with an imagined importance of their calling, and have such a contempt for ideas not emanating from themselves, that they will endeavor to argue away those of their employer, and indeed, often willfully ignore them altogether. In nine cases out of ten, such men are ignorant pretenders."[50]

Periam's designs were aimed at the middle class, with typical lot sizes of a half acre or less. He advocated winding paths with a few small side beds for flowers.

He suggested that less is more, and noted a common error of planting too many trees or flowers for the lot size. "Remember that a few well-kept flower beds are better than many unsightly ones," he warned. Lots were to be enclosed by arbor-vitae hedges, except in the front where deciduous trees or hedges were appropriate so as not to block the view. Periam cited the work of H. DeVry, then superintendent of Lincoln Park, as the "finest example in the United States of floral work on a large scale." [51] He opined that these flower designs, on a smaller scale, could be copied by the average homeowner. Round carpet-bedding schemes included *alternantheras, achemenas*, and *escheverias* with oxalis for a border. Ribbon beds used bright-colored geraniums in the middle, with dwarf geraniums as the next band, and coleus or alyssum on the outside.

The rampant growth of Chicago's suburbs, coupled with explosive population increases in the city, threatened the open space and unfettered views for which the region was famous. One of the most persuasive leaders in celebrating and protecting Chicago's indigenous landscape was Jens Jensen. Among the many threats to the prairie landscape, Jensen was concerned about unbridled real estate speculation. He observed, "Chicago was once called a garden city. What has become of the gardens? Great men and great women have grown out of that garden city, but why are there no leaders today? What has happened in Chicago has happened in many other large cities where speculation has been the guiding

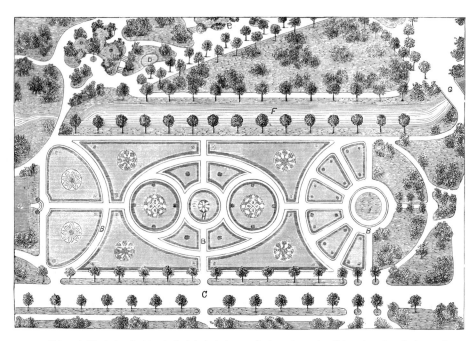

Although this design for Lincoln Park included space for bear pits and wolf dens, Jonathan Periam maintained it could be scaled to residential properties. Ca. 1870. From Periam, *Home and Farm Manual.*

force."[52] Among the solutions Jensen and his counterparts in Chicago's architectural and landscape fields recommended was a return to nature.

TWENTIETH-CENTURY DESIGN TRENDS

In Chicago, the early twentieth century was a period of many overlapping garden trends and specialty gardens. This period marked America's Progressive Era, when idealistic groups united to improve conditions ranging from child labor to impure food and drugs, to impoverished living conditions, to air and water pollution. In Chicago, the Hull-House settlement became famous for its programs to alleviate immigrants' crowded tenements and health conditions. Many of the solutions to these industrial problems included a garden component. The influx of European immigrants to Chicago also brought new ideas from the gardens of different nationalities. Complementing these trends, the City Beautiful movement promoted green space and gardens. Lastly, as the middle class grew, and with plant hybridization and transportation improvements such as motorized vehicles, a wealth of flower introductions made specialty gardens popular. In this "country-estate" era, wealthy homeowners in Lake Forest, Highland Park, and other northern suburbs hired talented landscape architects such as Jens Jensen, Warren Manning, the Olmsted firm, O. C. Simonds, and Ralph Rodney Root.[53] More modest gardeners adapted ideas from these estates on a smaller scale. Some of the prevailing trends included smaller gardens, perennials, water gardens, rock gardens, greenhouse and window gardens, Japanese gardens, and roof gardens.

City Beautiful

The Columbian Exposition, with its lofty committee of renowned architects, landscape architects, and sculptors, created an ideal of a sparkling White City. Visitors and Chicago residents alike were enamored of the possibilities of transforming their own urban spaces into the harmonious blend of buildings and green spaces epitomized at the 1893 fair. Following the fair, the City Beautiful movement spread throughout the nation.

In Chicago, the first steps in the movement were to clean up the city. As a major industrial center, air pollution was a major factor, with smoke darkening trees and houses alike. Trash and dirt in alleys and streets created a public outcry for more frequent garbage pickup and other municipal basic services. Billboard advertising became a major issue as automobiles and railroad traffic increased in the early twentieth century, and various groups fought to limit the amount and size of the so-called scars on nature. Local civic improvement groups sprang up

THE HOUSE BEAUTIFUL

COPYRIGHT, 1907, BY THE HOUSE BEAUTIFUL CO.   *EAST, WEST, HAME'S BEST*   PRICE, 25 CENTS A COPY, $2.50 A YEAR

VOLUME TWENTY-TWO   CHICAGO, AUGUST, 1907   NUMBER THREE

PERGOLA IN THE GARDEN OF J. B. HOOKER, ESQ., LOS ANGELES, CALIFORNIA
(See Page 16)

The City Beautiful movement spawned a number of ancillary efforts to beautify homes and gardens. One such project was Chicago publisher Herbert S. Stone's *The House Beautiful* magazine, which in 1896 featured artistic homes and gardens from around the world. From the author's collection.

throughout the city and suburbs to address such issues, and regional publications cheered their efforts.

One such group, the Municipal Art Society (later the Municipal Art League), founded in 1899, embraced the city's public spaces and addressed not only issues of smoke and dirt, but also the higher arts of public sculpture and art. Famed architect Louis Sullivan, speaking at the group's charter meeting, cautioned that the group must first focus on practical issues of the smoke nuisance, clean streets, and removal of unsightly structures. In its first early years, the league brought political and media focus on bridge beautification, changes in the smoke law, public comfort stations, street-lamp designs, automobile noise, and of course, parks and park adornment. Not all of their efforts were appreciated. In 1907, residents of the so-called ghetto district in the Ninth Ward appealed to Mayor

Fred Busse after the Municipal Art League allegedly vetoed a fountain, deemed "inartistic," in one of the district's small parks. Alderman Fick of the ward was quoted as saying "we don't care whether the blessed fountain is artistic or not. It is artistic enough for us people down in the Ninth ward. All we want is the fountain. If any artists would have their finer feelings hurt by looking at it, let them keep out of the ward."[54]

In 1900, Jens Jensen, recently fired from his position in the West Parks, wrote an article for the *Chicago Tribune*, "Parks, Boulevards, and Their Influences."[55] Jensen described public parks as "outdoor colleges open to all" where painters sketched, entomologists explored, amateur gardeners studied, and people played. Jensen also lobbied for small parks, woodlands surrounding the city (later to become the forest preserves), and a general cleaning up of city alleys. Jensen opined that the first step was to begin with one's own yard, which in the existing city was often distressingly bereft of beauty. "The first thoughts of beautifying the city are formed and executed at home in the little yard. That yard far oftener contains heaps of ashes and tin cans, with an occasional weed."

Among the city's many beautification efforts, the *Tribune's* garden contest was perhaps the most widespread and consistent. Many neighborhood civic groups and women's clubs also offered incentives for prettier gardens. These contests took the highbrow, esoteric aspects of City Beautiful and placed them in the hands of the city's workers. Except for occasional faux pas such as the Municipal Art League's fountain veto in the tenement district, middle-class Chicagoans increasingly were provided with more education and influence in making the city beautiful.

In 1905, after a two-year hiatus, the CHS decided to make their annual flower show more relevant and accessible to the typical Chicagoan. Moving the show to the Coliseum, a much larger and less intimidating venue than the Art Institute, the show attracted more of the city's general populace. As the *Tribune* reported, "Instead of catering to professional florists, or to 'society people,' the directors caught up the 'outdoor art' enthusiasm, and decided to direct their appeal to an amateur flower growing and flower loving public."[56] The flower show was the largest ever held in the city, with attendance over five days of thirty-five thousand. Subsequent shows in the next few years followed the educational trend, with exhibits on best plants, school gardening competitions, and model gardens.

Between 1906 and 1909, several actions were taken to further the City Beautiful goals and the role of horticulture therein. The Chicago Merchant's Club, later the Commercial Club of Chicago, commissioned famed architect Daniel H. Burnham to develop a comprehensive plan for the city. Completed in 1909, the *Plan of Chicago* included wide boulevards, an outer ring of green space, reclaimed lakefront, and more public parks. Mayor Busse appointed a plan commission

GARDEN CONTEST WINNERS

With the hope of inspiring a competitive spirit, a number of civic improvement organizations sponsored local garden contests. The *Chicago Tribune* began in 1901 what was to become a tradition—the citywide garden contest. Participants were invited to submit for review either their whole gardens or window-box gardens. Serving as judges were Jens Jensen and Mrs. Herman J. Hall, a leader in Chicago's art movement and the American Park and Outdoor Art Association. Garden entries were divided into the city's west, south, and north sides, with judges traveling more than eight hundred cumulative miles to visit each site. According to Jensen's summary, the South Side had the most contestants although, in his opinion, the North and West Side gardens were neater overall. The lawns, however, were better on the South and North Side than on the West, which Jensen attributed to the poor soil conditions there. Good taste was not correlated with social status. Jensen said, "The craving for loud colors was more evident in the gardens of the rich than of the poor; one garden on Drexel boulevard contained such a mass of vivid scarlet geraniums as to make one's eyes run."[1]

Throughout the contest period, the *Tribune* published various articles on proper gardening, including tree-planting tips, decorating porches, using vines, and so on. In a suggested design for the small city lot, plantings were simple and consisted largely of shrubs (e.g. lilacs, spirea, weigela, hydrangea) and two trees framing the front (American elm and Carolina poplar). Fences delineated the side boundaries, and sheds were relegated to the back. A well-tended lawn surrounded the entire property.

Contest winners exemplified contemporary definitions of fine landscaping. On the North Side, first prize went to John Whiteway, a varnish maker. His corner lot, 40 feet by 125 feet deep, featured a fenceless front yard, sodded parkway with "artistic" groupings of trees, a backyard with arborvitae hedge, and overall pleasing composition of continuous blooming shrubs without evidence of overplanting. Second prize went to Mrs. J. T. Rees, a janitor's wife. She kept her lawn neat and free of weeds, and used clumps of shrubbery and a grape arbor to block unsightly views of the outhouse and vegetable garden. Interestingly, first prize for the window-box garden was given to famed architect and landscape architect William Le Baron Jenney for his home on Bittersweet Place. Jenney, whom the *Tribune* described as "only an amateur gardener,"[2] was complimented for his taste in blending the window boxes in with the architectural style of the home.

On the South Side, first-place winner Louis Lawson was a teamster who lived with his wife and brother in four rooms above the barn on the rear of his lot near 7000 South May Street. Lawson planned to build a home on the lot, but until then tended a fine garden, despite poor soil and smoke and ashes in the air. A wooden pavilion adorned the center of the smooth lawn, which was accented "at artistic intervals" with beds of cannas. Lawson's use of his vacant lot was praised as a model of neatness and beauty. Second prize went to J. J. Curran, a janitor at Parkside School. He hauled in loads of topsoil to amend the sandy ground, and paid for seeds himself—showing citizens what a model schoolyard could look like.

The West Side prizewinners included language teacher Francis Brimblecom (first place), clerk A. N. Darling (second prize), and window-box winners Henry Tibbits and Mrs. L. A. Jowers. Brimblecom's house and summerhouse were clad in luxurious vines, while Darling's lot, although small (25 by 80 feet) was praised for its neatness, formal garden, and parklike garden. Tibbits, principal of the John Spry School, was applauded for the color harmony he brought to his house and window boxes. Mrs. Jowers's window boxes were praised for their thriftiness and variety. When the contest was over, the judges offered practical hints to all who participated and any prospective gardeners. Among the tips: nurture a lawn for a good garden foundation, remove weeds, don't plant too closely, don't plant large trees on small lots, group plantings instead of scattering them over the lawn, and use more perennials.

1. "South Side Prize Winners in the Tribune Garden Contest," *Chicago Daily Tribune*, September 8, 1901, 37.
2. "North Side Prize Winners in Tribune Garden Contest," *Chicago Daily Tribune*, September 1, 1901, 37.

comprising 328 men from a variety of business and social backgrounds to implement the ideas.

Concurrent with the development of the *Plan of Chicago*, a special committee was formed to protect the city's trees. No longer was it acceptable for individual property owners to select, plant, and care for their trees on public parkways. For better pruning, tree selection and siting, and overall health of the plants, a group meeting at the Art Institute began a campaign for city governance over public trees. Appointed to the new tree committee were thirty representatives from art, environmental, and other civic improvement groups, including the Chicago Park District, Outdoor Art League, CHS, Municipal Art League, Vacation Schools committee, the Chicago Women's Club, Commissioner of Health W. A. Evans, Hull-House, and prominent names in horticulture such as Bryan Lathrop, Jens Jensen, Mrs. P. S. Peterson, O. C. Simonds, and J. C. Vaughan. The group succeeded in convincing the city council and mayor to create the new position of city forester, ultimately filled by J. H. Prost.

Prost, an assistant landscape gardener in the West Parks who had worked with Jensen, did not limit his expertise to trees per his job title, but instead embraced the whole City Beautiful concept. Prior to his appointment, in 1908, Prost served as a judge in the *Tribune*'s garden contest. Traveling over a thousand miles by automobile, streetcars, railroad, and foot, Prost and his fellow judges became familiar with the gardens of the city and suburbs. The judges were pleasantly surprised by the variety of gardens.

Gardens in the poorer sections of the city nonetheless showed good taste in arranging rare plant specimens not even seen in wealthier areas. One of the city's pioneers, James P. Root on Washington Avenue, nurtured an acre of blooms including many old, beautiful shrubs such as a syringa that was planted in 1856. In awarding their prizes, the judges gave consideration to the special challenges in growing conditions unique to each locale; poorer sections that had to contend with soot, shade from tall buildings, and cramped yards were evaluated separately from suburban gardens, which enjoyed more favorable environments.

In October 1909, shortly after he was appointed, Prost continued to reach out to citizens to help him complete his mission. He urged homeowners to plant their front yards to harmonize with the trees in the parkways, keeping the property lines free of fences and hedges. Showing Jensen's influence, Prost proposed establishing backyard gardens in an informal style, noting, "Naturalistic gardens require the least attention . . . they cost the least to maintain and also give a pleasing winter effect."[57] Encouraging gardens in window boxes, porches, and dooryards, Prost wrote, "What a grand annual garden festival we might have and what an influence it would have toward civic betterment, if each improvement association in the city could be induced to have an annual garden contest in its district."[58] The notion of locally sponsored garden contests caught on: not only did improvement associations organize such contests, but private companies did, too. In 1910, for example, the Northwestern Elevated Railroad Company sponsored a garden contest for properties adjacent to the right-of-way. Window boxes, backyard porches, and small gardens were cultivated and gave train passengers a more beautiful view.

By 1913, the City Beautiful campaign was still under way, with many elements of Burnham's plan in progress. Nonetheless, the vision for the City Beautiful was not embraced by all—cynics believed these plans were for artistic niceties geared only for the city's elite. At a meeting of the national conference of city planners held in Chicago, various speakers emphasized the need to appeal to the common man. "Started under the handicap of the name 'City Beautiful,' the propaganda has lived down the sneer in that name and has become identified with the work of practical betterment of the living conditions of the great mass of residents of cities."[59] Hoping to inject new life in the city's revitalization, the *Chicago Tribune* made a five-year commitment to a column written by city booster Henry Hyde. Hyde's inaugural column on March 17, 1913, received many endorsements including that of University of Illinois landscape architecture teacher and author, Wilhelm Miller. Miller declared himself "delighted beyond words" about the new column and offered any help needed. He made note of the circular he was distributing, "The Illinois Way of Beautifying the Farm," and proposed to end each of his lantern-slide lectures with a "We will" slogan.[60]

More articles focused on how middle-class city dwellers, not privileged estate owners, could beautify their yards. Describing the current yard conditions of the typical "tired city dweller," J. Willard Bolte wrote, "A solid board fence about seven feet high shuts off the alley and there is a gate through which the ash man and the second story worker gain entrance. Maybe there is a shed backed up in the southwest corner. Two to one neither the fence nor the shed is painted . . . In most back yards there is a board walk running straight down the yard to the rear gate. It may be in the center or at one side, but no matter where it is it has no place in the garden spot we are going to make." For quick improvements, Bolte recommended painting the fence and sheds, putting an archway over the gate, using shrubs and trees to effectively screen out the eyesores, installing naturally curved pathways, and informal flower beds against the edges. A small lawn swing or hammock could be placed in the grassy lawn in the center of the yard. However, Bolte warned, "Don't let anybody persuade you to build a summer house or a long arbor in that little free space because it will look badly and you won't like it . . . Keep away from those winding paths, bordered by shells, or broken brick, or interlocking croquet wickets, and strive to make the whole thing as simple and naturally beautiful as you possibly can."[61]

Further extending the City Beautiful movement to the common man, egalitarian initiatives such as the creation of the Cook County Forest Preserves and demonstration gardens in the city's major parks took hold. The creation of the forest preserves involved years of lobbying and was intended not only to protect rapidly dwindling areas of green space, but also to provide oases of nature for anyone to enjoy. The Chicago parks' sample gardens—one each in the North, West, and South Parks—were to operate similar to the state's experimental farms. City Forester Prost hailed the effort, noting the current lack of city gardening—on average only one in one hundred backyards contained vegetable gardens. With the expertise gained from readily accessible demonstration gardens, Prost thought more city dwellers would start their own gardens. "The parks have done wonders for the amusement and recreation of the public," Prost said in 1913. "Now they have a chance to do something, at small cost, which in the long run will do as much for their profit, financially, physically, and morally."[62] Laura Dainty Pelham of the Better Gardens Association also supported the sample garden project, noting it would give her members better encouragement and support. Many of these members were Chicago's immigrants, whose numbers exploded in the late 1800s and early 1900s, and whose needs and traditions heavily influenced Chicago's gardens.

Ethnic Gardens Shape the City

As a port-of-call city, Chicago attracted hundreds of thousands of immigrants in its formative years between 1840 and 1920. In this time span, Cook County's total population increased from 10,200 to more than 3 million, largely due to the influx of foreign-born newcomers. In 1860, almost one-half of Cook County's population were foreign-born, and even in 1920 remained at just over 30 percent. Chicago's foreign-born population in 1920, at 28.4 percent, was almost twice the national percentage.[63] More than twenty nationalities had a significant base in Chicago, with many more present in smaller numbers.

Garden design, horticultural practices, and preferred plants of each of these immigrant groups brought great diversity to Chicago's landscape. The tapestry of different traditions, from plowing fields to cooking cabbages, enriched Chicago's horticultural knowledge and furthered its advancement. Unfortunately, documentation of ethnic gardens is tantalizingly scarce. Photographic and written evidence of such gardens, if available, awaits further discovery and research.

When given the opportunity, Chicago's immigrants leapt at the chance to garden. "Aliens in rush for city farms," headlined the *Chicago Tribune* in 1915. "Two hundred Italians, Bohemians, and Slavs crowded into [the] Hull house dining hall yesterday afternoon to participate in the drawing for free garden plots."[64] This program sponsored by the City Gardens Association offered Hull-House and other low-income city residents the opportunity to garden on vacant lots. To meet the demand, other community gardens sprouted at 31st and South

CHICAGO'S GERMAN AND IRISH GARDENS

German immigrants settled in pockets near the Chicago River in the 1840s. By the end of the 1860s, there were more than fifty thousand Germans in Chicago,[1] and their communities had spread far beyond the city limits. Reflecting the changing demographics of its customer base, the *Prairie Farmer* published its first German monthly edition in 1865. A large German community flourished on North Avenue between Halsted and State, and a number of smaller groups settled in today's northern suburbs. German farming communities also formed the nucleus of such outlying towns as Addison, Bensenville, Lombard, and Villa Park. Typical German farming areas just north of Chicago in 1890s were described as follows:

> They live and work the same here just outside of Chicago as they did in their faraway homes in the Fatherland. The men as a matter of course, are a little more conventionalized than the women, but they wear the odd coats, peculiar caps, the wide loose clumsy-looking trousers of the German peasantry . . . All day long during the spring, summer, and until the sharp frosts of autumn . . . these patient women work out of doors . . . they dig the ditches, plant the ground, plant the seed, watch it, tend it—everything . . . When the summer is over and all growing done they sit in long rows upon the damp ground and dig the vegetables they have grown, or they pass from the field to the cabin loaded down with shocks of corn or huge baskets of apples.[2]

German farms tended to be economical and market-driven versus ornamental. Design, other than that which would increase the farm's output, was a lesser consideration. "To own his own farm has always been the ambition of the typical German farmer of the Mississippi valley," according to the *Chicago Tribune* in 1909. "To have his market assured for him is one of the first considerations of the German farmer, after which he seeks the soil and applies to it the best energies that are in him . . . since the establishment of the old Haymarket on Chicago's great west side the proportion of Germans sending loaded wagons there has been enormous; and today in that widened section of west Randolph Street the German farmer is supreme."[3]

While farming was a common occupation for newly arrived German immigrants, others succeeded in a variety of careers, many garden-related. Some of Chicago's most prominent florists and gardeners were of German heritage, such as Charles Reissig, Peter Reinberg, Conrad Sulzer, and George Wittbold. Reissig, born in Hamburg in 1818, arrived in Chicago in 1845 as a boilermaker. He began growing flowers for his own pleasure in his household greenhouse, but soon expanded in 1869 to build large wholesale and retail greenhouses. He moved his business to Riverside, and expanded it to ten large greenhouses known locally as Reissig's Exotic Gardens. Reissig's love of flowers is never more evident than in the names of his four daughters; Flora, Rose, Lily, and Daisy. As a young boy, Peter Reinberg, son of a German immigrant, hauled produce to market from his father's farm near Rosehill Cemetery. He began experimenting with flowers, particularly roses and carnations, and successfully grew his business into a multimillion-dollar operation. Reinberg's love of nature was such that he was appointed the first president of the newly established Cook County Forest Preserves. Swiss-born Conrad Sulzer arrived in Chicago in 1836 and established one of the earliest nurseries in the Lake View area. George Wittbold, former gardener to the king of Hanover's estate, came to Chicago in 1857 and started what would become a multigeneration florist business.

Wealthier German Chicagoans expressed their love of horticulture in tastefully appointed landscapes. Wagon-maker Peter Schuettler built a mansion and exquisitely appointed grounds on Chicago's west side on Aberdeen and Adams streets. Schuettler, who served for a short time on the West Parks Board, shared a conservatory with his sister and brother-in-law next door. Palms, pineapples, and a wide assortment of other plants adorned this glasshouse of the man whose wagons helped Americans settle the frontier West. Wheelworks manufacturer Adolph Schoninger's Lake View estate in the 1890s featured broad lawns, winding sidewalks, and manicured floral beds in the decorative patterns of the day. Baden-born Edward G. Uihlein moved to Chicago in the 1860s and became an executive in the

Schlitz Brewing Company. An official with the West Chicago Park Commission, Uihlein also served as an officer with the 1890s reconstituted CHS. Uihlein's conservatory in Wicker Park housed a rock grotto with flowing water and a wealth of orchids and other exotics.

The Germans enjoyed nature as an integral part of their celebrations, and both outdoor beer gardens and picnics were popular. Ferdinand Haase,[4] recently arrived from Prussia, bought some acreage along the Des Plaines River in today's Forest Park western suburb and, in 1856, opened it for picnics and parties. German and non-German groups alike took the Galena and Chicago Union Railroad to the park to enjoy fishing, hunting, boat riding, and more. Typical was this scene on June 1, 1868, where crowds enjoyed Saengerfest: "The park was never in better order or more exquisite in the verdure of its trees, shrubbery and grass. The oaks which abound in every variety of size, were more than usually handsome in their virgin foliage and the grass was of so delicious and velvety a green as to court many to lie in repose by lying at length in its yielding softness."[5] In the 1870s, Haase began to sell off his land for cemeteries—the parcels' good drainage and location made them suitable and needed just as Chicago's City Cemetery closed. Concordia Cemetery (1872) was for German Lutherans; Waldheim (1873) was a German, nondenominational cemetery; and Forest Home (1876) was a nonsectarian cemetery that would appeal to non-Germans as well.[6] Haase and other civic leaders such as H. W. Austin traveled to Cincinnati's Spring Grove cemetery to study its design. Forest Park was laid out with curving paths and drives in the new cemetery design style.

Another cemetery, of a different ethnic group, time period, and design style, is that of St. James at Sag Bridge Church in Lemont. Founded as a mission church in 1833, St. James originally served immigrant Irish Catholics who worked as laborers in building the Illinois and Michigan Canal. In contrast to the "new" style of parklike landscaping, St. James's cemetery, with headstones dating back to the 1840s, is a traditional churchyard with neat rows of monuments surrounding the church itself. Although old-fashioned, the cemetery had (and has) undeniable charm. Writer and naturalist Louella Chapin wrote of an early 1900s visit, "When some one from the effete East tells you that Chicago is uninteresting because it has no history and no scenery, take him to Sag . . . set him on the edge of the pasture back of the churchyard . . . show him the long, placid, mirrored vistas of the old canal where once pioneers came . . . to try their fortunes. . . . . . Let him look across the miles of smiling bottom-lands to the purple bluffs beyond . . . Show him the flower-grown churchyard." In the churchyard itself, Chapin described peaceful paths, cut flowers placed near statues honoring saints, and "flowers and vines grow untrained over the graves and headstones."[7]

The area around Lemont, and other areas closer to the South Side of Chicago (such as Bridgeport), were heavily populated with Irish. Many Irish, like most newly arrived immigrants, worked hard at low-paying jobs such as digging the new canal. They lived in small shanties in the Sag Valley, or in tiny houses in the Bridgeport area, sometimes called Cabbagetown—for good reason. Writing as late as

St. James at the Sag Bridge. The gentle rolling hills and farms around the Sag valley were populated with Irish canal workers and other immigrants. Courtesy of Westchester Township History Museum, Chesterton, Indiana.

1892, a visitor noted, "Another sight was the vast prairies and cabbage fields which surrounded this vicinity [Bridgeport] for miles."[8]

Chapin describes these south Chicago Irish settlements as she rode the trolley between Bridgeport and the Sag: "Rows upon rows of tiny cottages line these side streets, telling of the ever-present longing in the heart of the homeless Irish peasant for a bit of the earth to call his own . . . Then comes the country, level as a floor. Acres and acres of vegetables, and fields tawny with squirrel-grass."[9] In Palos, where many Irish workers bought farms after the canal years, Chapin met a ninety-year-old Irishman whose garden was bedecked with portulaca and poppies and whose house was almost hidden by glorious crimson ramblers. Roses were not often missing from Chicago-Irish gardens.

Although some Irish, such as the aforementioned W. B. Egan, rapidly gained wealth and could afford well-appointed gardens, most worked hard to gradually gain their own homes and land. Typical is the story of Dublin-born Episcopalian Edward McConnell, who, at age twenty-five, arrived in Chicago by schooner in 1830. As a bachelor, McConnell worked variously as a land agent, farmer, and storekeeper. He traded heavily in the canal lot sales and, despite up and down years, did well overall in his investments. As an example of current real estate value, he swapped one of his lots of land in Joliet for a horse and what he termed "100 Morus Multitudinus" trees. Was this a quiet Irishman's wit about the ubiquitous mulberry tree, or a colloquial term? He ultimately made his home on land across from the Hastings' nursery on the south branch of the Chicago River.

Vaughan's Drumhead cabbage from Chicago's Bridgeport immigrant Irish neighborhood became a national favorite. Courtesy of Sterling Morton Library and Archives, Morton Arboretum, Lisle, Illinois.

In 1845, McConnell and his new wife, Charlotte (nee McGlashen), moved to this farm, and as they prospered, hired a multinational contingent of workers. There were at various times the (Irish) McNamara and his wife who cut the hay, a Scott named Hamilton, Irish Thomas Gannon and his wife to keep house; "Irish Catherine, and Anne Keating and Dutch Susan" also helped in the house.[10] McConnell's first year of crops in 1845 totaled $105 of oats and vegetables. By 1847, his revenue increased to $346 of vegetables, hay, corn shucks, and turkeys. In 1850 and 1851, he bought more trees, plants, and asparagus from Hastings' nursery, and sold $380 of crops. His young boys were also given a garden for fun and education. McConnell's farm grew in crop diversity and in prosperity. Like many upwardly mobile immigrants, however, any design efforts tended to the utilitarian and improved economy.

1. *Historic City: The Settlement of Chicago* (Chicago: City of Chicago Department of Development and Planning, 1976), 35.

2. "A Panorama of Nations Round About Chicago," *Chicago Daily Tribune*, November 3, 1895, 37.

3. "Practically 5,000,000 German Americans Have Found Homes in the Middle West," *Chicago Daily Tribune*, October 3, 1909, H5.

4. Sometimes spelled Haas.

5. "Sunday in the Country," *Chicago Daily Tribune*, June 1, 1868, 4.

6. Franzosenbusch Heritage Project, http://www.franzosenbuschheritagesociety.org/Histories/ (accessed September 10, 2006).

7. Louella Chapin, *Round About Chicago* (Chicago: Unity Publishing Co., 1907), 83–84, 72.

8. "Growth of St. Mary's," *Chicago Daily Tribune*, June 12, 1892, 34.

9. Chapin, *Round About Chicago*, 69–71.

10. Edward McConnell, "Reminiscence of an Emigrant from Dublin to America," 1860, Chicago Descriptions Collections, Chicago History Museum, 43–49.

California, on Bellevue Place, and on fifty acres at Foster Avenue. The yield from these gardens often met the needs of immigrant households, and sometimes enabled a small truck garden profit.

Early European immigrants to Chicago did not face the slum conditions of industrialized Chicago in the late 1800s, and were more likely to engage in agricultural pursuits either as tenant farmers or gardeners for large estates. Chicago's two earliest and largest immigrant populations, the Germans and the Irish, settled in enclaves both within the city and in outlying towns. Although their settlement patterns differed, both groups worked hard to own a piece of land large enough to raise vegetables and food for their own table or market. Theirs was a no-frills type of gardening, where landscape design was but a distant concern after cash crops and productive yields.

German and Irish gardening traditions melded with those of other nationalities between 1880 and 1900 as a large influx of new immigrants came to Chicago and other major cities. By 1900, although Germans (32.1 percent) and Irish (11.5 percent) still made up the largest groups, followed by Poland and Sweden and nine other countries that each contributed between 2 percent and 7 percent of the total population.[65] Italian immigrants, another major ethnic group, came to Chicago as early as the 1850s, but in much larger numbers after 1900. Each of these groups contributed to Chicago's horticultural experience. Early Swedes included pioneer nurserymen and landscapers Swain Nelson and P. S. Peterson. Italians and Greeks flourished as fruit peddlers in downtown Chicago. John Van Derwyk, a Dutchman, introduced the U-pick concept to roadside market customers and, along with his fair prices and excellent crops, revolutionized truck gardening in south Chicago.[66] Jewish immigrants celebrated Succoth, a harvest festival, featuring music, displays of Jewish farm produce, and a floral dance.[67]

Later waves of immigrants coincided with America's industrialization, and new arrivals in Chicago were often shuttled to the factories, stockyards, and steel mills. Chicago's so-called ghetto, the tenement district near Hull-House, was known for two occupational pursuits—sweatshops and small stores. The crowding of American cities became such a national problem that a "back to the soil" movement for immigrants was urged by church leaders, social workers, and civic improvement associations. In Chicago, this took the form of low-rent plots such as those offered by the City Gardens Association. "This is the greatest movement in the 'back to the soil' idea that has originated in Chicago," said Laura Pelham, president of the City Gardens Association. "The poor immigrant families will be impressed with the productivity of the soil and be encouraged to leave the crowded city for the simpler farm life."[68]

As immigrants assimilated, garden styles became more homogenized. Yet nearly all ethnic groups, whether newly arrived or several generations descended, yearned for reminders of home. After several florists turned him down, saying it would only grow on Irish soil, Alderman Michael McKenna persuaded German florist Gus Lange to grow shamrocks in his greenhouse. The experiment was successful, and McKenna delighted many with his three-leaved gifts on St. Patrick's Day in 1909.[69] By 1911, those Irish who harbored memories of the devastating potato famine of Ireland were importing Irish potatoes despite their cost, because of their preferred flavor.[70] Even harder to relinquish was a reliance on traditional plant-based remedies and medical "cures." The *Chicago Tribune* reported in 1909, "Thousands of foreigners in Chicago, according to druggists in foreign districts, come to the drug store with names of herbs that they have used in the old country and which they want to use here. Some of these herbs are easily identified. Others are hard to identify for the reason that they have no medical properties, and their value and cure ... consists only in the superstitious belief which the people have in them."[71] "Scare powders," to alleviate a child's fear, and "love drops," to attract beaux to a young lady, were popular requests. Local chemists such as the Murray and Nickell Company published pharmacopoeias in different languages to help druggists translate.

A wonderful celebration of Chicago's multicultural horticultural heritage was displayed in the Grandmother's Garden, near the Lincoln Park Conservatory. Nurtured by amateur botanist Julius Higgins in the 1890s, the garden paid homage to the old-fashioned favorite flowers of Chicago's immigrants and natives alike. The kind-hearted Higgins accepted plant donations from everyone longing for their homeland flowers. Thus, an Englishman offered a cluster of water

Grandmother's Garden included flowers favored by many of Chicago's immigrant populations, as well as humble plants from the prairie. Its casual design contrasted sharply with the formal gardens in other parts of Lincoln Park. Ca. 1900. From the author's collection.

dropwort, whose roots were used by druggists in medicinal preparations. With its patchwork of common flowers, rather than the newfangled varieties flaunted by some, this Grandmother's Garden was the "garden of the Illinois grandmother and yet garden of every other nationality which has remembrance of the old simplicities of home."[72] Yet, even as Chicago's ethnic population brought diversity and reminders of Europe to the landscape, a revolutionary regional design style, the prairie style of landscape architecture, made national headlines and local garden gems.

Chicago's Prairie Style of Landscape Architecture

Chicago's great prairie-style movement is most often associated with architecture titans Louis Sullivan, Frank Lloyd Wright, and Dwight Perkins, and others whose revolutionary, organic designs achieved international acclaim. Until fairly recently, Chicago's historic counterparts in landscape architecture had been overlooked. Prairie style celebrated Chicago's topography: its flat plains, open skies, and long views. Indigenous plants, such as hawthorns, crab apples, wild roses, and prairie flowers, were emphasized over earlier fads for exotics. Trees and shrubs were planted in natural groupings instead of highly formal arrangements. No longer did Chicagoans need to feel apologetic for the lack of hills and dales—flat terrain expressed limitless potential. No more did gardeners need to struggle with nonnative azalea, finicky firs, and rhododendron. Native plants could be just as lovely, and were much easier to maintain.

Jens Jensen and O. C. Simonds are perhaps the best known of Chicago's prairie-style landscape architects (or landscape gardener, as Simonds preferred). University of Illinois professor Wilhelm Miller was a much-publicized proponent of what he called the "prairie spirit in landscape gardening." In a 1915 address before the Illinois State Horticultural Society, Miller traced the movement back to the late 1870s when O. C. Simonds began working at Graceland. Miller observed that "some of the best-known work of Mr. Simonds is in Lincoln Park, Chicago, but the whole 'North Shore' shows his influence."[73]

Miller described Jens Jensen as "probably the first designer who consciously took the prairie as a leading motive." He singled out Jensen's work on the W. J. Chalmers estate at Lake Geneva in 1901 as the "first design in which prairie flowers were used in a large, impressive way," with hundreds of wild phlox, paniculata, purple flags, and swamp rose mallows. Jensen's 1901 design for Harry Rubens's Glencoe estate with its miniature spring, brook, waterfall, and lake was the "first attempt to epitomize the beauty of Illinois rivers."[74] Miller also praised Walter Burley Griffin for his prairie spirit in his work in Oak Park and Hubbard Woods.

Chicago's prairie-style landscape architects designed properties across the nation, but have been rediscovered only in the past few decades.[75] Signature prairie-style elements included the use of native plants, and massing of trees and shrubs in naturalistic groupings. Jensen and his protégé Alfred Caldwell also incorporated horizontally stratified rockwork, simulating the bluffs on prairie rivers. Geometric lawn cutouts of bright annuals were eschewed in favor of muted palettes and naturally curved bed lines. The prairie-style movement in gardening was the first major expression of approval for Chicago's natural topography and flora.

Many of Jensen's and Simonds's commissions were for the exclusive estates of Chicago's wealthy. But their public works, particularly Jensen's designs for the West Parks and Simonds's work in Graceland Cemetery, were very accessible to the average homeowner. Jensen, charismatic and outspoken, also formed conservation organizations such as the Prairie Club and Friends of Our Native Landscape. Through these nonsectarian groups, he promulgated the message of

prairie-style landscaping and celebration of the natural world. Jensen was also heavily involved in the Progressive movement that encouraged smaller, "pocket" neighborhood parks. Even in 1893, while the world's fair drew attention to Chicago's major parks, Warren Manning pointed out that, per capita, Chicago had less public green space than other comparable "developed" cities such as Washington, Paris, Vienna, and Tokyo.[76] The Progressive movement in Chicago helped bring neighborhood parks within close reach of where people actually lived.

Still, not everyone embraced the prairie landscaping style. Nurseryman Miles Bryant of Princeton wrote in 1917,

> A few years ago there was a great deal of agitation in the state over what was called "The Prairie Style in Landscape Gardening." Its chief motive was to increase the use as much as possible of native shrubs, and to use especially what were called

OPPOSITE TOP Jens Jensen, seen here cycling with his wife in his early neighborhood of Humboldt Park, was a passionate conservationist and formed a number of environmental groups. Ca. 1900. Courtesy of Sterling Morton Library and Archives, Morton Arboretum, Lisle, Illinois.

OPPOSITE BOTTOM LEFT Jens Jensen labeled this picture "our yard," depicting his own backyard in Humboldt Park. Since he was prone to taking pictures of properties prior to his new designs, it is likely a "before" picture. It is a good representation of the typical city backyard. Ca. 1900. Courtesy of Sterling Morton Library and Archives, Morton Arboretum, Lisle, Illinois.

OPPOSITE BOTTOM RIGHT The grounds surrounding Jensen's studio in Ravinia reflected his naturalistic design. Courtesy of Sterling Morton Library and Archives, Morton Arboretum, Lisle, Illinois.

TOP LEFT This image of the R. J. Dunham estate in Lake Forest was captured by Jens Jensen and reflects his design elements of stratified stonework and a signature council ring, a low circular seating wall. Ca. 1916. Courtesy of Sterling Morton Library and Archives, Morton Arboretum, Lisle, Illinois.

TOP RIGHT The Harold Ickes estate in Hubbard Woods shows a dappled clearing and trademark stonework. Ca. 1916. Photos by Jensen. Courtesy of Sterling Morton Library and Archives, Morton Arboretum, Lisle, Illinois.

"stratified" shrubs, that is shrubs whose branches ran horizontally or parallel to the ground, it being held that the horizontal line was symbolic of the broad horizons of the prairie. Its motto was to use in every planting as great a percentage of such stratified shrubs and shrubs native to Illinois as was consistent with the type of the planting.

The idea was very good, but in many of the places planted at that time the use of native shrubs was carried to the extreme. Too little consideration was given to the fact that the most of our native shrubs are of the coarser types, perfectly desirable for plantings distant from the house, but too coarse for the more formal settings in the vicinity of architecture, and especially too coarse for use upon the smaller suburban properties. In many instances the fact was over-looked that in the cities and towns the dominant line was changed from the dominant horizontal line of the prairies to a more vertical line, and that the use of many stratified shrubs was inconsistent in that they could not emphasize, as the intention was, a dominant line.[77]

Bryant observed the obvious but often overlooked reality—prairie was disappearing from Chicago, replaced by homes and suburban developments. Thus,

The grounds for this mansion are attributed to Jens Jensen, although never formally identified. Although prairie-school architect George Maher built the home using typical nuances of the genre, the landscaping, as shown here, shows ambivalence toward naturalistic design, with its formal fountain and exotic container plants. Courtesy of the Historical Society of Oak Park and River Forest.

FOREST PRESERVES AS NATURAL OASES

Coincident with the prairie-school landscape architecture movement, or perhaps an outgrowth of it, was the development of the Cook County Forest Preserves. Daniel Burnham's *Plan of Chicago* envisioned the city embraced by a circle of green spaces. In 1904, a report was prepared by Dwight Perkins and Jens Jensen for the City of Chicago's Special Park Commission, outlining natural areas deserving preservation for their beauty, historic, and economic value. Although legislation to create the forest preserve districts suffered setbacks, by 1914, citizens voted favorably on the Forest Preserve Act, and the district's first president, Peter Reinberg of florist fame, was appointed. Of the four citizen members of the Plan Committee for the forest preserves, three had distinct horticultural ties. Nurserymen J. C. Vaughan and William A. Peterson were plan members, as was Dwight Perkins, who, as an architect, was perhaps most fervent in dedication to saving Chicagoland green space. Ransom Kennicott, another well-known name in Chicago's early horticulture, was the district's first chief forester.

Although the original concept for the forest preserves was to include an outer belt of parks and boulevards, this emphasis shifted toward conservation of lands in their natural state. According to the Forest Preserve District of Cook County Web site, "The difference in philosophy of the two concepts is important. In one case the preservation of natural communities is proposed, the other proposal suggests the development of scenic highways through natural areas." The philosophy of preserving land in its natural state won, and it was that which was approved by vote from the local citizenry. The forest preserves are therefore examples of Chicago's conservationist efforts, rather than designed gardens.

Thoughts of including gardens and arboreta in the forest preserve charter persisted, however. A large tract in Palos was, at one point, destined to be a botanical garden. In 1920, a furor arose over whether the Skokie marsh should be forest preserve land, or be turned over to truck gardening. The

The brilliant colors of autumn are reflected in this quiet pool at the Forest Preserve.

Readers of the *North Shore Graphic* were encouraged to take advantage of the newly created forest preserves. 1927. From the author's collection.

debate turned into a class struggle, with some accusing the "golf fellows and dainty dilettantes of the aristocratic ridge towns" of claiming forest preserve land that would cost taxpayers money versus productive farmlands. Others, including Jens Jensen, O. C. Simonds, and various officials of ridge towns such as Wilmette, Kenilworth, Glencoe, and Winnetka, pointed out that the forest preserves would be open to everyone and offered a remarkable diversity of flora and fauna.[1]

The original staff of the Cook County Forest Preserves included scientific, dedicated men who were proud of their jobs. Ransom Kennicott in 1919 oversaw fifteen thousand acres of land and sixty-four foresters and caretakers. Rules governing the early use of the preserves had to be impressed on the public—fire pits must be extinguished and no naked swimming was allowed. According to the *Tribune*, "These men [the foresters] are zealots and they take a mighty pride in their work and in the preserve. They are constantly thinking up ways to make the public happy. Their policing is so thorough that they have cleaned out many a nest along the Desplaines that used to be the haunt of gunman from town, who infested abandoned barns in secluded tracts."[2] From gangsters to nude swimmers, to botanizers who stripped the woods of their flowers, the forest preserves had to be protected.

How to best use the wonderful resources of the forest preserves has always been a question. Peter Reinberg, in his 1918 status report to the public, noted that the recreational purposes of the preserves, for picnicking, hiking, and so forth, were obvious. Indeed, scores of people regularly flocked to the newly opened preserves to revel in nature. Less apparent were what he called the economic benefits of the preserves. Trees were vitally important in maintaining regional waterways, by preventing erosion and adding to the water cycle through transpiration. Nurseries should be established in the districts, according to Reinberg, to help reforest not only the preserves, but also other lands around the city. Experimental growing tracts could be maintained for scientific purposes. Since the forest preserves were opened at just about the same time as World War I started, emergency uses of the land were also considered. Damaged and diseased trees were cut for firewood to alleviate the coal shortage. About 525 acres of previously cultivated land was devoted to the production of oats, corn, and vegetables to meet the food crisis.[3]

1. "Skokie Marsh for All People or for Gardens?" *Chicago Daily Tribune*, February 5, 1920, 17.
2. "Custodians of Preserve Proud of Their Jobs," *Chicago Daily Tribune*, July 11, 1919, 11.
3. *The Forest Preserves of Cook County* (Chicago: Clohesey and Co., 1918), 45–65.

the prairie landscaping style, while bringing long-overdue attention to Chicago's regional beauty, was not appropriate for every dwelling, nor did it suit everyone's taste.[78] While many famous landscape designers came to work on Chicago's great estates and parks, most Chicagoans of moderate means relied on local newspapers, magazines, and books to get ideas for their gardens. As leisure time increased, homeowners had more time to dabble in ornamental gardens, and more time to read the growing body of garden literature aimed at amateurs.

Specialty Gardens

From 1910 through the 1920s, the type and variety of gardens expanded throughout Chicago and its suburbs. The middle class grew, creating more affluence and time to devote to gardens. Transportation improvements, particularly through motorized vehicles, enabled shipments of live plants from grower to homeowner. The sheer number of plant varieties exploded, thanks to improved hybridization techniques. Technology improvements, such as new chemicals to maintain lawns and plants, and new construction materials, including cement basins for water features, shaped the choice of landscape effects. With this smorgasbord of options available, Chicagoans sometimes specialized in particular plants or garden themes, or brought various ideas together in a pastiche of design, combining the newest ideas from around the world and redefining old traditions from the past.

Smaller Gardens

A common theme throughout the gardens of the early 1900s addressed strategies for smaller gardens. Homeowners no longer required large lots to grow food for personal consumption or for livelihood. J. A. Pettigrew, former Chicago Park District landscape artist, created designs targeted to smaller properties as early as the 1890s. His design for a "Cottage Garden" outlined in an 1893 Chicago-based *Gardening* magazine, presaged the later trends for smaller gardens. Pettigrew offered a suitable design for a typical city or suburban lot, with 50 feet of frontage and 150 feet deep. The first task, he said, was to fence out the chickens. The front yard was to have a walkway of porous material such as coal ashes between a front gate and front door. Lawn could be on either side of the walkway. A flower border of varying width, averaging seven feet, hugged the fence on the three sides of the yard surrounding the front of the home. Climbers such as *Clematis coccinea*, *C. paniculata*, and *C. Jackmanii*, or natives such as trumpet vine, moonseed, and bittersweet, could clamber over the fence. A variety of spring bulbs (snowdrops, crocus, narcissus) and hardy plants (delphinium, hollyhocks, perennial asters) were to be planted "to avoid a dotted repetition." Pettigrew advised planting in "as natural a manner as possible," with taller plants at the back of the border.

In 1914, landscape architect Ralph Rodney Root described the typical American home as a suburban residence with a lot size from sixty to two hundred feet wide, costing $4,000 to $15,000.[79] Lena McCauley described average city lots as seventy-five feet wide, and recommended gardens as solutions to those who live in "one of a score of little houses in a made-to-order subdivision."[80] Kate Brewster's entire 1924 book was dedicated to "Little Gardens," with an emphasis on getting the most out of a gardening investment. With smaller gardens, the delineation of outdoor space changed dramatically. Earlier, larger rural gardens had major areas dedicated to service areas and barns. Vegetable gardens were large and de rigueur to supply the family table. Early front yards were decorated with flowers aligned along walkways, or assembled in lawn cutouts. Leisure sports, such as croquet, were fair game in the front yard, perhaps as public displays of available leisure time associated with wealth.

In his 1914 book, Ralph Rodney Root identified three landscape zones for the modest home: the private yard for the family, the semipublic area, and utilitarian areas. Private areas, in the back, would contain the family flower-garden, "and that is the only place where flowers should appear except for accent purposes," according to Root. Private areas were to be screened by shrub masses, and were to contain the most "interesting and varied planting." Semipublic areas, visible to passersby, were to be simpler and more formal, with a few trees and masses of shrubs, but less "individual interest."[81]

In 1932, F. Cushing Smith, Chicago city planner and fellow of the American Society of Landscape Architects, published a book geared toward the smaller garden, *Landscape Designs for Your Garden*. Smith opined that every garden had a distinct personality and that homeowners should not slavishly try to imitate plans found in magazines. In the magazine he edited, *American Landscape Architect*, he proposed enclosing all gardens, by brick, fence, or living hedge. This departure from the ballyhooed American democratic ideal of continuous flowing lawns reflected the smaller spaces and need for privacy in Chicago's gardens. Smith also noted that "it is our belief that even on the smallest lot there is a place for both formal and informal landscape design."[82] According to Smith, the use of rockeries, water gardens, vegetable and cut-flower gardens, Japanese gardens, and a "host of others too numerous to mention" could be used to good effect if proper design principles were followed.

In smaller properties, the backyard became the new gardening frontier, and its borders provided the enclosure Smith mentioned as necessary for privacy. Changes in design fundamentals were needed to accommodate the fact that the wide-open spaces of the prairie were closing in. Siting the home became even more important as potentially unpleasant views had to be considered. As designer/nurseryman Miles Bryant noted in 1917, "On smaller properties, in city and town, orientation will of course be the most dominant factor, but even there the various outlooks from potential windows should largely determine the interior house plan. The plan for the small property should be internal in its conception rather than external, that is it should depend for the termination of its views on things within rather than beyond its own boundaries, for long views are beyond the control of the owner and may change at any time to the detriment of his out-look." Bryant suggested that, except for very large estates, most of the lawn be placed on the south and west sides of the house to maximize sun exposure, and that the public rooms of the home should be oriented in accordance. Bryant further recommended landscaping for the backyard since "the desire for privacy has lead to the development in the rear of private lawns and gardens shut off from the view of the street."[83]

TOP LEFT This sketch from the 1929 Chicago Garden and Flower Show program shows an increasing emphasis on enclosure and structure in the garden. Vaughan's "Formal Garden" recommended this style as being most complementary to a home's architecture, especially when gardens were closer to the homes as in small properties. From the author's collection.

BOTTOM LEFT The Men's Garden Club designed "A City Garden." "The architectural features of such gardens become of great importance" when plants face adverse growing conditions. From the author's collection. From 1929 Chicago Garden and Flower Show program.

RIGHT The author's mother, Gertrude Gallagher, is shown here with her paternal grandfather, Thomas Miller, in the backyard of a South Side home near 83rd and Halsted streets. This garden represents a wonderful fusion of current trends, including children's gardens, backyard enclosures, and the influence of the automobile. Typical of 1920s city properties, this garden adjoined a detached garage near the alley, and herbaceous perennials and annuals line the border of the property. Courtesy of Gertrude Gallagher.

Author and wealthy socialite Kate Brewster advocated an even stronger emphasis on private backyards. She suggested siting houses close to the street so as to maximize backyard space where families could enjoy their own gardens. Space should not be wasted on long driveways leading to detached garages in back; instead she suggested bringing the garage closer to the house. Brewster opined, "In America, we are hampered by preconceived notions as to ... stylish front-yards and messy back-ones." She approved of tidy, formal front yards with personalized gardens in the backyard.[84]

TOP The garage and automobile required design solutions. In some homes, such as this at Kenilworth Street and Touhy Avenue in the Rogers Park area, the garden extended and camouflaged the garage. 1922. Reproduced by permission of Rogers Park/West Ridge Historical Society.

BOTTOM Elsewhere on these grounds, the homeowner decorated a production grape and fruit orchard with a rustic arch. Ca. 1920. Reproduced by permission of Rogers Park/West Ridge Historical Society.

The layout of private grounds was also impacted by children's needs. Ralph Rodney Root called this era the "age of the child."[85] Creating outdoor space for children became a social imperative with Chicago settlements such as Hull-House. Active leaders in Progressive movements like the Playground Association of America, Jens Jensen, and others incorporated children's playgrounds into many Chicago parks, and popular literature stressed the healthful and moral benefits of outdoor play. School gardens became a staple of the Chicago

TOP LEFT In 1914, at the Chicago Teacher's College at 64th and Stuart streets, the premier school for certified teachers, gardening was an integral course topic. The teachers could then bring their gardening knowledge to their own students. From the author's collection.

BOTTOM LEFT This large gathering of school children and adults demonstrates the interest in the newly dedicated garden for Marshall School at Adams and Kedzie streets. The original photograph caption indicates that "Mr. Prost" was the gentleman delivering a speech—probably J. H. Prost, the city forester. 1917. Reproduced by permission of Wisconsin Historical Society WHi-11532.

RIGHT W. W. Barnard Seedsmen was one of many companies who promoted the value of gardening to children. 1924. From the author's collection.

Public Schools. In earlier years, children may have been relegated to working the fields—now they could play in them. Lena McCauley wrote that for "many college-bred parents . . . bringing up of children . . . hinged on the vital importance of play in the open air." McCauley observed that landscape planning must include space for children to "play ball and tumble about." She eschewed the overly designed children's gardens of landscape architects, and recommended simple beds with child-friendly flowers and nearby fields or pastures for romping about.[86]

Even as children claimed some of the space in the ever-shrinking lot size, so did the automobile. Chicago's position as a transportation hub had broad-reaching implications in the early twentieth century. Just as the city was a center of shipping and, later, railroad traffic, the invention of the automobile had both social and garden-design implications. Once again, Chicago figured prominently in the new age of the automobile. While not a major automobile manufacturing city, Chicago became a hub for highway roads. The American Automobile Association was formed in 1902 in Chicago, and became one of the largest promoters of automotive travel. Automobiles spawned detached garages, creating a new architectural element to incorporate in garden design. City gardeners with limited space found ways to grow plants on trellises near the garage, or simply used the garage as a backdrop to their plantings. Automobiles posed challenges, however, with their fumes and noise. A writer in the 1920s noted, "Growing perennial flowers in back yards just over the fence from Gasoline Alley is considered impracticable by those who have not had experience with city gardening." The writer described the garden of Mrs. H. M. Bowers as a success story—over the years she identified hardy plants for city garden conditions, and created shelter for tender plants with shrub masses. "On one side of [Mrs. Bowers's] garden is a public garage, besides many small ones, belching smoky fumes from early morning until late at night. Nature has apparently provided the flowers with gas masks, and the injury is hardly noticeable."[87]

Unfortunately, one of the less savory outgrowths of the automobile craze was outdoor billboards marring natural scenery. With sweeping prairie views now obsolete because of smaller lots and outdoor advertising, homeowners were encouraged to erect natural or architectural walls around their gardens to define their space and block offending views. This was a major shift in Chicago's gardens where, as in most of America, broad, sweeping lawns were viewed as the democratic ideal. In Chicago in particular, walled gardens had an even more dramatic impact. Heretofore, design styles were freely copied from local tastemakers' gardens, which were generally located in town and, thanks to Chicago's flat topography, were often in plain view. Now, with many gardens sequestered behind walls, middle-class homeowners had to rely more on books, newspapers, and recommendations from local nurserymen.

As the carpet-bedding fad waned at the turn of the century, and with England's Gertrude Jekyll and others touting the value of mixed borders, perennials began to assume greater importance in Chicago's vernacular gardens. John C. Moninger, local greenhouse builder, noted in 1908, "In the last few years the demand for hardy perennials has increased wonderfully, and this not at all because of a falling off in the demand for the class of plants commonly known as bedding plants, but rather because of . . . the merits of hardy plants."[88] The shift from annuals to perennials is nowhere more apparent than in Lena McCauley's book *The Joy of Gardens*, where she includes in her appendix a list of "old fashioned annuals." Garden makers could include annuals for a nostalgic touch, but McCauley proclaimed perennials the "crown jewels of gardens."[89] Her list of preferred perennials was organized for continuous bloom and winter interest—a significant achievement for Chicago's short growing season.

Planting perennials in Chicago gardens perhaps indicated a willingness to settle down. Frontier days were over, and real estate speculation, while never abated, became more the realm of suburban development consortiums rather than individuals. Chicagoans now viewed their land as "home," rather than an asset to be traded at ever-rising market prices. McCauley said, "Some have been heard to say, 'Decorate to-day, for tomorrow you move,' and they expend all their fancy on potted geraniums, palms and hastily sown annuals. To the winds with them! They well deserve to move often."[90] She opined that "permanence is a secret of the charm of old gardens," and perennials were essential for that sense of stability.

Highland Park resident W. C. Egan was tapped by New York publisher McBride, Nast and Company to write a book, *Making a Garden of Perennials* (1912). In addition to an extensive list of recommended perennials, Egan devoted considerable attention to bed preparation and mulching—cultural concerns that might not have been seriously considered with one-season annuals. He also dedicated a chapter to plant combinations—pairing such plants as shooting star (*Dodecatheon meadia*) and bellflower (*Campanula carpatica*), or forsythia with Virginia bluebell (*Mertensia virginica*). Design strategies for incorporating perennials evolved over the first decades of the twentieth century. McCauley's plans showed perennials mixed into outer garden borders, while preserving old-fashioned geometric cutouts in the lawn for annuals. By the 1930s, geometric cutouts were largely gone, and perennials were a mainstay of the backyard garden border along with flowering shrubs and hedges.

Water Gardens

Chicago's early gardeners were preoccupied with draining swampland, so it was a major turning point when gardeners started deliberately installing water gardens. "Water gardening is a fascinating modern development, for those interested in such pursuits," wrote gardener Sarah Frazier of Yorkville in 1879.[91] She recommended such a garden only if a natural water source were available to limit the expense. Hence, "if you have a boggy piece of ground do not have it drained unless proved that it influences malaria or the mosquito supply." Frazier appreciated water gardens for their romantic appeal—noting their historical associations dating back to ancient Egypt. In the 1880s, John Neltnor's *Fruit, Flower,*

TOP Most rock gardens of the early twentieth century attempted naturalistic settings, such as this garden in the Busse home of Mount Prospect. Courtesy of Mount Prospect Historical Society.

BOTTOM Some landscape gardeners specialized in rock gardens. Unfortunately, these were not often blended well into the landscape. From the author's collection.

and Vegetable Grower magazine had a regular column "Bog and Aquatic Garden," with seasonal suggestions for the "bog" garden. These articles contained a technical approach to water gardening, listing plants for deeper water (water lilies, pickerelweed, white water crowfoot) and those for the margin (rush, cattail hair, loosestrife, and water pitchers). Neltnor also proposed a "window bog garden," which allowed those with limited means or space to create a tabletop water garden.[92] Water gardens continued to be popular through the early twentieth century as city water and irrigation methods became more prevalent. The 1927 prizewinning entry in the *Chicago Tribune*'s garden contest showcased a major water feature as the focal point of a typical perennial border.

There is some irony in the concurrent rock garden fad in Chicago, where, unlike the rocky East Coast, stony soil is uncommon. Oak Park author Martha Rayne characterized Chicago's soil: "Especially in the prairie sections, . . . stones are so rare that a traveler, in recording their scarcity, said that when his dog saw one he stopped and barked at it."[93] Creating rock gardens on the prairie, versus a mountainous locale, seems a bit of an unnatural act. Yet the craze for rock gardens waxed and waned—from elaborate 1870 grottos constructed in public parks to backyard rock gardens of the 1920s and 1930s.

For beginning gardeners, Kate Brewster recommended rock gardens because they required only a bit of space and were relatively inexpensive. New Englanders, who built walls from castoff stones, would be amused that she identified inadequate stone supply as the chief obstacle for creating such a garden. For design guidance, she advised her readers, "A rock garden is not a flower bed with stone trimmings; it is a rock bed with flower trimmings."[94] For rock garden flowers, Brewster did not insist on true alpines; for our flat plains, she proposed some common garden substitutes. Pinks, alyssum, foxgloves, and columbine were just some of the plants she recommended. Chicago, like much of the country, suffered the effects of Plant Quarantine No. 37 which, until 1923, barred many imported bulbs that would have worked well in the rock garden.

"A rockery is not merely a pile of rocks or boulders thrown in a heap with a few starved plants set among them," wrote florist Fritz Bahr in 1922.[95] He suggested that rock gardens could be used to hide unsightly views, or be constructed as a backdrop to a water feature. Designing small rock gardens would be a good business for florists, Bahr opined, because there weren't sufficient guidelines available to the do-it-yourself homeowner. Bahr also suggested that rockeries were good places to showcase individual plants—an alternative to the massed borders then in vogue.

As with many fads, the Chicago flower shows were instrumental in bringing rock gardening to average homeowners. The *Chicago Tribune* reported the case of Mr. and Mrs. F. J. Koza, "amateur suburban" gardeners, who "wanted something new, something different." They got the idea of building a rock garden from the big Chicago flower show. Koza apparently liked the model garden so much he bought the rocks—ten tons of them—from the exhibitor and moved them to his Wilmette home. Koza, a railroad engineer, built his rock garden in the backyard corner of his 50-by-187-foot lot. Noted the reporter, "It will soon be like a favorite mountain scene cut out of the Rockies and transplanted within four blocks of Lake Michigan's shore line."[96]

Greenhouse and Window Gardens

Greenhouses were once available only to Chicago's wealthy. Being able to grow exotic plants, or even fruit and vegetables, in Chicago's harshest winters was a sign of wealth. Yet even modest Victorian homes often featured a large bay window on the south side, to capture the sun and warmth for parlor gardens. Even if adequate windows were not available, in 1871, Edgar Sanders offered advice on the management of ferneries, contained in Wardian cases or do-it-yourself glass-domed tabletop terrariums. Besides ferns, Sanders recommended foliaged begonias, spiderwort, dracaena, and other plants for the indoor glass garden.[97] By the 1880s, the *Fruit, Flower, and Vegetable Grower* magazine devoted a regular column to the "Green House and Window Garden." It gave seasonal advice, with tips such as whitewashing the outside of the glass in May to soften stronger sun rays.[98]

In addition to bringing the outdoors in via greenhouses, Chicagoans learned to enjoy the all-too-short summers by creating outdoor rooms. Philanthropist

The Potter Palmer residence on Lake Shore Drive was world-renowned for its castellated ornamentation outside and world-class collection of art inside. An infrequently seen side view of the home shows the extensive greenhouse. Bertha Palmer, the doyenne of Chicago's 1890s society, grew a century plant at the front entrance. Ca. 1895. From the author's collection.

Louise DeKoven Bowen recalls her early years: "Chicago was a delightful summer watering place. When hot weather came the drawing room was moved to the front steps. It was a poverty-stricken family that did not have its stoop rug put out every evening at dark, on which the younger members of the family sat and talked while the elders occupied rocking chairs in the vestibule."[99] Front porches became an essential part of Chicago architecture, and homeowners flocked to the porch in summertime to catch a cool breeze. Elaborate outdoor furnishings became part of the outdoor room. Hanging baskets—for both indoors and porch decoration—were popular, and made of terra cotta, rustic work, or wire.[100]

TOP Porches were essential in hot Chicago summers. Chicagoans moved to front porches in the late 1890s and early 1900s, before the backyard became the private family zone. Marshall Field's, Chicago's premier department store, displayed the latest in porch furniture in this postcard advertisement. Ca. 1910. From the author's collection.

LEFT This garden in the 4000 block of Greenview Avenue demonstrates the effort taken in even the smallest garden space. Here, cannas and elephant ears decorate lawn cutouts, vines are trained up the front entryway, and flower beds adorn borders around the garden apartments. Ca. 1914. From the author's collection.

By the 1920s, porches moved to the back of the house, to complement the new "private" backyard, and to provide sleeping porches for hot summer nights. Multistory "flats" in the city commonly featured open back porches for ventilation and egress. "Chicago's back porches can be brightened up by spending a few cents for annual flower seeds or plants which are easily grown in boxes,"[101] recommended a *Tribune* writer, noting trailing plants such as petunia hid any clutter stored on the porch.

Window boxes regained popularity as the "poor man's garden" in the years of the Depression. "No one has a monopoly on the window box garden—it is everybody's garden, the rich and poor alike," according to an article in the *Chicago Tribune*. Thousands of visitors saw "the largest collection of window boxes ever assembled in one place," at Chicago's 1929 garden and flower show. Created by local garden clubs and nurseries, the show included plant combinations for all seasons of the year.[102]

Roof Gardens

Chicago is often called the birthplace of the American skyscraper, since William Le Baron Jenney created the metal-framed Home Insurance Building in 1885. No wonder that roof gardens became popular with both rich and poor in the city. Roof gardens, as a public entertainment venue, made their debut in Chicago in 1894. Taking cues from a temporary attraction on the Midway, the roof garden atop the Masonic Temple included two theaters with a total seating capacity of two hundred. More than three hundred feet above the sidewalk, this roof garden, like many to follow, was established mainly as a novel place to view plays or musicals near the skies.

Roof gardens created mainly for their green-space value, whether ornamental or practical, became popular in the early 1900s. Necessity initially drove the use of the rooftop in the so-called ghetto area of the city. While there were many instances of kitchen gardening in available porches or pots, purely decorative gardens were also grown—a testament to the love of flowers in every station in life. One well-publicized roof garden was that of Mr. F. Burkhart, who lived on West Twelfth Street. Burkhart's garden was enclosed by fences trained with morning glories. He laid boards over the gravel-topped roof to deflect the sun, and planted dozens of flowers in pots, including verbenas, mignonettes, pinks, hollyhocks, and castor beans.[103]

Roof gardens were also employed as a poor-man's summer resort. In the early 1900s, many working-class girls were drawn to the city and found themselves working in the clothing factories, as domestics, or in other low-paying jobs. The young women stayed in boarding houses or in low-rent flats. Fortunate girls got a bit of fresh air in the summer through philanthropic country homes such as Wildwood. Others, such as Carrie Cuyler, devised a roof garden curtained

Roof gardens were the province of rich and poor alike in the 1900s. Here, a very modest roof garden is created with a simple potted plant and wooden crate. From the author's collection.

with vines and adorned with flowers. Cuyler's garden was featured in the *Chicago Tribune's* garden contest, and the newspaper applauded her ingenuity and observed that such a garden might help a young woman attract and retain an eligible bachelor.[104] In that same year of 1901, the *Tribune* published hints specifically related to roof gardens for amateur gardeners. Readers, particularly those in crowded districts, were urged to utilize roof space. Planting vegetables and flowers in wooden boxes was best accomplished by selecting varieties that could tolerate high winds and scorching sun. This included poppies, phlox, marigold, pot marigold, larkspur, snapdragon, portulaca, sweet alyssum, petunia, bachelor's button, strawflowers, candytuft, calliopsis, asters, tuberoses, gladiolus, and tigridia lilies.

In 1907, the Cliff Dwellers, a newly formed club of creative cognoscenti including artists, architects, writers, and patrons of culture, designed their private clubhouse atop a Chicago skyscraper. Plans for the new meeting space included a roof garden designed by member landscape gardeners (O. C. Simonds and Jens Jensen were charter members). In the next year, architect Dwight Perkins brought attention to the possibilities of roof gardens in his design for a West Side school. The most exciting element of the design was the "aerial garden," which included flowers and plants as well as a playground.[105]

The Cliff Dwellers and other socially prominent individuals or groups brought cachet to the roof-garden concept, and "elevated" it from the lower and middle classes. As more of Chicago's social elite adopted a downtown pied-à-terre in conjunction with their country estates, the roof garden became a necessary accoutrement. Among the first was Herbert Bradley, whose 60-by-100-foot roof-top garden on South Hyde Park Boulevard yielded blooms and views galore.

Roof gardens sprang up on Sheridan Road, on the Gold Coast, and points be-tween.[106]

Japanese Gardens

Victorian America's fascination with Japanese culture extended to gardening, and the Japanese garden exhibit at the 1893 Columbian Exposition intensified interest. Located prominently on the fair's famed Wooded Island, the Japanese teahouse and garden drew thousands of visitors. Although the Japanese exhibit was not a particularly faithful reproduction of a true Japanese garden, it caught the mind and spirit of many. Visionary artists and cultural leaders were caught up in Japan mania. Frank Lloyd Wright's work was influenced by his study of Japanese design, and many of Chicago's elite became enamored of the Eastern aesthetic. The interest soon spread to the middle class, who found ways to in-corporate Japanese elements into their gardens. Local nurseries offered design advice, such as the D. Hill nursery whose 1927 catalog advised, "For the conve-nience of our customers desiring to make gardens in accurate Japanese style we have engaged the noted Japanese landscape artist, Mr. T. R. Otsuka."

Flower shows, including Chicago's 1933 World's Fair, continued to showcase the Japanese art of gardening until World War II. At the 1933 fair, two Japanese gardens were exhibited by Charles Fiore of Prairie View and the Fleming Land-

Smaller city properties such as this home on Touhy Avenue sported Japanese gardens. While not as exten-sive as its suburban counterparts, this garden nonetheless included water features, ornaments, and plants that expressed Japanese design. Reproduced by permission of Rogers Park/West Ridge Historical Society.

scape Company. The horticulture program book said, "He [Fiore] has shown in this garden how running water, a bridge, tea house, and American grown plants in Japanese variety, beautify the American home." The Fleming company's garden exhibit used authentic Japanese ornaments, imported by a Chicago company. Fleming asserted that "this garden as a whole is hardy and well adapted to local climate conditions."[107]

The 1933 World's Fair represented a major milestone in Chicago's landscape design evolution, and encapsulated all the disparate ideas above. The main landscape architect of the fair, Ferruccio Vitale, along with influential members of the

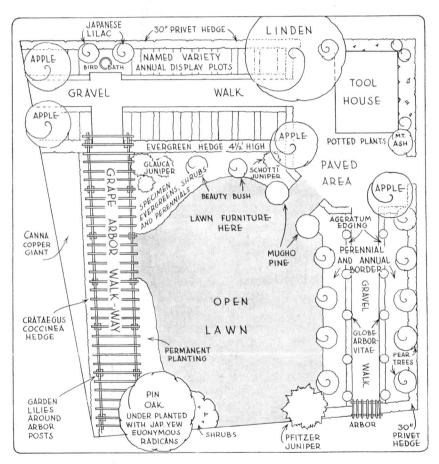

While the overall fairgrounds of the 1933 World's Fair were designed by a New York landscape architect, and many gardens from around the world were represented, Chicago's own nurserymen combined many ideas in their displays. Here Vaughan's "Home Garden" shows a simple combination of formal and informal plantings in the "new" backyard. Structure is provided through the parallel grape-arbor walkway and allee of shrubs and trees around the gravel walk. An open lawn offers a play area for children, and a tool house is discreetly tucked in the corner. A privet hedge borders two sides of the outdoor room. Shrubs included forsythia 'Golden Bell', oak-leaved hydrangea, mock orange, privet, Japanese lilac, and evergreens. Nearly one hundred annuals and perennials were displayed and interchanged during their flowering seasons. From the author's collection.

Lake Forest Garden Club, promoted Lake Forest's Foundation for Architecture and Landscape Architecture, which existed from 1926 to 1931.[108] Through this foundation, and the new Chicago-based magazine *American Landscape Architect*, Chicagoans raised their profile in the design community. The 1933 World's Fair was testament to how Chicago had matured in landscape design. No longer chained to European or New England ideals, Chicagoans felt sufficiently comfortable with their own landscape identity to invite classicist Vitale (and, upon his premature death, his successor, Alfred Geiffert) to design the fair. Yet a broad spectrum of design influences came together at this fair. Its horticultural advisory committee included a legacy of Chicago's early horticulturists such as Swain Nelson; Miles Bryant of Princeton's Bryant family; Hal Kennicott, descendent of John Kennicott; Roy West, partner of O. C. Simonds; and Leonard H. Vaughan, descendent of J. C. Vaughan. Jens Jensen, who ignited the prairie school of landscape architecture, was on the committee, as was Franz Lipp, who would take Chicago's landscaping to modernist ideals.

Thus, Chicago combined traditional homegrown talent with expansive ideas from abroad. Despite the turmoil surrounding the impending World War II, many foreign nations participated in this world's fair. There were Japanese gardens, naturalistic gardens, Italianate gardens, nostalgic landscapes, desert and tropical plantings, and "modern" flower gardens. But the garden that perhaps best encapsulated Chicago's own style, having selected the best from all these influences, was Vaughan's exhibit Home Garden. Chicago's landscapers had finally found a place to call home.

LEGACY LANDSCAPES: DESIGNED GARDENS

Very few, if any, gardens remain with their original design. Naturally, plants die and are replaced, styles change, and properties themselves are divided. Even fewer properties are open to the public. Those that are described below contain bits and pieces of heritage landscapes, both residential and public.

Riverbank

1925 Batavia Avenue
Geneva, Illinois
http://www.co.kane.il.us/Forest/fp/fabyan.htm

Where else would you find a bear's cage, alligator pool, Japanese garden, windmill, greenhouses full of flowers, grottoes, acres of gardens, deer paddock, Indian statue, working quarry, cannons, and Greco-Roman swimming pool? At Riverbank, the riverside estate of George and Nelle Fabyan, all this and more once existed. Scientific farms were a trend in the early 1900s. In the first decade of the twentieth century, moneyed gentlemen farmers enjoyed experimenting with new ways of raising livestock and growing crops. Their estates offered interesting examples of landscape design as well as insight into the food markets and technology of the day. On Chicago's North Shore, a number of estates featured experimental farming or animal husbandry, including Samuel Insull's Hawthorne Farms, J. Ogden Armour's Mellody Farm, and Arthur Meeker's model dairy at Arcady Farm. George and Nelle Fabyan, with highly eclectic interests, had perhaps one of the most diverse landscapes, and one of the most interesting, too.

TOP The Fabyans created gardens and farm acreage on both sides of the Fox River. Drawn by George L. Adams from reconstruction to 1920s by Darlene Larson and John Marshall Butler. Reproduced by permission of Friends of Fabyan.

LEFT Riverbank gardens included traditional planted beds, fields of peonies, and ornamentation, including raised sundials, stone walls, and a patriotic flagpole. Ca. 1920. Reproduced by permission of Friends of Fabyan.

BOTTOM An elaborate arbor covered with roses extended from the house grounds to the river. At left is Nelle Fabyan. Ca. 1920s. Reproduced by permission of Friends of Fabyan.

Businessman George Fabyan and his wife, Nelle, bought ten acres of farmland along the Fox River in 1905, and expanded the property over the next three decades to nearly six hundred acres, one of the largest self-sufficient estates west of Chicago. Riverbank includes the family home, Fabyan Villa, its grounds and outbuildings, and a research center across the street. Colonel Fabyan was born into a wealthy Boston family and, in 1903, moved to Chicago to head the Midwest branch of the family cotton textile business.[109] He and Nelle Wright Fabyan, married in 1890, each brought passion and dedication to many individual pursuits, and their far-flung interests are equally represented in this crazy-quilt landscape of curiosities and wonders.

Fabyan Villa, the original farmhouse remodeled over the years with input from Frank Lloyd Wright, sits near the road at the top of a long, sloping hill that descends to the Fox River. Stretching from north to south along both sides of the river, individual gardens, outbuildings, and follies dotted the landscape. Unlike the manicured estates of the North Shore, which typically unfolded under the master planning of a prestigious individual landscape architect or firm, Riverbank grew through the years, with additions and influences from its many employees, and the exotic interests of its owners.

Gardening traditions from many countries were represented at Riverbank. George Fabyan hired U.S. Department of Agriculture staff member Charles McCauley, an Irishman, to oversee the estate's horticultural interests; native Czechoslovakian Louis Kostel to manage the south greenhouse; and Susumu Kobayashi, Japanese gardener. Also in this mini-league of nations were a resident sculptor, Italian Silvio Silvestri, who created many of the garden ornaments; German Theodore Matthews, who headed the farm operations; and Norwegian Jack Wilhemson, who, in addition to managing the boathouse and water needs of the estate, used his nautical expertise to fashion a garden folly: a fourteen-foot rope spiderweb between two large elm trees.[110]

There are so many points of interest at Riverbank, it's perhaps useful to study the estate in sections, as suggested by the map printed here. This map shows Riverbank as it might have been in the 1920s, as recollected by former workers and local villagers. The east portion of the estate (above the river), is bisected from east to west by Middle Road. The upper (westernmost) part of the estate near the road contained the Fabyan residence called the Villa. Near the home, Nelle Fabyan planted flowers in old-fashioned lawn cut-outs of stars, crescents, and other shapes. One of the most impressive surviving landscape features, the rose arbor led from the bottom of the slope at the rear of the home down to the river. The arbor was planted with roses and featured a large sundial mounted on a pedestal. North of this arbor was a series of greenhouses. Torn down in the 1920s, this was but one of two sets of production glasshouses Fabyan had built. The greenhouses, vegetable gardens, and peony fields provided flowers and food for the family, and commercial crops sold in Chicago.

Through an elaborate gate, the arbor leads to a bridge that crosses the river to an island where once a large Greco-Roman swimming pool stood. Across the river on the eastern portion of the estate, Fabyan built a Bavarian beer garden during Prohibition for the entertainment of his workmen and local community, and installed an octagonal, sixty-eight-foot-high windmill. Fabyan milled a variety of grains farmed on his estate. Listed on the National Register of Historic Places in 1977, the windmill was relocated from nearby York Center to Riverbank in 1914, and reconstructed at great cost.

In the middle of the estate, the Fabyans' Japanese garden takes center stage. Built between 1910 and 1914, the restored garden was originally designed by Taro Otsuka from D. Hill nursery in Dundee. Surrounded by a simple wooden fence, the garden then, as now, had four garden gates, one at each point of the compass. The western gate is a Shinto-inspired Torii gateway, the traditional symbol of a nearby shrine. Concrete lanterns were created for the garden by Silvio Silvestri. The garden includes plantings, an arched bridge, and water features characteristic of Japanese gardens. The inspiration for the garden may have been Colonel Fabyan's role in a 1905 peace negotiating team during the Russo-Japanese War, coupled with the novelty of Japanese gardens sweeping America during the period.

Near the Japanese garden are several smaller gardens and follies of note. A bear cage remains just southwest of the garden, one of many animal displays that the Fabyans favored. At one time, alligators were kept in a pool just east of the bear cage, and other cages or pens on the estate contained exotic birds, pheasants, a kangaroo, monkeys, ostriches, and more. Slightly northwest of the Japanese garden near the restored flagpole were two formal gardens, the Centennial and Susquehanna gardens, which showcased a wide variety of flowers. Two grotto gardens once graced the landscape. One, whose remains can still be seen, was built north of the estate's boathouse and used fossil stone imported from Florida. Another was built into the hill close to the villa and was planted with ferns and shade plants.

Across the main road (now Route 31) is Riverbank Laboratories. This was George Fabyan's think tank. Fabyan was interested in such diverse subjects as plant genetics, animal husbandry, health, and physical fitness. Especially during wartime, his patriotism and interest in military strategies, acoustics, and encryption employed scientists full time at the laboratories.

Riverbank was often the height of social activity. The Fabyans were quite generous with their property, and allowed local residents to stroll the grounds. Fabyan, a patriotic man, permitted military practice maneuvers during the war. Although many structures are missing, it is quite easy to imagine the amazing landscape diversity this property once held.

Landscape Integrity

Several groups are actively involved in maintaining Riverbank today. The Kane County Forest Preserve District acquired the property in 1939, and is responsible for maintaining the grounds. The Friends of Fabyan, a volunteer group, is dedicated to preservation, restoration, and education on the legacy of George and Nelle Fabyan.

Although many of the outbuildings have been removed (greenhouses, the alligator pit, swimming pools, etc.), the remnant landscapes still hint at the former glory of this estate. The Japanese garden has been rehabilitated by the county and Geneva Garden Club, and more work is planned. The rose arbor is also being restored and still runs from below the villa to the river. Various outbuildings have been repurposed into shelters (the bear cage and boathouse). The Japanese garden and some areas around the house are planted, but other vast formal gardens no longer remain. Yet 245 acres purchased by the county in 1939 remains, and the spatial relationships among important garden features still suggest the original extent of the landscape.

Frank Lloyd Wright Home and Garden

951 Chicago Avenue
Oak Park, Illinois
http://www.wrightplus.org/homestudio/homestudio.html

How did the master of prairie-school architecture landscape his own home? Frank Lloyd Wright, whose architecture used natural forms and blended with the environment, built his own home and studio in Oak Park around a ginkgo tree. Early photos show Wright enjoying the prairie vistas from his front porch. Today, virtually the entire surrounding prairie has been developed, but the Wright home landscaping has been redesigned to include native and simple plantings that he might have espoused.

Frank Lloyd Wright, seated here, extended his home into the outdoors through a terrace. Nearby forested views no longer remain, but at the time helped create a sense of being immersed in nature. Reproduced by permission of Frank Lloyd Wright Home and Studio Archives.

In 1889, Wright bought a plot of land at Chicago and Forest avenues in Oak Park. He was a young man of twenty-one, about to be married. The site had previously been owned by John Blair, an early landscape gardener. Blair was famous as a prominent landscaper in Chicago and had planted a nursery around his own home. A number of trees and shrubs had been planted there for propagation and resale. The existing greenery attracted Wright to the spot: "In regard to Oak Park," Wright said in an early address to the Fellowship Club of Oak Park, "a certain profusion of foliage characterized the village. I remember well that I came to Oak Park to live for no other reason than that, and the remarkable character of the foliage on the old Blair lot."[111]

In the 1890s, Oak Park was experiencing a growth boom. Oak Park developers and city fathers like H. W. Austin tried to improve the flat prairie views through extensive plantings. Wright thought many of these landscaping efforts were unnatural and did not blend into their surroundings. We get a clue for Wright's preference toward natural landscaping in his comments about Oak Park greenery: "Speaking of Oak Park's feeling for nature," he lectured to the Fellowship Club, "I presume we might say that Oak Park . . . likes the country but prefers it as the lady thought she would the noble savage, more dressed." Wright continued his tongue-in-cheek comments on the formality of Oak Park landscaping:

> Disorderly natures may sing the charm of the rugged oak, the spreading chest-nut, or the waving elm; to Oak Parkers, all such, with their willful untidy ways, are eyesores. The Lombardy poplar grows where it is planted, and as it is planted. It has no improper, rugged ideas of its own. It does not want to wave or spread itself. It just grows straight and upright as a proper Oak Park tree should grow, and so let us root out all other trees and replace them with Lombardy poplars. Let us talk no more nonsense about untrammeled nature. In Oak Park, nature has got to behave herself and not set a bad example to the children.

Discussing the prevailing taste in rose planting, Wright further chided the "sober citi-zen who hates fuss," who ties his roses to sticks and plants them in straight rows so that they behave themselves. Perhaps, he continued, his neighbors should fence around their grass plots and put china dogs smack dab in the middle. "A china dog never digs holes in the lawn, nor scatters the flower beds. He too will stay where you put him . . . It is truly a tidy little land—is Oak Park."

Early photos of the Wright home and studio show a heavily wooded lot. Blair's house, which still exists, was behind Wright's house and was occupied by Wright's mother. Mag-inel Wright Barney, Wright's sister, recalls their shared yard in her book, *Valley of the God-Almighty Joneses*: "There were great oaks in the yard with trunks wrapped in woodbine; there was a beautiful tulip tree, as well; its fancy flowers smelled like cinnamon . . . We had every variety of lilac snowball and spirea. And in the spring, flowers sprang up along the fence: violets, white and blue, wild ginger, and lilies of the valley. Mother made a garden that was our joy."[112]

The original house that Wright designed and built in the shingle style includes wide verandas that brought the outdoors right to his front door. His trademark planting urns at the entryway provided built-in greenery. The garden walls of his home were covered with ivy. Wright built a passageway around an existing willow tree when he added his studio in 1898. The tree was half inside and half outside the house. Although the willow did not survive, the Frank Lloyd Wright Home and Studio Foundation has recreated the built-in tree with a honey locust. The south side of the home was used as a utility/service area. Chicken coops lined the south driveway, and a shed for the children's ponies was maintained at the back. Delivery men had access to the kitchen along this driveway, and brought coal and ice.

Landscape Integrity

When Wright first purchased his lot, Forest Avenue had just been paved, and his was one of the first homes built. The surrounding area, especially north of Chicago Avenue, was prairie and farmland. It is much easier to see how his home blended into the prairie then, than in the decidedly urban Oak Park of today. Nonetheless, the Frank Lloyd Wright

Home and Studio Foundation has made a significant effort to restore the property to how it may have looked when Wright was there. The house itself has been restored to the year 1909—the last year Wright lived in Oak Park. Restoration of the gardens has been an ongoing effort since the early 1980s.

Using historic photos and consulting landscape architects specializing in prairie gardening, the team developed a plan for the grounds. The front (west) yard has been restored as closely as possible to what might have existed then. Other gardens (e.g. the Book Shop Garden in the courtyard behind the studio, and plantings on the south) are interpretations of general planting trends of the era, to allow for the building's new use as a public museum. Plants are chosen to resemble those in pictures—for example, to replicate the trailing plants in Wright's signature built-in urns, heliotrope, green and black sweet potato vine, red fountain grass, and bacopa are planted.[113]

Some of the original plantings remain. The ginkgo tree, the centerpiece of the east courtyard garden, predates even Wright. Wright remembered this tree in a letter to the *Oak Leaves* newspaper in 1940: "It was planted by an old Scotch gardener, a Mr. Blair, who laid out Humboldt Park. It was growing before we got there. When I built the house the ginkgo was young and slender, about four inches in diameter."[114] A hawthorn tree thought to have been planted by Wright during a 1911 remodeling survives in the front yard. In the restoration project, other shrubs and trees were planted based on historic photos or written descriptions. A small white pine tree was planted near the entrance, based on early pictures. Snowball bushes, viburnum, and lilacs, while not necessarily native, were planted as indicative of the area.

The lot has not been subdivided or changed in size since Wright lived there; walkways and paths are essentially the same, with minor changes to the driveway and bookstore area to accommodate the public. Since this was Wright's place of business as well as his home, public use of the grounds was expected (although not nearly as heavily trafficked as today). The sense of nature Wright may have felt while sitting on his front porch most definitely changed with the proliferation of nearby apartments and housing. Yet the organic architecture of his home, thanks to meticulous restoration, continues to reach into the garden beyond.

Ragdale

1260 North Green Bay Road
Lake Forest, Illinois
http://www.ragdale.org

Ragdale, an elegant country house on the North Shore, is surrounded by exquisite gardens, woodland trails, and expansive prairie views. Built as a family summer home by one of the nation's leading Arts and Crafts architects, Howard Van Doren Shaw, Ragdale has inspired generations of artists. Today, it is a retreat for artists, and the grounds are undergoing a remarkable rehabilitation.

Ragdale is a romance, a story of a gifted artist and his talented wife and family. This Arts and Crafts home and landscape, which embraced generations of the Van Doren

Tall locust trees grace the front of Ragdale. Howard Van Doren Shaw enhanced the exterior of his summer home with handmade accents, including outdoor benches and signature window boxes. Ca. 1899. Courtesy of Ragdale Foundation and Lake Forest College Library Special Collections.

BOTTOM LEFT Old-fashioned perennials and rustic garden ornaments, such as these informal stepping stones leading to the wellhead, added casual charm to the garden at Ragdale. Courtesy of Ragdale Foundation and Lake Forest College Library Special Collections.

BOTTOM RIGHT Sylvia and "Sister Bill," as noted on the handwritten caption, gaze at the view behind Ragdale, then an expansive prairie. Courtesy of Ragdale Foundation and Lake Forest College Library Special Collections.

Shaw children, was owned by the same family for nearly a hundred years. Ragdale's gardens evolved with sensitivity and respect for the original design. Created in 1898 by local architect Howard Van Doren Shaw as a family retreat, Ragdale, with its quiet, country setting, was muse to generations of the artistic Shaw family.

A Yale and MIT graduate, Howard Van Doren Shaw became the architect of choice for many of Lake Forest's country estates, including those for the Swift, Ryerson, and Donnelley families. Shaw designed clients' homes to blend seamlessly with the landscape. Circulation patterns around the home, and garden views from within the home were all carefully planned to integrate indoors with outdoors.

Shaw imagined a simple, country place at Ragdale, which he named after an estate in

England. In her book, *Ragdale: A History and Guide*, Shaw's granddaughter, Alice Hayes, recalls, "To him, Ragdale meant meadows and woods and hollow apple trees and country vistas. The raggedy look of the shrubbery, the low hanging branches of trees, and the invasion of the lawn by violets were all deliberate effects."[115] Shaw brought outdoor views into virtually every room of his house, and extended outdoor living spaces with comfortable screened porches or covered porticos. Taken together, the home and garden are remarkable surviving examples of an Arts and Crafts home and garden.

Shaw was a romantic, and lover of nature. A vivid picture of the romantic Shaw is this love letter penned to Frances, eight years before they married. Not only is his love of flowers evident, but also his predilection for outdoor pursuits in the bygone, pastoral days of early Chicago.

> May 27, 1885
> Dear Fanny:
> I am very glad that you liked those little violets.
>
> I went out to South Park on my bicycle, with some other boys, & we wandered off toward Woodlawn where the Sunday School picnic-grounds are, & found lots of wild flowers, but couldn't bring many home because it is so hard to carry anything when riding.
>
> The other boys said they intended to give their flowers to the girl they liked best, so I did the same & sent mine to you.
>
> I think you treated them very well & I guess the flowers felt highly honored to be carried by such a *pretty* girl.
> I'm very sorry dancing school has closed because I see you so seldom now.
> With love,
> Howard[116]

From his travels in Europe, particularly England, Shaw became enamored of the Arts and Crafts ideal that favored handmade craftsmanship and simple, "honest" design. Ragdale is replete with original touches, from heart motifs on shutters and inglenook, to handmade benches, to the signature Ragdale blue paint on window boxes and trim—a proprietary hue melding tints of robin's egg and weathered copper.

Shaw was very much a plantsman. He collaborated with many leading landscape architects on Lake Forest estates, including O. C. Simonds, Warren Manning, Rose Standish Nichols, and Jens Jensen. From a simple farm field and orchard, Shaw created a bucolic, peaceful retreat: an American interpretation of English countryside. Key to this vision was a compelling vista of virgin prairie seen from Ragdale's living room, dining room, and enclosed porch. This prairie, once part of Ragdale, is now managed by Lake Forest Open Lands through a cooperative arrangement and, in recognition of its unique plant and animal life, is designated an Illinois nature preserve. Whereas Shaw's contemporaries Jensen and Simonds might have waxed eloquent over the native flora in the prairie, for Shaw, it simply meant a beautiful backdrop, as one might have experienced an English meadow.

Thanks to Shaw's benign neglect, the virgin prairie and its precious plants survived unscathed by plow or garden hoe. Sunsets over the prairie create an unforgettable vignette, one forever immortalized by Shaw's grandson, cartoonist John T. McCutcheon Jr., in his evocative 1912 rendering "Injun Summer." A green lawn separates the house from the prairie beyond, and in that greensward, Shaw built a bowling green and later a lawn tennis court. Both are now gone. The open vista remains, however, framed by woods on both sides and made even more precious by today's overcrowded world.

Trees and wooded areas were important parts of the Shaw landscape. Elegant American elms were used to frame views and define space. Only one or two of these elms remain, the rest having succumbed to Dutch elm disease. But charming garden remnants survive in the shady wooded lanes that Shaw designed along Ragdale's north and south borders. Over a mile and a half in total length, they were designed "to look like 18th-century English country lanes—sun-dappled tunnels, grassy and narrow, between green walls of native shrubbery planted by nature or by Shaw."[117]

One of the most charming landscape features at Ragdale is the vegetable and flower garden on the property's north side. Visually separated from the home with an arborvitae hedge, this garden features planting beds of fruits, vegetables, and ornamental flowers. Thought to be planned largely by Shaw himself, the garden is accented with charming ornaments such as a sundial, dovecote, wellhead, and picnic benches made by Shaw. Evelyn, Shaw's oldest daughter, kept a list of flowers planted in this garden. Full of old-fashioned favorites, the list describes rectangular beds of alyssum, ageratum, white zinnia, candytuft, snapdragon, dahlias, stocks, nicotiana, and cosmos, which formed a large portion of the garden. Largely planted with annuals, the garden also included lilies, delphiniums, chrysanthemums, iris, peonies, and other hardy favorites. Apple trees, a grape arbor, and vegetable beds also adorned the original garden.

Landscape Integrity

Today, thanks to the stewardship of Shaw's poet-writer granddaughter, Alice Hayes, Ragdale is an artists' retreat. Through a juried selection process, writers, painters, musicians, and other artists from around the world are chosen to live and work at Ragdale. The Ragdale home, and its relationship to major garden areas, is well preserved, although necessary changes were made to accommodate its new public use. One significant garden feature, the so-called Ragdale Ring, an outdoor theater enclosed by low stone walls, was lost to subdivision north of the estate. Frances Shaw, Howard's wife, used to write plays that were performed in Ragdale Ring.

In 2001, the Ragdale Board established a Landscape Restoration Advisory Committee to develop plans for rehabilitating the grounds. A "dream team" of landscape architects, historians, gardeners, and board members served on the committee and, after more than a year of research and planning, produced a multiphase plan to rehabilitate the gardens. Because Shaw was constantly experimenting with the garden, the plan calls for rehabilitating the garden to the 1920s period. The first part of the three-phase plan included restoration of Shaw's handmade posts and split-rail fence bordering Green Bay Road, and accommodation of the parking lot for visitors. Roadside undergrowth and shrubbery were thinned, and young shade trees given more room to grow to maturity. The parking lot, a necessity for Ragdale's public, is better screened with vegetation.

The landscape immediately surrounding the home and the formal garden area are to be restored. The fountain and St. Martin sculpture areas will be transformed to their original shady-glade appearance. Plantings around the house's foundation will be simplified, using historic photos as a guide. Walkways in the formal garden will be reestablished and paths recreated to restore the connection between the home and formal garden. Flower borders will be replanned, keeping the original design in mind while accommodating today's shadier conditions.

Much of Ragdale's charm, typical of the Arts and Crafts ideal, lies in handmade or personalized garden accents. Many of these ornaments remain today as part of the landscape legacy. There is a story behind each ornament, bringing visitors closer with the Shaw family and their lifestyle.

> *Bird Girl*: In 1938, Sylvia Shaw Judson, Shaw's daughter, created the original Bird Girl statue, popularized on the 1994 book cover and movie *Midnight in the Garden of Good and Evil*. This reproduction was commissioned by Alice Hayes. The original is in Bonaventure Cemetery in Savannah, Georgia.
>
> *Purling Fountain*: Just off the home's south porch is a simple concrete fountain and pool designed by Shaw in 1905. Family history notes that children and adults alike would cool off by dipping their toes in the little wading pool. Water sprays from the mouths of three fanciful fish decorating the fountain's stone basin. Frances Shaw composed the inscribed poem: "Purling Fountain Cool and Gray / Tinkling Music in Thy Spray / Singing of a Summer's Day."
>
> *Garden Gate*: The flower and vegetable garden entrance is marked by two concrete gateposts topped with floral-basket sculptures, trademark Shaw motifs. A wrought-iron banner inscribed with *R* for Ragdale connects the posts. The path ultimately led to Ragdale Ring, long gone.
>
> *Wartime Lambs*: These two gentle lambs, works of Sylvia Shaw Judson, greet visitors upon entering the Ragdale compound. During World War II, the family raised sheep, and these peaceful lambs are said to have been inspired by the flock.
>
> *Sundial*: This simple Arts and Crafts sundial was designed by Shaw and includes an inscribed verse.

Many landscape views have been preserved at Ragdale—from the open prairie vista to the wooded, shady lanes bordering the property. The cutting garden remains a delightful nod to the past with many of the garden ornaments remaining from Shaw's day. Despite the necessary changes, the overall ambience at Ragdale is that of a restful country estate.

Graceland Cemetery

4001 North Clark Street
Chicago, Illinois
http://www.gracelandcemetery.org/

Graceland Cemetery brought together some of Chicago's best architects and landscape gardeners in this celebrated "garden of the dead." Representing Chicago's foray into the landscaped cemetery, Graceland's roads, pathways, ponds, and naturalistic plantings surround monuments to some of Chicago's leading figures. The cemetery design was influenced by a host of landscape luminaries, with O. C. Simonds having the longest tenure.

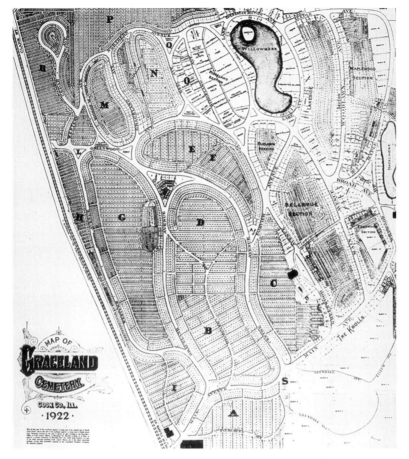

This map of Graceland shows the curving streets and open spaces in the "modern" American cemetery. 1922. Reproduced by permission of the Frances Loeb Library, Harvard Graduate School of Design.

Today, with landscape-design ideas readily accessible through the plethora of consumer magazines, television shows, and Web sites, it is hard to appreciate that cemeteries were once major showcases of design talent. Yet in Chicago, as well as other major cities in America and Europe, cemeteries predated city parks as sites for pastoral strolls and quiet contemplation. According to early historian Andreas Simon, until 1835, Chicago had no designated cemetery; families buried their deceased relatives in a "convenient spot near their home."[118] In 1835, two cemeteries were mapped out, one at South 23rd Street and the lake, and one on the near North Side. In the 1850s and 1860s, more cemeteries were created on the far outskirts of the city, in response to increasing public health concerns. Cemeteries were good business for Chicago's nursery and greenhouse industry, which supplied cut and potted flowers as funeral decorations.

Graceland Cemetery is a fascinating time capsule of historic landscaping. Here, on about 120 acres of land surrounded by the bustling city, saint and sinner alike rest in perpetual silence. In life, they were friends or foes, artists or patrons; in death, fellow tenants of God's green acre. Here is the monument to railroad-car magnate George Pullman, whose labor practices prompted widespread unrest and strikes. Here is the marker for John Altgeld, Illinois's governor during the Pullman strike, whose sympathetic ear labeled him an anarchist. Exquisite monuments were crafted for Chicago's leading family names such as Ryerson, Getty, and Goodman, by equally renowned architects such as Louis Sullivan and Howard Van Doren Shaw, who later were themselves laid to rest at Graceland.

Thomas B. Bryan, a wealthy Chicago businessman and avid horticulturist, was the major force behind the creation of Graceland in 1860. Along with other investors including Chicago's first mayor, William B. Ogden, his brother-in-law Edwin H. Sheldon, Sidney Sawyer, and portraitist G. P. A. Healy, Bryan formed the Graceland Cemetery Company. Graceland's location was ideal: readily accessible from Green Bay Road (now Clark Street) and later the Chicago and Evanston Railroad, yet far enough removed from the city to avoid health and sanitation issues. The company chose the high ridge area along what is now Clark Street, which was once an old Indian trail. The site offered good drainage with its strong drop-off to the east and slightly to the west. In the sandy soil here, plants thrive better than in Chicago's typical clay soil.

Site location was an important element in Bryan's vision for Graceland, but landscaping also played an early and continuing role. Bryan's journal of Graceland expenses prominently enumerated the landscape gardener's salary, trees, and plants. A star-studded cast of landscape gardeners helped configure the space of the early cemetery. Philadelphian William Saunders and local landscaper Swain Nelson developed some of the first plans in the early 1860s.[119] Swain Nelson, recently arrived from Sweden and hungry for work, told how he approached Bryan about Graceland's landscaping. Nelson had been asked by his client, James Waller, a concerned neighbor near the proposed Graceland site, to scout out Bryan's plans for the cemetery—a project that did not enthuse Waller. "I called on Mr. Bryan the next day and gave him my card, and told him that I had heard that he was intending to lay out a cemetary [sic] and I came to solicit the work . . . He was much astonished he said he had not bought the land yet, he kept me in his office talking on various subjects."[120]

Nelson apparently made a good impression on Bryan, for he not only loaned him some money but also called him back once the Graceland property was purchased. Bryan and Nelson together rode around the property on horseback, and Bryan indicated where he wanted the entrance, office, and a main straight road. The next day, Nelson returned with a worker and team of horses, and marked the road, then the center road in the cemetery. After reviewing and approving the work, Bryan asked Nelson for his bid to grub out the oak trees, grade, and make the road bed. "I made a low bide [sic]," Nelson recounted, "for I needed the work and labor was very cheap. And to be paid in gold."[121] Bryan then gave instructions for a second road, which Nelson completed, and was about to suggest a third, when Nelson said, "I could not work any longer without a plan, well he said make a plan."

Nelson then commenced to design the grounds. "I started to work out the whole ground in fifty feet squares and marked out the roads I had already made on the plan, and also the one I suggested to make." Bryan, a very hands-on manager, approved the plan and then had Nelson mark out the main features on the property. Again with Bryan's oversight, Nelson implemented the plan over the course of the next two to three years.

In the 1870s, H. W. S. Cleveland was retained to work on the design, as was William Le Baron Jenney. These early landscapers improved the major drives and pathways that can be seen today in Graceland. Both Jenney and Cleveland had associations with Frederick Law Olmsted and shared many of his design principles. Among other affiliations, Cleveland worked with Olmsted on Brooklyn's Prospect Park, and Jenney collaborated with Olmsted in the Chicago suburb of Riverside. In fact, in comparing the 1869 general plan of Riverside and early maps of Graceland, the curving drives and elegantly fitted paramecium-shaped plots are not all that dissimilar.

Jenney brought O. C. Simonds to Graceland, and therein began an affiliation that was to last several decades. Simonds flourished under two mentors—Jenney and businessman and horticulture aficionado Bryan Lathrop. Simonds had studied architecture and civil engineering under Jenney at the University of Michigan, and later joined his architectural practice. Bryan Lathrop, Thomas B. Bryan's nephew and president of the cemetery's board of managers, also strongly influenced Simonds. Lathrop had both the means and interest to study landscape design, and was active in such far-flung interests as a bamboo plantation in the south and the parks movement in Chicago. As Simonds's employer, he was a strong proponent of ongoing landscape improvements.

With this support, Simonds was able to study national examples of cemetery landscaping, such as Cincinnati's Spring Grove, and develop his own design style. Cemetery design was challenging, however, as Simonds himself noted: "The great diversity of tastes, opinions, superstitions and prejudices that must be consulted or controlled make cemetery landscape-gardening the most difficult branch of the art."[122] The design conundrum is readily apparent: how to create a cohesive design when multiple property owners have autonomy over small, contiguous plots. Although Graceland Cemetery Company did have rules governing plot design, individual taste could not be completely legislated.

Simonds developed his own set of principles governing cemetery design. Examining two of his writings on cemetery landscaping, dated several decades apart, it is remarkable how consistent his design precepts remained. He emphasized[123]

The contour of the land and arrangement of shrubs should afford privacy for visitors and mourners.

The site should be gently graded with slopes not exceeding a rise of 6 feet per 100 feet.

There should be a minimal number of driveways. Roads should be made of macadam, gently curved, and should pass within 150–200 feet of each lot. (His later 1920 writing, modified for the increase in automobile traffic and new technologies, suggested the use of bitulithic concrete, or bitulithic macadam.)

Water features such as lakes and ponds provide restful vistas.

Open lawns offer "cheerful effects," and can be achieved by grouping large plots together.

Much has been written about Simonds's penchant for native plants. His early prescription for cemetery planting notes, "In making selection for planting, we should seek those things which give cheerfulness." For example, he recommended deciduous blossoming trees over somber Norway spruces. He also noted, "Natural beauty is not expensive. Usually in country places, all the trees, shrubs and herbaceous plants really necessary to produce the effects desired can be had for the labor of digging them. The best things supplied by nurseries—that is, the things that are hardy and will usually take care of themselves—can be had for very little money."[124] Nonetheless, Simonds was not a purist when it came to native plants. In an 1887 essay entitled "Mr. Simonds' List of Shrubs," Simonds listed the shrubs and vines successfully tried at Graceland—many of which were nonnative species (all spellings below as originally written).[125]

> Serviceberry (*Amelanchier vulgaris*, *A. Botryapium*), Aralia Japonica, Japan Quince (*Prunus Japonica flore albo pleno* and *P. Japonica flore rubro pleno*), Witch Hazel (*Hamamelis Virginica*), Hydrangea (*Hydrangea paniculata grandiflora*), Honeysuckle (*Lonicera*), Mock Orange Syringa (*Philadelphus coronaries*, *P. coronaries flore pleno*, *P. foliis aureis*, *P. Gordonianus* and *P. grandiflorus*), Current (*Ribes sureum*, *R. gordonianum* and *R. sanguineum*), Sassafras, Spirea, Snowberry (*Symphoricarpus racemosus* and *S. vulgaris*), Lilac (*Syringa Josekoea*, *S. Persica*, *S. rothomagensis*, *S. Siberica*, *S. Sinensis*, *S. vulgaris*), Arrow-wood (*Viburnum*), Wild Barberry (*Berberis Canadensis*) and Common Barberry (*B. vulgaris* and *B. vulgaris purpurea*), Wild Carolina Allspice (*Calycanthus floridus*), Button-bush (*Cephalanthus occidentalis*), Dogwood (*Cornus Siberica foliis albo-marginatis*), American Hazelnut/Filbert (*Corylus Americana*), Cotoneaster, American Strawberry bush (*Euonymus Americanus*), St. John's wort (*Hypericum*).

Perhaps more important, Simonds espoused a naturalistic planting effect whereby plants were arranged artfully, yet in a way that imitated nature. He suggested that common bedding plants (typically brightly colored annuals) were less attractive than naturally growing trees and shrubs, which also provided winter interest in our bleak Chicago January and February months. He eschewed "dreary" spruce and weeping willows in favor of "bright cheerful effects [achieved] by the selection of all kinds of flowering happy-looking plants." Noting that "cemeteries, indeed, rank with parks in preserving open spaces and in the growth of foliage which purifies the air," Simonds even went so far as to suggest that "the modern cemetery becomes, in fact, a sort of arboretum."[126]

Simonds, and other cemetery designers of his time, preferred that manmade buildings or structures be sublimated to natural views. Office buildings were to be discreetly located, and other structures should be covered by vines. Grave markers and monuments were to be low to the ground and subtle. Simonds, in fact, supported cremation and helped design the first oil-burning crematorium, making Graceland a pioneer in this field.[127] In the aesthetic aim for discreet monuments, Simonds and other cemetery designers had their largest struggle with individual plot owners. Particularly in the western portion of Graceland, the older part, elaborate monuments reflect Victorian fashion with ornamentation, grandeur, and exotic art such as Egyptian symbols and statues. Interestingly, two of Chicago's major architects' monuments are most sympathetic to the scaled-down ideal: Daniel Burnham's monument is a simple, rough-hewn stone placed on a little peninsula surrounded by Lake Willowmere, and Ludwig Mies van der Rohe proves less is more with his simple, rectangular stone set flush into the ground.

Landscape Integrity

As the city has grown around Graceland, adjustments have been necessary. Many of the shrubs and understory plantings had been removed over the years, possibly due to maintenance costs, the effects of the great Depression, the lack of maintenance personnel during World War II, and personal safety and security issues in the postwar period, when massed plantings were considered potential hiding places for unwelcome intruders.[128] The cemetery is currently bounded by walls, whereas Simonds would have preferred a naturalistic massing of trees and shrubs or, at least, a fence disguised by plantings or vines. (Given the three-foot right-of-way adjacent to the perimeter walls and the clearance requirements of modern maintenance equipment, it is a challenge to include camouflage plantings to hide the walls.)

Nonetheless, the landscape retains intriguing glimpses into Simonds's work and that of his predecessors. The curving drives and overall layout reflect the thinking of Jenney, Cleveland, and Nelson. The western part of the cemetery is older, and even in 1894, author Andreas Simon noted, "It was still the fashion to surround the family-lots with low stone walls or fence them in with iron railings or natural hedges and then to adorn them with monuments and grave-stones, more or less gorgeous, as the means of the owners would permit. About 50 acres of the grounds were disfigured in this way. Of course at that time this ancient system had not as yet been recognized as a mistake."[129] Today, we can admire the older headstones as works of outdoor art.

The landscaping of the southeast portion of the cemetery has been described as being in the style of the American lawn.[130] It is a low area, with few plantings, and feels rather empty. Andreas Simon identified the area in 1894 as a recent advancement, and "one worthy of imitation." Apparently many of the plots in this area had been neglected, and so "these graves, forgotten by the living ... have now been cleared of the weeds and grass covering them by the management; the mounds have been leveled and the whole has been changed into a beautiful lawn, on which appear here and there the tops of small numbered stones, marking the resting-places of the dead."[131]

The best area to see Simonds's aesthetic ideal realized, both in 1894 and today, is around Lake Willowmere. Willowmere was one of two lakes created both for artistic effect and to improve the site's drainage. (The other lake, near the property's east side, has been filled in.) Large family plots of five thousand to twelve thousand square feet surround the lake, and were given evocative names such as Lakeside, Bellevue, and Fair Lawn. Priced at $1.00 to $1.25 per square foot in the 1890s, Andreas Simon observed that "only persons blessed abundantly with this world's goods can think of buying." Simon temporized that this privileged spot, with its natural beauty, nonetheless benefits all who visit Graceland. As he describes this eastern portion of the cemetery, "The principal charm of 'new Graceland' is found in the large rolling lawns, which appear as grand velvety green carpets, from which the blooming decorations of the low mounds dotting the lawns here and there stand out like many-colored embroideries."[132]

The Wolff-Clements firm rehabilitated the landscape around Lake Willowmere, particularly in the section known as Ridgeland, just west of the lake, which includes the Marshall Field, McCormick, and Armour family plots. Research using old photographs and planting plans helped ensure historical accuracy, although some changes were needed, such as substituting shade-loving plants for sun lovers in recognition of changed sun patterns, or preserving old trees that, while not on original planting plans, added maturity to the landscape.

The Lily Pool

Fullerton Parkway between Cannon and Stockton drives
http://www.chicagoparkdistrict.com/index.cfm/fuseaction/custom.natureOasis06

Shortly before he died in 1998, Alfred Caldwell returned to the Lily Pool, his masterpiece of prairie-school landscape architecture nestled in Chicago's Lincoln Park. Some sixty years had passed since he first created this sylvan haven, and time had not been kind to the site. Weed trees and shrubs choked sunlight from the clearing, stonework was broken, and the lagoon was murky and filled with debris. This was the place for which Caldwell had once cashed in his own life insurance policy to pay for needed perennials. Now, no flowers grew.

The recently restored Lily Pool emphasizes native plantings and includes the exquisite stonework favored by Alfred Caldwell and Jens Jensen. Courtesy of Brook Collins, Chicago Park District.

"It's a dead world," Caldwell said, upon revisiting his masterpiece in the 1990s. But Caldwell would be pleased today upon seeing the spectacular rehabilitation of his Lily Pool, wrought through the combined efforts of the Chicago Park District and local community groups. Interested parties included birdwatchers, preservation advocates, local governments, and common citizens.

Alfred Caldwell designed the Lily Pool in the late 1930s as a refuge from the city. With his signature prairie style, acquired through mentorships with Jens Jensen and Frank Lloyd Wright, Caldwell redeveloped an old Victorian pond to one and one-half acres of naturalistic sanctuary. In the years between Caldwell's creation and his visit sixty years later, the original Lily Pool design had degraded. Historically insensitive updates to the landscape and a general lack of maintenance caused overgrown trees, inappropriate architecture, and poor water quality in the pond. The bones of Caldwell's original design, however, held together over the years. His plan included a lagoon, made to look like a prairie river cut through limestone bedrock. Inviting stonework paths circled the lagoon, and a council ring was sited on a hill, providing views to both the Lily Pool below and glimpses of Lake Michigan to the east. The strong inward orientation of the site encourages personal reflection and relaxation, despite the hectic pace of its urban surroundings.

Caldwell's characteristically detailed drawing called for groupings of crab apple, sumac, serviceberry, and hawthorns underplanted with native shrub roses, viburnum, and literally tens of thousands of woodland perennials. "He knew plants upside down and backwards," says Dennis Domer, author of the biography *Alfred Caldwell*. Having studied with both Jensen and Chicago botanist H. S. Pepoon, Caldwell was not only a master stonemason, but a plantsman as well.

When the Lily Pool first opened in the 1930s, the public responded enthusiastically to the new space. So enthusiastically, in fact, that the effect of human traffic caused significant erosion and damage to the plantings. Compounding the problem, the Lily Pool had evolved into a favorite stopping-over place for migratory birds in the 1950s. Renamed the Rookery, the site was host to the birds who came from far and wide on their lakeshore migrations and turned Caldwell's quiet sanctuary into an avian O'Hare.

Significant historic research and community input took place before the rehabilitation began. Key points of agreement included a commitment to Caldwell's original design, access for people with disabilities, removal of 1960s limestone, additional plantings of various heights and forms, and continued maintenance after the rehabilitation. The concept plan therefore specified significant restoration of the stonework on paths, ledges, waterfall, and council ring, reconstruction of two prairie-style pavilions, reopening of the eastern path, and extensive replanting.

The first phase of the work, begun in the fall of 2000, was more a process of subtraction than addition. Over four hundred weed trees like mulberry, box elder, and buckthorn were removed. Using old photos and Caldwell's notes, historically significant trees were identified and preserved. When old trees had to be removed because of disease, the decision was not made lightly, nor were the trees unceremoniously dumped. A big old cottonwood, for example, diseased beyond repair, was recycled as a climbing tree for the nearby bear habitat in the Lincoln Park Zoo. In deference to birder advocates, all work was scheduled so as not to conflict with prime migration seasons. The lagoon was dredged with utmost care not to disturb the wildlife, a key concern of some constituents. A family of turtles, for example, were gently carried to a neighboring pond.

In the spring of 2001, work began on the stonework and replanting. Although the original craftsmanship of the stone was superb, inevitable wear and tear had damaged it. Matching the unique weathered edge stone was a particular challenge. All the stone paths had to be removed and reset, and about 1,600 rocks were removed from the site with a few hundred wall stones removed and replaced. Stone that was inappropriate to the design was reused elsewhere in the park district.

Now that the invasive trees have been removed, sunlight pours into the clearing, just as Caldwell intended. It is a dead world no more.

Landscape Integrity

The Lily Pool is one of the best examples of restored prairie-style landscape architecture. Through a public and private cooperative effort, the lagoon was dredged and replanted with appropriate shoreline plantings. Stonework was relaid in the original circulation patterns. Inappropriate structures were removed, a council ring restored, invasive vegetation eradicated, and native plantings reintroduced. Some modifications were made to accommodate the Americans with Disabilities Act (ADA) requirements and other regulations. But the tranquility and "other-worldliness" present in this very regional landscape remain as Alfred Caldwell would have wanted.

FIVE

SUBURBAN SOJOURN
Gardens in the Country

Metropolitan Chicago has grown exponentially since the city's original lakefront plat. Agricultural villages and suburban developments, linked by transportation and commerce, form the larger Chicago region. Gardening traditions in the suburbs often followed those of the city; however, because suburban properties tended to be much larger than city lots, privacy fences were not as big a concern (although backyard fences were used to keep out livestock). Without fences, as in Chicago's earliest days, suburban passers-by could readily view a homeowner's front garden, and neighbors tended to follow the lead of the local tastemakers.

Chicago's outlying communities extend over enormous territory and showcase great geographic diversity. Each suburb's unique topography contributed to markedly different garden styles and a unique gardening ethos. Helen Oakes, a Garden Club of Illinois member, noted on a WGN radio broadcast in 1933, "Suburbs vary much in the style of their gardens and grow to have certain characteristics." She cited Wheaton as being famous for peonies and thought it "rather English in its gardening." Barrington's rolling land offered "unusual effects in terraces and arbors." Riverside "makes use of the winding Des Plaines in the landscaping of some

233

of its loveliest gardens," and "all the north shore suburbs make use of the lake where possible as a garden setting."[1] Sometimes charismatic individuals shaped the trends and styles of local gardens. While gardens changed over time, the individual personalities of many communities' gardens remained intact.

One way to understand the changes in Chicago's suburban gardens is to trace the evolution of land use from farms to suburbs to city. While all three forms of rural and urban land always coexisted in some form, the shift from agrarian to industrial land use—so dramatic during Chicago's major period of growth in the late nineteenth century—greatly affected garden design and the plants used. Many outlying towns around early Chicago began as agricultural communities. Farm gardens, while frequently the source of new plant varieties and plant culture experiments, seemed persistently under fire from self-appointed tastemakers for perceived unsightliness.

TOP & BOTTOM Farmer Thrifty (*top*) had a nicely fenced-in vegetable garden and carefully tended trees, whereas Farmer Slack's garden (*bottom*) was overrun by animals and his trees seemed anemic. 1884. From Periam, *Home and Farm Manual*.

CHICAGO'S PRAIRIE FARMERS

Most of the Chicago area was heavily agrarian from the 1830s to the 1870s, with government land sales encouraging early settlers to stake their claims and farm the land. The task was not easy in the early decades—unfamiliar conditions, inadequate transportation, and rudimentary machinery challenged even the most experienced grower.

Although the *Prairie Farmer* and other publications understood the harsh realities of agrarian life, well-meaning horticulture writers constantly exhorted farmers to improve their gardens. Farmers were accused of neglecting the landscapes around their own homes. Ornamental and even home vegetable gardens were often afterthoughts when the farmer finished planting his cash crops. Yet

TOP These views of the farms of Captain T. S. Rogers and his neighbor and brother, J. Warren Rogers, show front yards heavily populated with evergreen and deciduous trees. The *DuPage County Atlas* of 1874 reports that J. Warren Rogers sold his farm in 1864 and moved into the village. Note the outbuildings removed to the back, what appears to be a straight hedged walkway on J. Warren's house (*right*), and an attempt at a curved drive on his brother's house (*left*). Parkway trees have wooden structures surrounding them, presumably to stabilize the new plantings. Fences surround the properties, and the farmland and outbuildings can be seen in the background. Reproduced by permission of DuPage County Historical Museum.

The photo postcard of T. S. Rogers's house, taken about thirty years later than the etching, shows mature parkway trees and a dooryard planting within the curved walkway—although the curve is inverted! Ca. 1900. From the author's collection.

writers maintained that farmers should plant non-income-producing flower gardens and shade trees for the delight of farmers' wives and for the moral values of their children. The rhetoric increased in the 1870s and 1880s as technology, railroads, and other improvements made the farmer's job a bit easier. With more leisure time, writers opined, the farmer no longer had to worry about survival and could afford to plant for aesthetics. Jonathan Periam, like many writers of the day, equated a neat, tidy farm with moral righteousness. His diagram of Farmer Slack and Farmer Thrifty implied that a sloppy yard meant sloppy virtue.[2]

Proselytizing for better farm landscapes continued among members of the Horticultural Society of Northern Illinois. Member B. O'Neil of Elgin compared American farm landscapes unfavorably with those of European farmers, noting, "The French and Belgian farmers take as much interest in the cultivation of fruits, vegetables and flowers as they do in general farming."[3] O'Neil went on to assert that "it is a melancholy fact that farmers and their families, as a class, supply the largest percentage for our insane asylums, and our best scientists tell us this is due to the lack of ennobling influences among this class." He suggested that farmers subscribe to agricultural and floricultural journals, put south bay windows in their homes for houseplants, keep farm animals and stock away from the house, and other ennobling landscape schemes.

In the late 1800s, Chicago-area residents, like the rest of the nation, were torn between city and country. This dynamic tension pitted the excitement of the city against the healthfulness of the farm. In 1872, debunking the myth of an idyllic country life, the *Chicago Tribune* observed that the typical man who built country homes after succumbing to the "best arts of the landscape gardener and the most ingenious devices of the advertising agent" ultimately regretted his decision. "There is a solemn silence as to the prevalence of the ague, a virtuous indignation is exhibited at the mention of mosquitoes, . . . the running brook has been transformed into stagnant water, the dairy has proved to be a failure, and now you must go back to town if you desire to secure the choice cuts of beef and the early vegetables of the season."[4] Yet the same newspaper in 1892 headlined an article "Our Rich Farmers," and listed hundreds of individuals who had reaped a fortune in farming near Chicago. "The newcomer in Chicago who may have received the impression that Cook County is taken up by Chicago and its multifarious suburbs will be surprised to learn that there are some great farms just without the borders of the city."[5] The secret to living happily on a farm, according to years of earnest advice, was to beautify the personal grounds.

We can only speculate as to whether the typical hardworking farmer appreciated this unsolicited advice. Regardless, entreaties to spruce up farm properties continued well into the 1930s when, for example, philanthropic leaders such as Myrtle Walgreen made farm beautification a key goal of the local 4H program.[6] Even as suburban developments began to consume nearby farmland, members

of such influential groups as the Garden Club of Illinois entreated rural neighbors to spruce up their yards.

SUBURBAN DEVELOPMENTS

Beginning in the 1870s, Chicago's suburbs experienced rampant growth as outlying settlements changed from agriculture to other industries, and as railroad
commuting became easier. In her book *Building Chicago*, Ann Durkin Keating
notes that the economic base of newly incorporated Cook County settlements
declined from about 83 percent agricultural processing in the 1830s to only 9
percent in the 1880s. In these latter decades, the rest of the economic base was
fairly evenly split among market, industrial, and other businesses.[7] Chicago's
outlying communities began take on one of two forms: suburban commuter developments or self-contained villages.

Suburban commuter developments, leveraging new railroad access to Chicago, exploded in the 1870s. Fueled by an exodus of Chicagoans after the Chicago
Fire of 1871 and encouraged by speculative real estate developers, suburbs grew
to accommodate upper-middle-class workers. Railroads were particularly instrumental in suburban development by offering more frequent and flexible schedules. In 1866, the *Chicago Tribune* identified railroad suburbs on the Chicago and
Milwaukee Railroad (Hyde Park, Winnetka, Evanston, Glencoe, Highland Park,
Lake Forest) and on the Chicago and Northwestern (Harlem, Cottage Hill [now
Elmhurst], Babcock's Grove [now Lombard], Danby [now Glen Ellyn], Wheaton,
Geneva, Elgin). Railroad executives not only saw the potential in suburbia—
many were principal investors in outlying villages—but often built their own
homes in the fledgling towns. Planned landscape improvements were important
for new suburbs, as the *Tribune* reported that "in the suburbs of Chicago improvements are going on with more than usual rapidity. Chicago architects, and
especially those who pay attention to landscapes, and the arranging of grounds
generally, have their hands full of business."[8]

Indeed, the outlook for Chicago looked rosy in 1870, with the city being called
a "Paris of America." One year and one day before the Chicago Fire, the *Tribune*
observed that solidity, strength, and endurance were the new watchwords of construction. With the city's accumulated wealth, grand buildings of substance were
erected, and the days of quick and cheap balloon frames and shanties were over.
Boldly, but tragically proved wrong, the *Tribune* predicted in 1870, "There could
be no such thing as 'fire limits' in a city whose improvements were running from
east to west, and from north to south with the rapidity of a prairie fire."[9]

The burned district resulting from the Chicago Fire extended from about 12th
Street on the south, almost to Fullerton Avenue (2400 north), and west from
Lake Michigan almost to Halsted Street. Not only did thousands of refugees

escape to existing suburbs to seek temporary shelter, but many more suburbs were created to permanently house the displaced. Suburbs once inaccessible became more attractive as building costs in the city skyrocketed. As development marched relentlessly outward, areas once considered suburbs became more densely populated, and their rural character disappeared. Lake View, the township adjacent to the northern limits of the city, underwent massive change typical of nearby suburbs. Everett Chamberlin noted that the township had, until the years following the fire, existed happily with its natural attractions such as

TOP William C. Goudy, a politician and attorney, built this home fronting Clark Street, just north of Fullerton in Lake View. According to Chamberlin, "the grounds are ornamented with shrubbery, flowers and fountains, and covered with native trees." This etching shows Goudy's landscape as an elegant suburban villa of the times. Ca. 1873. From Chamberlin, *Chicago and Its Suburbs*.

BOTTOM James B. Waller, a real estate agent, owned approximately sixty acres surrounding his own homestead, a local landmark since the Civil War. Noted poet Eugene Field, who lived nearby, immortalized the property in his 1894 poem, "The Delectable Ballad of Waller Lot," which began "Up yonder in Buena Park / There is a famous spot, / In legend and in history / Yclept the Waller Lot." In 1914, after the inevitable subdivisions, the house itself was torn down for an apartment building. Ca, 1873. From Chamberlin, *Chicago and Its Suburbs*.

wooded groves, lakeside views, and gently sloping terrain. With the influx of fire emigrants, Lake View began an expansion program that included new brick sewers and improved roadbeds of cinder and gravel, and that, according to Chamberlin, "awakened them [Lake View residents] to the necessity of throwing off their partial inertia and contend for all the auxiliary advantages which could be secured."[10] Owners of multiacre estates in these inner-ring suburbs found their property value escalated as subdivision became the norm.

Where earlier planned suburbs such as Lake Forest by Almerin Hotchkiss (1857) and Riverside by Olmsted and Vaux (1869) were developed for wealthy families who might have both city and "country" homes, these post-1871 suburbs were intended for year-round daily living. Some suburbs were created from scratch and laid out by recognized landscape designers, for example Highland Park, designed in 1872 by H. W. S. Cleveland. These designs capitalized on existing landscape features and incorporated extensive areas of greenery in the form of planned parks, generous house setbacks, and parkways, among other features. Other suburbs were created as model, self-contained towns, such as Zion or Pullman.

In 1883, Rosalie Buckingham, a member of one of Chicago's prominent grain-dealer families, purchased land between 57th and 59th streets on Harper Avenue. In one of the very few woman-planned subdivisions, Buckingham hired the team of Solon Beman and Nathan Barrett to design the subdivision. It is a fine example of a planned suburb on a small scale. Even within this relatively small space, the landscape architect and architect team who designed Pullman created courts and other open spaces. Alphonse Park was adorned with a rustic basket filled with flowers. The entire subdivision, now subsumed by Hyde Park, enjoyed views of Jackson Park, before the railroad interfered with the sightlines. From the author's collection.

Still other suburbs grew organically from existing villages that were once self-sufficient towns. Arlington Heights, for example, founded in 1854 as the farming community of Dunton, was incorporated in 1887. In his 1874 book, *Chicago and Its Suburbs*, Everett Chamberlin noted, "Arlington Heights has not until recently assumed any considerable suburban importance, but Chicago parties, in connection with the Messrs. Dunton, have laid out a new subdivision, on which they have graded streets, constructed sidewalks, set out large shade trees, etc." Like Arlington Heights, many other sleepy towns were discovered by enterprising businessmen who bought land and added improvements for the new homeowner—a commuter, not a farmer. Similarly, Downers Grove, founded in 1834 and described by Chamberlin as "the old and familiar town" whose land was "rich and peculiarly adapted to fruit and vegetable culture" was a prime target for developers. Three hundred fifty acres were purchased by Chicago businessmen "who propose to convert it into an attractive residence park."[11]

Often, the natural beauty of the surroundings beckoned early settlers. New Englanders who longed for the hilly topography of their birth villages secured Lake Forest, with its exquisite lakefront and intriguing ravines. Palos Park, with its equally attractive ravines and wooded hills, remained a well-kept secret for years, but ultimately was discovered by developers. The *Encyclopedia of Chicago* identifies another suburban form, the automobile suburb. Created in the 1920s and later, as automobiles lessened a commuter's dependency on railroads, these suburbs included Buffalo Grove, Rolling Meadows, and Lincolnshire in the northwest; Darien, Bollingbrook, and Woodridge on the west; and Hickory Hills, Palos Heights, and Park Forest in the southern suburbs.[12]

Upwardly mobile suburbanites, who often commuted to work in downtown Chicago, now had more leisure to enjoy their gardens. In keeping with prevailing health and exercise trends that advocated outdoor exercise, children and adults alike could often be found outdoors enjoying the healthful benefits of nature. Enabling outdoor enjoyment, the lawn made its debut in Chicago and environs, with homeowners using rotary reel lawn mowers to clip grass to military precision. The burgeoning nursery industry, growing from a handful of Chicago-area nurseries in the 1850s to hundreds listed in the 1900 census, provided homeowners with a wealth of options in seeds, plants, shrubs, and trees (plate 9). Suburban nurseries catered to their local clients with vegetables and flowers suited for the smaller spaces of suburban living. Where early settlers harbored seeds from old homesteads or from eastern mail-order companies, now flowers could be obtained locally and used generously throughout the yard. Nurserymen recognized the opportunity and expanded their stock to include more ornamental trees rather than productive fruit-bearing plants.

Suburban garden design changed from the 1870s to the 1930s, reflecting new technologies in garden structures such as water pools and fountains, new

LEFT Peterson Nursery, one of the oldest in the Chicago area, successfully made the transition to serve suburban residents. From the author's collection.

RIGHT The D. Hill nursery changed its emphasis from evergreens as windbreak trees to evergreens as foundation plantings in this design for a suburban home. The recommended selection included one Blackhill spruce, one Douglas golden arborvitae, two pyramidal arborvitae, two Woodward arborvitae, three American arborvitae, and five Pfitzer junipers. Ca. 1920s. From advertisement appearing in *Garden Glories*.

hybridized plants, and trends from leading tastemakers. Suburban landscaping evolved from carpet bedding in the front yards to more secluded, individualized private gardens in the backyard. Early showcase suburban properties featured curving drives, large expanses of lawn, fountains, and trees. Lot sizes were sufficiently ample to provide for outbuildings and turnarounds for horse-drawn carriages.

Tastemakers who had access to the latest design ideas from popular landscape architects or gardeners influenced many local homeowners' yards. These local landmarks were often the estates of the horticulturally savvy and wealthy individuals, nurserymen displaying their wares in situ, or real estate developers who often employed landscape gardeners for their own estates as well as for their subdivisions. Chicago's suburban showcase gardens were well known within the community, and often served as the site for social functions and ceremonial occasions. But as architect William A. Radford wrote in 1926, "Everywhere we go we see the signs of the real estate sub-dividers. The paved highways north and northwest out of Chicago are platted and staked out almost to the Wisconsin line . . . The new house of today is being planned and built much smaller than the style of forty years ago; in fact, just about half as large."[13] Suburban showcase gardens succumbed to subdivision, and few are left today.

The early twentieth century was a transitional time for suburban garden planners. Many homeowners were caught in the change from horse to automobile,

TOP Nurseryman and newspaper editor John Neltnor's home in Turner Junction, now suburban West Chicago. We can infer from the rendering that there was a nice mix of deciduous and evergreen trees, and winding pathways around the house. Hedges around the property and an elaborate gate and fountain would have been consistent with Neltnor's expertise as editor of the *Fruit, Flower, and Vegetable Grower*. On the left, we see what might have been cut-out beds of flowers. But this drawing is a good example of the artistic license engravers took with their subjects: the trees would not have been so heavily pruned or uniform as is depicted here to better show off the architecture of the home. Ca. 1874. Reproduced by permission of DuPage County Historical Museum.

BOTTOM Vaughan's nursery in Western Springs is an example of extensive acreage that once included display gardens that greatly influenced suburban homeowners in their garden design. The land was subdivided for homes. Courtesy of Western Springs Historical Society.

and the shift from farms to commuter suburbs. Well-to-do clients often maintained a stable for their horse yet needed a place for their newfangled automobile. Those garden designers with forethought recognized the opportunity for redesigning the driveway and approach to the house when horses were not needed. As better roads were constructed to the suburbs, homeowners not only adopted cars as the preferred mode of transport, but also converted summer homes to year-round homes. The four-season home was landscaped with winter interest and summer harvests in mind.

As suburban property sizes gradually decreased in the early twentieth century, homeowners turned to the privacy of their backyards for the enjoyment of an outdoor room. The shift to backyard gardening effectively obscured the ability of neighbors to see new garden trends as adopted by their neighbors. Where once garden fashion was prominently displayed in the front yard, now one had to be invited into a neighbor's backyard to see any new flowers or landscape designs. As home designs began including attached garages about the 1930s, neighbors

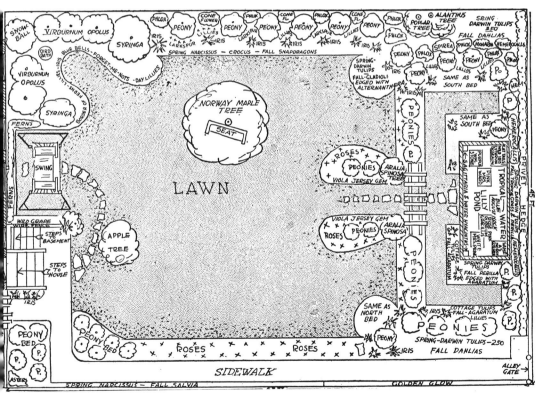

This plan of Leo W. Nack's garden won the *Chicago Tribune*'s 1927 grand prize. Now part of Chicago, Nack's then suburban 45-by-63-foot backyard got this headline in the *Tribune*: "Any suburbanite can duplicate this prize garden." Nack included a birdbath, focal lily pond, shrub borders, perennial beds, lawn, and swing set for children's play. Copyright 1929 by Chicago Tribune Company. All rights reserved. Used with permission.

TOP Rogers Park evolved from a farming community to an incorporated village in 1878. It was annexed to Chicago in 1893. Homes such as this of A. B. Bramaon on West Touhy Avenue were once landscaped as small suburban plots. Even though Rogers Park adjoins the suburb of Evanston, it is a Chicago neighborhood, and large landscaped lots such as this are virtually nonexistent. Reproduced by permission of Rogers Park/West Ridge Historical Society.

MIDDLE Even as suburbs attained their own identity, and unique landscaping, in the 1920s, rural scenes such as this one in Wilmette were not far away. From the author's collection.

RIGHT Residents of River Forest, an upscale suburb west of Chicago, adjoining Oak Park, enjoyed suburban garden style yet at the beginning of the twentieth century harbored rural characteristics. Here, the grounds of the R. H. Pierce residence include a manicured lawn, gracefully curved wooden sidewalk, artistically placed mature trees near the home, and casual lawn furniture. And a cow. Reproduced by permission of the Historical Society of Oak Park and River Forest.

no longer walked along driveways to their backyards, further losing opportunities to espy their neighbors' gardening prowess. Backyards thus became highly individualized and privatized, and gardeners had to pick up cues from local garden clubs, newspapers, and, later, radio talk shows.

Particularly in the late 1800s and early 1900s, lines blurred among Chicago's outlying villages, suburbs, and agrarian hamlets. In 1889, the city of Chicago annexed over 125 square miles of land, more than tripling the size of the old city.[14] With this expansion, Chicago absorbed many erstwhile self-sufficient communities and transformed them from rural hamlets to urban conclaves. Suburbs were carved from agrarian fields, yet rural life remained a romanticized fantasy. The *North Shore Graphic*, for instance, ran a picture of a simple farmhouse surrounded by cornfields and noted, "One of the charms of the North Shore lies in the fact that such rural scenes as this are easily within reach of its residents as are the more sophisticated scenes of urban life."[15] Barns and outbuildings disappeared from city residences as automobiles replaced the horse and as plumbing became more affordable, and excess land was sold off to developers. With influxes of new residents, some former suburbs took on more citified appearances (plate 10).

NATURAL BEAUTY DRAWS THE CROWDS

With Chicago's landscape much maligned as being flat and featureless, early settlers, particularly those with wealth, gravitated to those areas with beautiful scenery that might have been reminiscent of their homesteads on the East Coast. Hilly areas were preferred, and Lake Michigan shoreline was the next best thing to the ocean. North Shore communities such as Winnetka, Lake Forest, and Glencoe readily met these criteria. The South Side, while enjoying the lake view, lacked the varied terrain. Hence, although Prairie Avenue on the South Side enjoyed a brief heyday as an enclave for the wealthy, the homes were built as city mansions rather than country getaways. Hyde Park, another well-to-do early south community, later annexed to Chicago, lacked the hills that would buffer it from the stockyards and other encroaching industrial influences.

Lacking hills or a lake view, prairie rivers or mineral springs were another draw for early settlers seeking a claim. Rivers offered the possibility of dams to power mills or other forms of industry, and provided scenic vistas and recreational opportunities. Suburbs formed around these natural lures often incorporated the beauty into their gardens. Lake Forest and Palos Park are two such examples.

Lake Forest: Behind the Scenery

In 1857, a group of men, bound by a common Presbyterian religion and affluence, hired professional landscaper Almerin Hotchkiss to plan a new suburb, Lake

Forest. The area's unique features beckoned these former New Englanders, reminding them perhaps of home, where the ground was not uniformly flat, where the prairie was not all-encompassing. Lake Forest boasted deep ravines, thickly wooded forests, and bluffs overlooking sparkling views of Lake Michigan. It was a decidedly un-Chicago site topographically speaking, and a departure from the perceived social ills of the growing urban center.

With the Hotchkiss commission, the Lake Forest community demonstrated its early and enduring commitment to fine landscaping. Hotchkiss laid out the streets in winding curves to complement the natural contours of ravines, bluffs, and wooded hills. Lots were generous, with multiacre sites the norm. An elegant hotel accommodated visitors to the gracious summer or permanent villas in the village, and a university was planned as the literal and symbolic center of the community.

Lake Forest residents, through their international travels, Ivy League schooling, and vacation homes around the country, frequently transplanted trends from afar to their own properties. While local gardeners were employed to maintain the grounds, Lake Forest estates were often designed by some of the nation's most preeminent landscaping practitioners. Frederick Law Olmsted, the succeeding Olmsted firm, Warren Manning, Ellen Biddle Shipman, Rose Standish Nichols, James L. Greanleaf, and Ferruccio Vitale were among high-

The Cyrus McCormick Jr. estate in Lake Forest was "the greatest Lake Forest landscape plan of the era," according to the authors of *Classic Country Estates of Lake Forest*. McCormick's Walden estate, initially designed by Warren Manning in a naturalistic style, included wooded areas of native trees, ravines filled with wildflowers, and elegant gardens. Ca. 1913. Courtesy of Westchester Township History Museum, Chesterton, Indiana.

profile landscape artists who designed Lake Forest landscapes. Local landscape architects were also patronized—O. C. Simonds's firm was popular, Ralph Rodney Root's work was prominent, and Jens Jensen developed plans for about forty residential properties, ranking Lake Forest one of the Chicago suburbs with the most Jensen designs.

Many overlapping landscape trends were implemented in Lake Forest, attesting to the varied architecture and interests of the homeowners. The penchant for Italian gardens was embodied in H. F. McCormick's Villa Turicum (ca. 1908–18). The idyllic, naturalistic Walden (ca. 1890s–1930s), Cyrus McCormick Jr.'s estate, was inspired by the open country fields of his ancestral Virginia home. Plant material for this estate was sourced from nurseries, arboreta, and private estates from California to New York, a typical example of the transcontinental reach of Lake Forest properties. Albert D. Lasker's French Provincial manor house, Mill Road Farm (ca. 1920–40), was complemented by a lush landscape dubbed "An American Chantilly" by *Country Life* magazine.[16] Rose Standish Nichols inserted her English-inspired plantings into many Lake Forest gardens. Add to these Italian, French, English, and naturalistic American landscapes, the "gentlemen farms" where affluent hobbyists experimented with dairies and crops, and Lake Forest properties represented a world of new ideas and design styles.

Trends in Lake Forest landscapes were not quickly assimilated by rank-and-file Chicago-area residents, however. Lake Forest estates were rarely viewed first-hand by nonresidents. The exclusive suburb, twenty-five miles from Chicago, wasn't readily accessible in preautomobile days. The hilly terrain obscured cultivated gardens from passersby, and the homeowner's natural disposition to privacy further precluded casual sightseeing. Furthermore, the scale of these grand landscapes was hardly reproducible on a workingman's budget. Still, the area was habitually frequented by outdoor enthusiasts for its natural beauty. Early public health advocate Dr. John H. Rauch led a field trip for the Chicago Academy of Sciences to Lake Forest in 1860. Nearly one hundred participants boarded a Chicago and Milwaukee train for Rauch's botany lecture and ravine ramble. Hundreds of native plants were gathered and scientifically identified, including sumac, milk vetch, New Jersey tea, butterfly weed, hepatica, and maidenhair fern. Jens Jensen, a powerful link between Chicago's working class and the city's elite, sponsored trips to his clients' landscapes.

Lake Forest's contributions to Chicago regional landscaping and horticulture were more directly achieved through its citizens' extensive philanthropic and cultural efforts. From the outset, Lake Foresters were among the "wealthy amateurs" involved in Chicago's early horticultural societies. H. M. Thompson exhibited floral specimens in the 1862 Illinois State Horticultural Fair. Among his prized displays was a date palm from the collection of the late Andrew J. Downing. Early Chicago gardener Frank Calvert was employed by Thompson and was his emissary at this and other fairs, as were other gardeners from Lake Forest. In horticulture and flower shows, conducted under the auspices of the reconstituted CHS in the 1890s, or later the Garden Club of Illinois, Lake Forest individuals appeared prominently as patrons. The names of Armour, Brewster, Cudahy, McCormick, Pirie, Ryerson, and Van Doren Shaw were ably represented throughout the years. The Lake Forest Garden Club, founded in 1912, was a charter member of the Garden Club of Illinois, which managed the Chicago flower shows from 1927 through 1977. While their influence was not overt, Lake Forest gardeners shared their landscaping and horticultural knowledge behind the scenes, and behind their beautiful scenery.[17]

Palos Park: Forests, Farmers, and Artists

Palos Park was incorporated as a village in 1914, but European settlers had been drawn to the area for decades. As workers completed the Illinois and Michigan Canal in the late 1840s, they were drawn to the beauty of the area and the possibility of good farmland. Copious belts of forests dotted with ponds, streams, and natural springs promised fuel, shelter, and plentiful water.

In his book, *Flora of the Chicago Region*, botanist H. S. Pepoon described Palos Park as being on the edge of an ancient beach—Glenwood Beach—formed through glacial activity.[18] When the waters receded, the hills and ravines characteristic of this Valparaiso Moraine area formed, creating a topography unusual in Chicago's typically flat land. Home gardens often featured plants native to this diverse habitat, as well as incorporated natural features such as hills, terraces, streams, and ponds.

The natural beauty of Palos Park drew hikers and summer vacationers. In the early 1900s, groups such as the Prairie Club adopted Palos Park as one of their favorite destinations, boarding the train from Chicago for an all-day hike amid the forests and streams. For overnights, the club established a permanent camp in Palos in an old farmhouse near a natural spring and ravine. Individuals and families also enjoyed excursions to Palos. Louella Chapin devoted two out of sixteen chapters on Palos Park in her nature book, *Round About Chicago*. She described the wildflowers, crab-apple blooms, and autumnal colors of native hawthorns and oaks as they existed circa 1907. On one trip, Chapin recounted, "This day we had elected to go to Palos Park, perhaps the favorite of all our resorts. It is real country, with high, rolling wooded hills and a babbly creek; real country, for its beauty is still unknown to the multitude."[19]

Artists were beginning to discover Palos Park as an inspirational haven, and Chapin described "the long-haired artist who carves his face in the clay of the bluffs, and attracts to himself numerous beauty-lovers from the city, so that the artist colony of Chicago is beginning to know Palos."[20] Whether this long-haired artist was Lorado Taft is unclear, but Taft was known to have frequented the area.

Protecting Palos's natural beauty has been a long-standing effort. Palos Park is ringed by forest preserves, among the oldest in Cook County. Over the years, the preserves have accommodated many uses—giant toboggan slides at Swallow Cliff, bridle trails for equestrians, and fishing in the many sloughs and ponds. In the 1920s, several novel uses of the forest preserves were proposed: forty acres were roped off as a corral for a herd of buffalo and elk donated from Yellowstone Park, a ski course was recommended by a champion Norwegian ski jumper, and a committee including professors H. C. Cowles and Charles B. Atwell of Northwestern University identified two thousand of Palos's four thousand forest preserve acres for the "world's finest flower garden." The site was to have greenhouses, a museum and a library. "Already a garden spot . . . the Palos district offers a nucleus, in the opinion of the board, from which may be made a scenic marvel where the scientist may revel in botanical wonders while the layman revels in its beauty," reported the *Chicago Tribune* in 1919.[21]

The Palos Improvement Club and Palos Park Woman's Club often joined forces for civic beautification. In 1913, the Improvement Club hosted its ninth annual flower show. Prizes were awarded for the best six ears of corn grown by a boy, the best bouquet or basket of wildflowers arranged by a girl, and best photographs and sketches of Palos Park views.[22] Children's gardens were encouraged by the Woman's Club's inclusion of school garden produce in weekly open-air markets. This nonprofit venture became a local tradition each Friday morning as the *Tribune* reported, "Down the driveways of country estates and on every road in Palos Park women and children were seen yesterday with large bundles. They were going to 'market' at the Palos Woman's clubhouse. Some were afoot, others in carriages, but the majority went in automobiles, surrounded by their garden products."[23] The Friday farmer's market has endured and is thriving today.

Traditional suburban development came late to Palos Park. Most homes before the 1920s were farmsteads, country villas, or small summer cottages. Trains were infrequent, perhaps delaying the inevitable suburban boom. It wasn't until 1926 that the first real estate office opened. As the *Tribune* noted, "It doesn't seem possible that in these days of whoop 'er up realty markets going in high throughout Chicago's suburbs that there's a town which can point to a brand new building and say it's the first and only real estate office in the community. Palos Park, just recently awakened by subdividers from doing a long Rip Van Winkle, at last has a real estate office."[24] Where previously Palos Park properties consisted of many acres in farmland or forests, now more typical suburban lot sizes prevailed. Gardens became more manicured, although many incorporated natural features of the land. Typical was the garden of the "Norman French and Cliff Dweller"–style home that won a prize from the *Tribune* in 1929. The white

Palos Park was initially populated by farmers, many of whom had settled in the Sag Valley after working on the Illinois and Michigan Canal. Groups such as the Prairie Club enjoyed hikes through the region's hills and forests. Ca. 1915. Courtesy of Westchester Township History Museum, Chesterton, Indiana.

stucco residence had a sloped slate roof and was sited in a stream valley on a 100-by-200-foot lot. A winding driveway mirrored the curves of the stream, and flagstone walks echoed the natural materials of the home. The heavily wooded lot included a number of sinuous footpaths to enjoy the trees and stream and swimming pool at the rear of the lot. The home was built so that views of the stream were visible from the terraces and many rooms.[25] Today, many of Palos Park homes continue the tradition of incorporating precious natural features into their gardens.

RAILROAD SUBURBS

The railroad completely transformed Chicago's outlying communities. Where once residents depended on local general stores and village blacksmiths to provide for their needs, now the railroad offered opportunities to trade and work in downtown Chicago. Many self-sustained villages became bedroom communities almost overnight as the railroad exponentially improved access. Oak Park is one suburb that began as a self-contained village and became transformed with the railroad, whereas Wheaton was originally planned around the iron horse.

Oak Park: The Thinking Man and Woman's Gardens

Oak Park is quintessentially a Chicago suburb, founded on flat plains, bordered by a slow prairie river, and having wooded forests. Platted in utilitarian grids, Oak Park grew slowly from the 1850s until the influx of Chicago Fire refugees in the 1870s. As the railroad emerged, Oak Park quickly evolved from a standalone village to a bedroom suburb in close proximity to Chicago. While this may seem the typical story of many a Chicago suburb, Oak Park, like its diverse population, is a charming study of contrasts.

In his 1874 book, *Chicago and Its Suburbs*, Everett Chamberlin observed that "Oak Park has always been a favorite resort for literary and religious people." This tradition of intellectual curiosity is reflected in the village landscapes and gardens. Oak Park has always had an eclectic mix of garden styles, with elaborately landscaped estate grounds easily sharing fences with humble dooryard gardens. Garden styles were not slavishly imitated among neighbors, just as the architecture of their homes varied from Victorian Queen Annes to native son Frank Lloyd Wright's prairie-style creations.

Oak Park's groves of trees, clustered densely along the Des Plaines River and scattered over the prairie, attracted early settlers. A sawmill on the river drew the village's earliest settler, Joseph Kettlestrings, in the 1830s. Even as Kettlestrings and others found employment in milling timber, Oak Parkers lined their streets with new trees and reveled in their leafy shade. Controversy arose when the Oak Park tree tradition was threatened by new utilities in 1902—the *Oak Leaves* newspaper editorial warned against the risk of "marring Oak Park's chief adornment—its shade trees." The editors continued, "If it comes to a question of pavements or trees, give us the trees."[26] Frank Lloyd Wright, then an emerging architect, also voiced disapproval against tree pruning for utility wires: "Rows of our beautiful trees are yearly assassinated in the interest of the wires of the [telephone] franchise."[27]

Civic improvement and social groups flourished in Oak Park and became driving forces behind village beautification efforts. In the early 1900s, the Nineteenth Century Club promoted planting vines along unsightly fences, the Association for the Beautifying of Oak Park encouraged alley cleanups and tree planting, and the Oak Park Improvement Association planted shrubbery on railroad right-of-ways. Typical of Chicago's prairie suburbs, vacant lots remained among developed land, and became patches of prairie for local youngsters. Unfortunately, weeds and trash also accumulated, so in a movement dubbed "Every lot a Garden," the Citizens' Association aimed in 1912 to convert each plot to a cultivated garden using youth groups and private homeowners.

As early as 1903, efforts for public parks were spearheaded by realtor Anson T. Hemingway (Ernest Hemingway's paternal grandfather). At that time, sufficient green space was available in the undeveloped prairies north and south of the village, and in the vacant lots interspersed throughout. By 1912, however, with its burgeoning population, Oak Parkers foresaw the need for open space, and passed a referendum creating a park district. By 1927, Oak Park possessed eight parks and four playgrounds, encompassing over seventy acres.[28]

The Oak Park Garden Club took a hands-on approach to civic beautification, but also demonstrated a healthy academic interest in plants and horticulture. The club, which evolved from a 1913 group called the Applied Arts Circle of Unity Church, became the Oak Park Garden Club in 1917.[29] Club activities included flower exchanges, flower shows, the establishment of a bird sanctuary, public tree plantings, and public garden lectures. Roof garden boxes were planted at Cook County Hospital, an Alley Flower Walk Parade was hosted to beautify the alleys, and members appeared on the radio as garden experts.[30] In the 1920s, "members engaged a Boston landscape firm to draw a plan for the beautification of the plantings in the Village."[31] Member programs and lectures were provocative and topical; Jens Jensen spoke on landscape gardening, Ransom Kennicott on the forest preserves, and H. S. Pepoon on various garden topics. Field trips were sponsored to the North Shore gardens of W. C. Egan, Kate Brewster, and Charles Hubbard. No mere social gatherings, the 1919 program included an extensive list of over twenty-five books and several periodicals to be read in connection with the club lectures and outings. Books included Gertrude Jekyll's artistic *Color Schemes for a Flower Garden*, the nuts and bolts in L. R. Taft's *Greenhouse Management*, and regional author Wilhelm Miller's landscape gardening treatise, *What England Can Teach Us about Gardening*.

The Freer home, now the site of the Cheney mansion, once included many greenhouse ranges. While rural in appearance, it was not far from the downtown center of Oak Park. Reproduced by permission of the Historical Society of Oak Park and River Forest.

Oak Park home gardens were as individual as their residents. Affluent residents often employed landscape architects. Town leader H. W. Austin not only introduced the Chicago area to the talents of landscape gardener John Blair in the 1860s, but also commissioned him to design his private residence. The heavily wooded property, with rustic garden furnishings, included tennis courts and separate utilitarian areas for livestock and kitchen gardens. An oak fence, wrought in rustic fashion from the town's namesake trees, enclosed the property and became a local landmark. Edwin O. Gale, another affluent Oak Park resident, provided an ode to John Blair in his *Reminiscences of Early Chicago* (1910). The Gale family was said to have planted trees descending from Fort Dearborn in their own suburban development, Galewood. Jens Jensen drew plans for thirteen residential properties and two public parks in Oak Park.[32] Attesting to Oak Parkers' love of plants, successful nurseries and florists were many, including the 1880s greenhouses of Albert Schneider near Harlem and Chicago avenues.

Most telling of Oak Parkers' early and sustained commitment to enlightened horticulture was the movement to build one of suburban Chicago's very few public conservatories. In addition to its aesthetic qualities, however, the conservatory would serve a key educational mission. As one supporter wrote to the *Oak Leaves* in 1927, "It is strange to say that while we bandy names of trees and shrubs about, very few people can walk down one of our village streets and name a half dozen different species."[33] Oak Park taxpayers voted with their wallets in favor of horticultural education. In April 1929, the Oak Park Conservatory opened with plants from around the world, and over 1,500 residents in attendance. It was a wonderful blend of local and worldly horticulture, a typical Oak Park mix.

Wheaton: Wheels of Fortune

Unlike many suburbs that jumped on the railroad bandwagon as an afterthought, Wheaton actively embraced the new technology. Wheaton founders and brothers Jesse and Warren Wheaton donated a right-of-way to the Galena and Chicago Union Railroad in 1848 to actively encourage railroad owners and customers to patronize the fledgling village. Farmers of the community greatly benefited from this more efficient way to ship goods to market.

The village's good relationship with the railroad also helped ignite a civic beautification program. The Rural Improvement Society, one of many civic groups in Wheaton, formed in 1884 and had an objective of "planting of trees, grading and tiling of streets, sowing grass and seed for lawns along the street and the general beautifying and improving of the physical aspect of our village."[34] A town meeting was called to encourage citizens to participate, with the inducement that the railroad would reward such enterprise with a new depot and other improvements. The society hoped commuters would be attracted to the beautified village and decide to build a home and stay.

Civic leaders took a similarly assertive stance in 1867 to change the county seat from Naperville to Wheaton. As the center of county government, Wheaton became the regular host for agricultural fairs and shows. Typical was the Thirty-first Annual Fair of the DuPage County Agricultural and Mechanical Society, held in September 1885 in Wheaton. In addition to the usual display of animals, implements, and products of the soil, the fair featured the Grand Exhibition of Handsome Babies, with categories such as girl babies, boy babies, and twins. A glass-ball shooting tournament, grand donkey race, egg race, and great wheelbarrow race—with blindfolded contestants—added to the festive atmosphere. Despite this frivolity, entrants in the horticulture-related classes were judged quite seriously. In the "Fruits, Flowers and Plants" category, displays included hanging flower baskets, floral table ornaments, cut-flower arrangements, greenhouse plants, apple collections, dried apples, and baskets of grapes. Winners were awarded premiums of floral chromolithographs by Vicks Seed Company of Rochester.[35]

Local newspapers offered a good deal of gardening advice. The *Wheaton Illinoisian* featured gardening columns written locally and gleaned from national sources. Articles in the late nineteenth century included cultivating rhubarb, hilling strawberries, and using innovative fruit-picking contraptions. The *DuPage County Tribune*, which circulated not only in Wheaton but in many surrounding communities, included these imponderable horticultural notes in 1911: "Dehorning makes old trees become as new ... A strawberry plant is naturally an evergreen ... Not everyone who farms has a good family orchard."[36]

A real estate brochure published in the 1890s emphasized the advantages of green space in Wheaton. The builders stated that Wheaton wanted people who would build and stay rather than real estate speculators. With a few hundred dollars and a small mortgage, the promoters promised a family could have "a home in a good community with the best educational advantages the state affords, and a garden spot from which they can raise all the vegetables and small fruits they require during the year. The average family can save a hundred dollars a year by devoting their spare hours during the spring to their garden."[37]

Not only were homeowners encouraged to garden for their own benefit, but prospective homeowners were urged to embark on a market gardening career. The brochure cited the railroad and fortuitous growing conditions: "There is an opening here for people who understand market gardening, not equaled anywhere else near Chicago. The transit facilities and low express and freight rates make it possible to deliver garden produce in Chicago every day at a profitable rate. The soil here is so wonderfully fertile that with but comparatively little expense the finest of garden produce may be raised ... Land around Wheaton is cheap, and a thousand persons might find profitable employment to-day in market gardening."[38]

The city's attention to progressive infrastructure improvements aided city beautification and affected landscape design. Realtors declared that Wheaton's landscape effects would attract the discerning buyer. "When water works are constructed . . . so that lawns may be kept sprinkled, we will have a city that is a park in itself . . . Much care has been taken by the citizens in the cultivation of lawns and shrubbery." The brochure also noted that with newly laid sidewalks, "many fences were abolished and now open lawns between the walks and houses is the rule. The effect of this is so pleasing that eventually there will be no fences in Wheaton except the numerous, tasteful hedges."[39]

In the Progressive Era, Wheaton found opportunities not only to showcase its advantages but also to continue to improve its landscape efforts. The Home Town Club of Wheaton sponsored lectures aimed toward civic improvement, including one by Chicago city forester J. H. Prost, who spoke on "Town Beautification by the Planting of Trees and Shrubs" in 1911.[40] A spring-cleaning day was also encouraged wherein every homeowner examined his backyard to remove rubbish and tidy any nearby vacant lots. In the same year, Wheaton staffed a booth at the Chicago Land Show, a realtor's convention. The village promoted Wheaton as a great place to build a residence, and displayed samples of crops grown in the vicinity.[41]

Wheaton citizens were open to new ideas about how to use crops. In one article in the *DuPage County Tribune*, readers were encouraged to try edible weeds. Roots of golden thistle were found to taste like salsify. Leek was elevated from a weed to a proper food staple. The common mallow root was thought to contain valuable minerals and offer a mild lettuce flavor. Milkweed was touted as tasting like asparagus, and "lamb's quarter" (probably *Chenopodium album*) was said to taste like spinach. Receptive as Wheatonites were to these potential food uses, they adhered to their strict temperance heritage. When druggists in town were found to have marketed witch hazel with alcoholic content, the local newspapers were outraged. "To flaunt in the faces of the good citizens of Wheaton . . . that the witch hazel Mr. Dollinger sells contains nearly one-sixth alcohol is taken by many as an affront to a Christian community."[42]

Transportation improvements continued to shape Wheaton's landscape. In the first decades of the 1900s, Wheaton consisted of farms, town gardens, and summer homes. As new boulevards were created connecting Chicago to Wheaton and suburbs farther west, many summer residents decided to convert their homes to year-round residences. Like some other western suburbs, Wheaton sported a sizable summer colony. Individuals such as Joseph Medill, owner and publisher of the *Chicago Tribune*, built his summer home, Red Oak, in 1896 in Wheaton. Colonel Robert R. McCormick, who inherited the estate, renamed

Cantigny, made it a year-round home after World War I. Cantigny, a public garden and museum today, became an experimental farm and gracious permanent estate for the McCormicks.

In the 1920s, Wheaton demonstrated its patriotism as well as its love of horticulture by planting four miles of "memory trees" honoring World War I soldiers. The trees, planted along Roosevelt Road, were primarily white elm. The project was a community effort marshaled by the women's auxiliary of the American Legion. The 1920s also brought Wheaton women together through their newly formed garden club, one of the charter members of the Garden Club of Illinois. The club, which ultimately grew to five chapters, sponsored many flower shows and included guest judges and visitors such as Mrs. Charles L. Hutchinson, Mrs. Walter Brewster, and other notables in Chicago's gardening lore.

MODEL TOWNS

Chicago's model towns offer excellent insights into the evolution of landscaping. Two of Chicagoland's model towns, Riverside and Pullman, are designated national landmark districts, and many of their city-planning concepts, including the landscaping, are fairly well preserved. These two suburbs (Pullman is now a neighborhood in the city of Chicago) are presented at the end of this chapter as living legacies. Zion, once a self-contained village, has undergone significant industrial development, but still retains its central green village park. Kenilworth, which enjoys the beauty of Lake Michigan but shares the flat topography of many other suburbs, is presented here.

Kenilworth: One Vision, Community Action

> If this were my last day I'm almost sure
> I'd spend it working in my garden. I
> Would dig about my little plants, and try
> To make them happy, so they would endure
> Long after me.
>
> ANNE HIGGINSON SPICER

Envisioning an idyllic suburb where nature could impart health and happiness to its residents, wealthy businessman Joseph Sears bought a tract of farmland on the North Shore in 1889. Dubbing the community Kenilworth after Sir Walter Scott's romantic novel of love and nature, Sears sought to achieve the ideals of the emerging City Beautiful movement through town planning and restrictive covenants governing everything from physical layout to community demographics.

Kenilworth's physical layout is a charming example of ingenuity and creativity in prairie village design. Unlike other older North Shore suburbs such as Lake

Forest or Highland Park, Kenilworth lacked ravines and bluffs. Like the city of Chicago itself, Kenilworth is basically flat with its chief natural asset the Lake Michigan shoreline. Designing a village for this prototypical Chicago site could honor the region's level topography and demonstrate how perfection could be made from the plain.

Sears initially achieved beauty in his suburb through the use of light and greenery. As a practical businessman, he approved the ubiquitous grid layout of streets. But unlike the neighboring suburbs of Winnetka and Wilmette, the grid was set on an angle—its streets platted on a northeast-southwest axis to maximize a home's exposure to sunlight. Sears's other early major investment, beyond streets and utilities, was in planting trees. Ledgers for the first ground-breaking months of November 1889 through January 1890 show payments made to "clear woods." Immediately thereafter, when presumably the minimum number of trees was removed to allow building, Sears began to plant trees. In fact, in one ledger for February, 1890, the largest expense was paid to one J. Nellesen for "Planting Trees."[43]

Trees were an important aesthetic and unifying element in Kenilworth. Colleen Browne Kilner, Kenilworth resident and wife of horticulture publisher Frederic Kilner, wrote a book, *Kenilworth Tree Stories*, memorializing the importance of trees to the community. There were tales of residents who chose their lots specifically for the trees, memorial trees for weddings, births, and engagements, landmark trees, Indian trail trees, and more. Kilner recalled fond memories of "Sassafras Island," a small tree oasis on the south of town, churchyard yews bracketing a stone archway, "avenues of gold" sugar maple trees, and "cathedral aisle of elms" on Kenilworth Avenue.

Achieving a city beautiful required more than trees. By design, there were no disfiguring alleys in the village, and streets were paved to reduce dust and mud. Lot size minimums were specified, and citizens were encouraged to contribute to the overall aesthetics of the village. Kenilworth both attracted and cultivated individuals who appreciated natural beauty. Mary Babcock, widow of renowned local botanist Henry H. Babcock, relocated in the early 1890s from Chicago to become headmistress of Kenilworth Hall for girls. Her daughter Mabel Keyes Babcock grew up to become one of America's pioneer female landscape architects. Roy West, partner of landscape gardener O. C. Simonds, also attended school in Kenilworth.

Anne Higginson Spicer, poet, artist, and gardener extraordinaire, perhaps influenced local landscapes the most. A leader in local and national garden clubs, Spicer's hands-on horticultural example inspired many Kenilworth women. Spicer's own garden was a charming example of wildflower and woodland plant groupings on a 200-by-175-foot suburban lot. One garden club member recalled the Spicer garden as an inspiration: "Whereas it covered only a small area, it seemed large as [Spicer] had narrow paths around an irregular shaped pool."[44] The paths meandered through primroses, flowering bulbs, trillium, marsh marigolds, mayapples, and Spicer's favorite mertensia. The latter was so prevalent that a garden club member supposed "that all the mertensia in Kenilworth came from the mother plant in Ann's [sic] garden." In her memory, the Anne Higginson Spicer Memorial Path, a winding foot trail through crab apple trees and Virginia bluebells, was created near the railroad station.

More powerful than individual contributions, however, were the collective efforts of Kenilworth residents to create a beautiful harmonious neighborhood. From the outset, civic pride was an essential trait in Kenilworth. A variety of groups formed to ensure the village attained and retained the idealized collective vision of beauty. Through the efforts of organizations such as the Kenilworth Club (formed in 1894), Neighbors of Kenilworth (1895), and Kenilworth Improvement Company (1898), public parks were established, trees and flowers planted, and natural areas near the village protected. A list of "Kenilworth Don'ts" codified by one of these groups spelled out proper civic etiquette, including:

> Don't be indifferent to the appearance of the village.
> Don't forget that the village can be made much more beautiful with a little more thoughtfulness.
> Don't expose an unsightly spot when a vine or shrub from the woods would hide it.
> Don't think that your rights or duties stop at your lot boundary.[45]

Littering, cutting across public lawns, and dumping refuse in the Skokie Creek were also deemed unacceptable. Given the pervasively positive public spirit, it hardly seemed necessary to post such rules. Kenilworth's continuous improvement was fueled by a shared sense of place.

The Kenilworth Garden Club, founded by Anne Spicer in 1915, brought garden fairs and more civic beautification projects to the village. According to one garden club member, "In the early days every bit of work was done with our own hands. We had cold frames and hot beds and raised every plant that we finally transplanted into our garden."[46] The club's garden lectures, such as "The Little Garden," were deemed "very valuable to our small gardens in Kenilworth."[47] Committees in the club included wildflower preservation, conservation, billboards, and garden fairs. The club's June 1930 fair was modeled after the famous Chelsea Flower Show, in keeping with the village's English ambience.[48] The Kenilworth Home and Garden Club also sponsored parkway plantings of crab apples, hawthorns, and other flowering trees.

Architect and Kenilworth resident George Maher and landscape architect Jens Jensen arguably had the most impact on Kenilworth green space between

Assembly Hall, Kenilworth, Ills.

The prairie-style architecture of the Kenilworth Assembly Hall is evident, yet the landscaping, while carefully tended, conveys a traditional, formal style. From the author's collection.

1910 and 1930. Maher, a contemporary of Frank Lloyd Wright, designed homes in his own interpretation of the prairie style. Maher, who also served as park commissioner, designed public spaces such as the village's annexed western addition and the railroad depot environs, and helped implement what came to be known as the 1922 Kenilworth Beautiful plan. Jensen's projects in the village began in 1906 and continued through 1933. He designed six residential properties and two public spaces, including a wildflower preserve and bird sanctuary in his 1933 Mahoney Park. The Kenilworth Assembly Hall, built in 1908, combined the talents of both men. Maher designed the building, including its living centerpiece, an existing elm tree, and incorporated the elm motif throughout his design. Jensen was subsequently commissioned to develop a plan for naturalistic landscaping around the building.[49]

Kenilworth properties were not as expansive as, say, their neighbors in Lake Forest—which is not to say that their gardens or homes were inexpensive. Theirs was a suburban village rather than a collection of country estates. As a village, community effort in civic beautification resulted in a cohesive landscape that truly showcased the beauty of the natural prairie and woods.

TAKING THE HIGH GROUND

With much of the Chicago environs a swamp, it was natural for settlers to seek out the high ground. Many early village names allude to their greater altitude:

Mount Prospect, Summit, Westmont, Clarendon Hills, Park Ridge, and Hillside are just a few. Suburbs built on higher ground often did not remain agricultural communities—a flat plain would be better for farming. But the higher elevation warded off fears of fevers and malaria—often associated with low-lying plains and swamps. Hinsdale and Elmhurst, two early western suburbs, capitalized on their elevations.

Hinsdale: Hills and Rails

Strolling around today's Hinsdale, visitors are struck by the majestic, mature trees. Surely they have been here for centuries? Instead, thanks goes to the foresight and nature-loving attitude of Hinsdale's early planners, who planted hundreds of shade trees. The gently rolling prairie hills, however, came with the territory and, along with the convenience of newly laid railroad track, jump-started Hinsdale's popularity as a suburban retreat in the 1870s.

At 150 feet above the level of Lake Michigan, Hinsdale was said to be the highest point nearest Chicago on the West Side. In 1897, the Chicago-based *Campbell's Illustrated Journal* reported, "The surface is of wooded hill and dale; romantic, rolling, billowing land, lacking the dull flatness of all other Chicago suburban towns, and inviting in its changing aspect. Its magnificent trees, beautiful flowers and foliage are unsurpassed outside of the New England States. It is the New England town of the West."[50] Reflecting both the City Beautiful movement and persistent infatuation with East Coast landscape ideals, Hinsdale's hilly and leafy environs were hailed as the region's gold standard.

"Father of Hinsdale" William Robbins literally paved the way for landscaping elegance when he commissioned H. W. S. Cleveland in the early 1870s to lay out a portion of the suburb. The so-called Robbins Park addition on the east side of the village, with its gently curving streets and green spaces, is a clear departure from the grid pattern of most of the suburb. Cleveland decried the prevailing grid system in the West: "The monotonous character of their rectangular streets, which on level ground is simply tedious in its persistent uniformity, becomes actually hideous when it sets at defiance the . . . natural topography." About Hinsdale, Cleveland wrote, "The adjacent portion of the town had previously been laid out in squares, but owing to the inequalities of surface of this tract, much of its beauty and convenience would have been sacrificed by continuing the streets in straight lines across it."[51] Indeed, Cleveland placed building sites on the highest points of the site, and wove streets and green spaces delicately around.

The 1874 *Atlas of DuPage County* shows more evidence of early horticultural pursuits. O. J. Stough, another Hinsdale founder, retained twelve acres for his own residence in the center of town. Stough's grounds were improved with evergreens and deciduous trees and, to further validate Hinsdale as a great land investment, a thriving orchard. According to contemporary author Everett Chamberlin, "Mr. Stough has proved the soil to be well adapted to fruit growing. As an evidence of the success that has attended his experiments, it may be stated that 15,000 pounds of grapes were raised on a one-acre patch in 1872."[52] The same atlas shows the holdings of Alfred Walker on the village's east side. Walker was reputedly a progressive farmer whose agricultural experiments earned accolades from the federal government. Near Walker's addition was a tree-studded picnic park, conveniently located next to the railroad, for local enjoyment of the outdoors and perhaps to entertain prospective investors visiting via the railroad.

A perusal of 1890s Hinsdale photographs shows how the village evolved from a patchwork of farms to a cluster of stately suburban residences. Most estates of the era are defined with well-groomed lawns, wide lots, and stately trees. Pennsylvania transplant and newspaper publisher O. P. Bassett, for example, purchased the Robbins's homestead, which he rechristened Bonnie Heights. Lawn dominated the front of Bonnie Heights, uninterrupted by flower beds or shrubbery groups. Bassett and neighbor C. L. Washborn built a greenhouse and grew plants as a hobby. Local lore has it that their experimentation produced the exquisite rose known as the American Beauty. Bassett displayed a collection of American Beauties at the World's Fair.[53] Bassett and Washborn's success as rosarians was such that they expanded their greenhouses and opened a wholesale flower store.[54]

Other early landscapes certainly reflected a love of gardens. Englishman John Hemshell sited his home, Sylvan Castle, on the northeast corner of a fifty-acre land tract (like many, Hemshell positioned his home with an eye toward future subdividing). The ivy-covered brick residence was framed by tall evergreens and featured a hand-built rustic twig trellis sheltering the front entryway. Oaklawn, the estate of grain trader Vermont native W. L. Blackman, featured curved drives and the stately trees of its namesake. Fellow Vermonter Dr. D. K. Pearson reportedly made his fortune in real estate rather than the medical profession. His turreted home featured vine-clad porches, curved drives, specimen urns and planters filled with exotics, and a whimsical garden folly.

While many of the homes of this period were constructed of wood, William D. Gates's home was impressively fashioned of terra cotta. Not surprisingly, Gates was founder of the American Terra Cotta Company, where, with the prairie's own clay, he developed functional building materials used by many of Chicago's prairie-school architects, and where the highly desirable Teco pottery was ultimately created. Gates's residential landscaping included a broad lawn ringed by trees and pasture, and an elegant urn (presumably terra cotta) marking the front entryway. While each property's landscaping was unique, the general themes in the 1890s included wide, unobstructed lawns, vine-clad porches, minimal foundation plants, specimen and boundary shade trees, and isolated garden ornaments or furniture such as planted urns, hammocks, rustic swings, and benches.

As more acreage was subdivided for private homes, Hinsdale's public spaces became especially valued for their natural beauty. The wooded Salt Creek, which traversed Hinsdale, was prized for its water lilies and recreational opportunities including canoeing, fishing, and hiking. The village's drives supported the healthful exercise found in riding horses and driving carriages, and later, the new bicycle sport taken up by Chicagoland's "wheelmen." The 1897 *Campbell's Illustrated Journal* declared, "In no other suburban town are the streets so wide, clean and inviting, and bordered with endless foliage, vines and flowers." A corporation was formed in the 1890s to create Oak Forest Cemetery (now Bronswood Cemetery) in Hinsdale, with grounds including winding walks, natural stream, and native oaks. *Campbell's* asserted, "All the systems of the cities [*sic*] great cemeteries were carefully studied and the most commendable features were selected and adopted, thereby giving to the patrons of Oak Forest the benefit of the years of experience."[55] Hinsdaleans were hospitable about sharing their healthful natural resources; several social organizations were formed to benefit underprivileged or health-impaired individuals. The Fresh Air Home, created in 1892, offered countryside immersion experiences to underprivileged women

GARDENER'S PLANTING PLAN

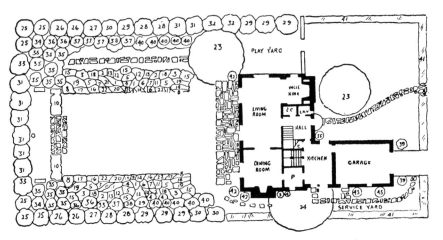

This 1920s garden plan from the Garden Club of Hinsdale featured a wide lawn in the front and a backyard ringed by shrub and perennial borders. As shown by the plot numbers, trees and shrubs included elm (*24*), Lombardy poplar (*25*), high-bush cranberry (*26*), wayfaring tree (*27*), snowball (*28*), lilacs (*29, 32, 33*), syringa (*30*), and honeysuckle (*31*). Copyright 1929 by Chicago Tribune Company. All rights reserved. Used with permission.

and children. Dr. David Paulson purchased Judge Beckwith's picturesque land in
1903 to create Hinsdale Sanitarium, which emphasized the outdoors and vegetarian menus. Alexander Legge, president of International Harvester, purchased
land as a summer home in Hinsdale in the 1920s. When his wife died, Legge created Katherine Legge Memorial Park and Lodge as a retreat for women employed
at International Harvester. (The lodge is now a park on fifty-two acres of woods
and gentle hills in south Hinsdale.)

Private gardens in the early twentieth century were greatly influenced by the
efforts of the Garden Club of Hinsdale, and of local nurseries. The Morris family
established a florist and greenhouse business in 1894. Their greenhouses, originally at Vine and Hickory streets, were sold for a subdivision. Their later location
at Monroe Street and Ogden Avenue was later sold to Amlings. Littleford Nurseries (later Hinsdale Nurseries) on Madison Road started under the pioneering
work of Lauren Havens in 1837, who brought apple grafts from his native New
York. By the 1930s, F. Jason Littleford, Havens's great-grandson, cultivated more
than one hundred acres of trees, shrubs, roses, and more.[56] Littleford, an officer of the Illinois State Nurserymen's Association, was a strong supporter of
the Garden Club of Illinois, and not only gave lectures to its members but also
was featured on WGN radio broadcasts. The Garden Club of Hinsdale, a charter
member of the Garden Club of Illinois, included a number of influential ladies.
Mary Dynes was elected to serve on the statewide board and, with fellow Hinsdale resident Mrs. Euclid Snow, served for years as editor of the club's national
newsletter, *Garden Glories*. They also gave talks on WGN's radio broadcasts. Both
Dynes and Snow created beautiful gardens for their own homes, praised by the
national society as "a mecca for gardeners in iris and peony season, and Mrs.
Dynes' roses are exceptional."[57] Other Hinsdale enterprises and garden club
supporters included Fanny L. Brent of Windy Hill Gardens, who promoted her
services as a rockery garden consultant, and S. T. Collins, who advertised in *Garden Glories* as a gladiolus grower.

Hinsdale ladies were at the helm of the Garden Club of Illinois during the tu-
multuous years of the Great Depression. In 1934, president Mary Dynes wrote,
"Although the past three years have proven the worst ever experienced in both
a business and a horticultural way—with the alarming era of unemployment—
The Garden Club of Illinois has weathered the storm and finds itself in a strong
healthy condition."[58] The respected stature of the Garden Club of Hinsdale was
such that it was selected to host the first Central District Conference—representing clubs from Illinois, Indiana, Iowa, Kentucky, Michigan, and Wisconsin—in August 1933. The chief topic of the conference was conservation, and, as
part of the schedule, the group was escorted by Mrs. Snow, Dynes, and others to
Chicago's 1933 World's Fair. The gardens of this hilly railroad village had become
known worldwide.

Elmhurst: Town of Trees and Hills

The two names of this suburb, originally Cottage Hill and then Elmhurst, indicate the topographical reasons many early settlers staked their claims here. West of Chicago by about sixteen miles, Elmhurst enjoys an elevated site above the prairie. The ground proved fertile farmland. Transportation aided market gardeners. In the 1830s, pioneers settled there, and Cottage Hill became an important stagecoach stop. Later, when the Chicago Great Western Railroad came through town in the 1880s, both commuters and freight traffic helped Elmhurst grow.

But it was the interest in horticulture that is still reflected in Elmhurst's name, which was changed from Cottage Hill in 1869 at the suggestion of one of the town's most influential citizens, Thomas B. Bryan. Bryan moved to Cottage Hill in the 1860s, when most of his peers maintained homes either in downtown Chicago or on the North Shore. Bryan, businessman and horticulture patron, was one of the first to build a gracious home in the then-wild prairie town of Cottage Hill. A Virginian and Harvard graduate, Bryan and his wife, Jennie Byrd Page, moved to Chicago in 1852. Their first home was in an exclusive district on Michigan Avenue near Madison Street. Following his purchase of one thousand acres at Cottage Hill, Bryan set out to protect his investment and prove that there was life on the prairie. Trees were deemed critical to tasteful landscaping, and Bryan began correcting the "flaws" of natural prairie scenery. The *Chicago Tribune* reported, "He began at once making extensive landscape improvements, including the planting of many trees, chiefly elms. The name of Cottage Hill seemed to him inappropriate, for there was neither cottage nor hill in the vicinity, and he invented the name Elmhurst, the last syllable 'hurst' being the old Saxon word for 'planting' or 'seedtime.'"[59]

Landscaping at Byrd's Nest evolved over the decades. Bryan's early gardeners were European immigrants, including Thomas Wakefield, an early member of the Chicago Gardeners' Society, and Mr. Swenson, a local nurseryman. Byrd's Nest in 1860 included curved carriageways, some midsized trees, a grapery, and greenhouses. Later photos show very mature landscaping replete with a pond, formal gardens, and broad vistas. Byrd's Nest was the epicenter of high society and culture, and the Bryans hosted many parties and meetings, including an international gathering of officials from the Columbian Exposition in 1893.

Other wealthy industrialists followed suit and built estates in Elmhurst. Many of the estate names reflected an interest in nature or horticulture. John R. Case built Cherry Farm, named after his extensive cherry orchard; Seth Wadhams created White Birch; Henry W. King (father-in-law of garden writer Louisa King) settled in Clover Lawn. Later, after the Chicago Fire of 1871 brought more permanent settlers to Elmhurst, additional estates were created, including Hawthorne for Lucian Hagans, Sweet Briar for George Runsey, and Shadeland for Lee Sturges.[60]

It is said that the first elm trees to be planted in Elmhurst were planted by Jedediah H. Lathrop, Thomas Bryan's brother-in-law, in 1867. He planted a double row of elms along Cottage Hill Avenue from Park Avenue to St. Charles Road.[61] Indeed, many of the estate owners, wanting mature landscapes on this flat prairie, transplanted large trees from nearby woods. The technology of digging and transporting these trees became the topic of many a horticultural publication at the time.

Elmhurst was home to many individuals influential in regional and national horticulture. Louisa King created her first garden at her mother-in-law's home in Elmhurst, now Wilder Park. King was one of the early twentieth century's most influential women garden writers. Bryan Lathrop, son of Jedediah, was an early and avid horticultural patron, sponsoring O. C. Simonds in his naturalistic garden designs in Graceland Cemetery. Bryan Lathrop also investigated a number of horticultural experiments, including a bamboo plantation in the southern states. Later local gardeners included Emma Glos, whose property in the 1930s was famous for its trees such as beeches, blue spruce, and ash and colorful flowers such as zinnias. P. M. Keast was well known as a gardener and rosarian; he also served as president of the Chicago Rose Society. His garden was featured in *Better Homes and Gardens* magazine. These gardeners, while not attaining national notoriety, helped showcase the potential of beautiful gardens.

In the 1920s and early 1930s, following a population boom resulting from the end of World War I, Elmhurst residents intensified their interest in gardens and green space. The Park District was formed, obtaining space such as Wilder Park for public greenery. In addition to hosting horticultural fairs, the Elmhurst Garden Club became an important catalyst for civic beautification. Working with the town officials, the club planted flowers and shrubs in common areas such as railroad embankments. In 1929, the club inaugurated a sale of elm trees to the public. According to *Garden Glories* newsletter, "Elmhurst is known for its stately and beautiful elms; Cottage Hill Avenue is spanned by a veritable Gothic arch formed by specimens planted by far sighted city founders."[62] The club's tree sale coincided with Arbor Day celebrations and encouraged citizens to use elms in their street plantings. The Elmhurst Men's Garden Club was formed in 1932, and hosted annual flower shows for over thirty years, published a newsletter, joined in civic projects, and in 1939, hosted the national convention of the Men's Garden Clubs of America.

Although the scourge of Dutch elm disease took its toll on the Elmhurst namesake, the city fought back with innovative horticultural practices. Today, Elmhurst continues to offer a mix of shade trees along its graceful streets, with disease-resistant elms offering testimony to past and future.

SUMMER RESORT SUBURBS

Some suburbs, due to their natural attractions or pleasant climate, became sum-
mering destinations for Chicagoans seeking a breath of fresh air. Whether the
attraction was a local lake, river, deep woods, or intriguing rock formations, the
nearby town made the most of its seasonal tourist business. As the railroad improved daily commutes to the city, quite often the summer resort became a year-
round community. Lake Geneva was a consummate example of a summer resort
that ultimately became a year-round town. Its residents, mostly wealthy Chicagoans and Milwaukee residents, created elaborate summer estates with exquisite
gardens. Although on a railroad line, Lake Geneva did not become a commuter
suburb of Chicago because of its distance. Other towns, such as Glen Ellyn, also
initially considered summer resorts far removed from Chicago, became commuter suburbs as railroad and automobile improvements reduced the commute
time to a manageable distance.

Glen Ellyn: A Lake Resort

Hospitality was the watchword for Glen Ellyn, from its earliest days in 1846 at
Stacy's Corners where Moses Stacy's stagecoach tavern offered respite for weary
travelers to and from Chicago. Farmers driving their wagons to market were
charged fifty cents for dinner and breakfast, lodging, and hay for two horses.[63]
Stacy's Tavern was sited at the intersection of several plank roads. When the
Galena and Chicago Union Railroad came through town, the central business
district was laid out in the predictable grid pattern, several blocks south of the
tavern. The town underwent several name changes, from Danby to Prospect Park
in the 1870s to Glen Ellyn in the 1890s.

The town was rechristened Glen Ellyn in honor of the new subdivision and
resort designed and funded by a group including Thomas E. Hill, a resident,
teacher, and author of numerous books. Hill and his group took advantage of
the village's natural resources—the heavily wooded hills and a newly created
lake made to drain nearby marshy land. Hill's own home was surrounded by
gracious grounds, orchards, and vineyards where he and his wife entertained
prospective investors and friends in style.[64] Glen Ellyn became a summer resort
town with city dwellers arriving by train to enjoy the manmade lake, nearby medicinal springs, and first-rate accommodations at the new Glen Ellyn Hotel.

"By careful analysis it has been demonstrated that nature has supplied Glen
Ellyn with mud and mineral water which excels that of the famous 'Indiana Mud
Springs,'" declared an advertisement in the 1894 *Wheaton Illinois* newspaper.[65]
The mud baths, according to the ad, were beneficial for "rheumatism, gout, lum-
bago, kidney complaint, and nervous prostration."

This design for home and grounds for a suburban residence was featured in the "Landscape Architecture" section of the August 1887 *Hill's National Builder*, published by Thomas E. Hill, who developed the Lake Ellyn summer resort. According to the accompanying article, like the Lake Ellyn area, this home was sited in "hilly country," and a "charming little lake" was added. Carriageways and paths curved gently, and large existing trees were preserved. This home included a greenhouse and was surrounded by native plants. "In lawn planting there must be a catholic taste shown in selecting plants," Hill advised. While diversity in plants was encouraged, homeowners were not to follow every passing fancy. "Skylines" (or sightlines, as termed today) were to be preserved. This sketch likely showed Hill's vision for suburban Glen Ellyn. From the author's collection.

Although the hotel enjoyed only a relatively short lifespan, the curative powers of Lake Ellyn's natural scenery endured. The *Chicago Tribune* opened a summer hospital for Chicago's indigent women and children in 1905. A bulletin was sent to hospitals, physicians, settlement houses, and other social agencies announcing "you are invited to make use of The Chicago Tribune Summer hospital, at Glen Ellyn . . . for the benefit of convalescent women and children. The hospital is intended for those recovering from acute diseases . . . and whose recovery would be accelerated by removal to the bright, pleasant rooms and verandas."[66] Wagonloads of ice were brought to the sanitarium daily, to provide cool air and refreshments. While many patients were recuperating from illness, some suffered from sheer exhaustion, such as the mother who could no longer stand while doing her ironing, but ironed with a board on her lap. "All day long the children at the hospital played under the trees or swung in the hammocks, with only the thought of having to return to their tenement homes in Chicago— sooner or later—to trouble them."[67]

Little wonder then, with such pastoral surroundings, that Glen Ellyn's permanent residents were inclined to gardening and the outdoors. Their honor roll of horticulturists is impressive. In the early 1900s, Isaac A. Poole, a botanist, worked with Ruskin College, then located at Glen Ellyn, and developed a white iris with blue borders. This stately flower soon adorned many Glen Ellyn gardens.[68] In 1910, George Ball, of internationally known Ball Seed, began his seed and nursery business as a wholesaler with seven greenhouses on Hawthorne Street.[69] Ada Douglas Harmon, whose ancestor was Elijah D. Harmon, Fort Dearborn doctor and inveterate horticulturist, maintained a bountiful garden and created 175 watercolor paintings of native DuPage County wildflowers.[70] Benjamin Gault, an internationally known ornithologist, lived in Glen Ellyn from 1890 on, and directed the planting of a bird sanctuary and wildflower preserve (named for him) in the village.

Frank Johnson, who became the village's first forester, earned a moniker as Glen Ellyn's "Johnny Appleseed" for his donations of saplings to the village.[71] Johnson, a Glen Ellyn resident since the 1890s, was a lilac connoisseur who grew over three hundred varieties on his grounds. His collection, said to have been larger than that of the Arnold Arboretum, easily rivaled that of his friend and fellow lilac hobbyist, William Plum of Lombard. Johnson donated his lilac collection to the village of Glen Ellyn in 1927 to adorn its new Memorial Park. Johnson was commended for public service by Gifford Pinchot, U.S. forester.[72]

Glen Ellyn residents formed a number of civic organizations to beautify the village or contribute to social causes through horticulture. Improvement committees developed to work on everything from roads to a fly-swatting campaign wherein grocers were required to screen their fruit and vegetables against the pesky insects.[73] The Glen Ellyn Garden Club was a charter member of the Garden Club of Illinois in 1927. In fact, the club was so popular that its charter group included three chapters (the Rose, Scilla, and Iris), and later expanded to five groups with the Aster and Poppy chapters. Among their many local beautification efforts, the club chapters worked to improve roadside planting and the village-entrance landscaping. In 1933, working with proceeds from their flower show, and with advice from Wheaton landscape architect John Brown, the club screened unsightly poles and other utility structures with decidedly naturalistic landscaping. As reported in the *Garden Glories* newsletter, "Glen Ellyn is hilly; its streets have many sweeping curves. Hawthorne, flowering crab, elms, maples and oaks are native. It is, therefore, considered most appropriate that our roadside planting be informal and naturalistic."[74] The summer resort village, now a thriving suburb, recaptured its natural beauty.

LEGACY LANDSCAPES: SUBURBAN BY DESIGN

Many Chicago suburbs grew organically around a natural physical feature of the land. An enterprising pioneer built a grain or sawmill near a river, and towns like Hinsdale, Oak Park, and Elgin sprouted. A grove of trees provided shelter or fuel, and villages like Downers Grove evolved. Higher land brought relief from prairie wetlands, and thus Mount Prospect and Summit were born. Little towns grew around these natural nuclei, and farms radiated outward. The railroad came, bringing more settlers, and changing growth patterns. The automobile and suburban developer soon followed, and the original suburban gardens in most cases were long gone.

Some Chicago suburbs were planned outright in their entirety. The land patterns and relationships among buildings and green space have remained relatively intact in Riverside and Pullman. Private gardens and public spaces in these very different communities reflected the social conditions of the times and the goals and priorities of their designers and investors. Riverside, designed in 1869, exemplified the idealized suburban village, designed by Frederick Law Olmsted and funded by a consortium of businessmen. Railroad barons were just making their mark in Chicago, and Riverside was built to capitalize on this modern form of transportation. Pullman, built almost twenty years later, a company town based on the vision of a man whose fortune was made through the railroad, offered homes for the workingman. Both communities have adopted modern improvements, yet retain the design intent of their makers. Brookfield Zoo, on the other hand, is a unique example of a world-class attraction that brings crowds of tourists daily to a onetime sleepy workingmen's suburb. The landscaping in the zoo is almost as interesting as its inhabitants. The Cheney Mansion, one of the few early-twentieth-century suburban properties close to the city to remain virtually intact, contains excellent remnant garden elements.

Riverside, Illinois

Between 26th Street and Ogden Avenue, and Harlem and First avenues
http://www.riverside-illinois.com

A masterpiece of community planning, the general plan for Riverside was developed by Olmsted and Vaux in 1869. Frederick Law Olmsted, Father of American Landscape Architecture, created this village with his partner, Calvert Vaux, as a prototype for healthful, suburban living. Today a national historic landmark village, visitors can readily see the key elements of Olmsted's design: curving streets, wide green spaces, natural vistas, and an abundance of trees.

Riverside is an unlikely oasis of greenery amid encroaching suburban sprawl. Sandwiched between strip malls, major shopping centers, a large quarry, and an industrial

Riverside's curving paths through wooded river walks and open, grassy views offered a respite from the city, and an alternative to grid street plans. Photo by the author.

district, it is miraculous that the village has survived basically intact and true to Olmsted's vision. It is perhaps the genius of that vision that draws multiple generations of families back to settle in Riverside, and to fight passionately to preserve the village's historic landscape.

When Olmsted and Vaux planned the village, suburban communities were just coming into vogue in Chicago, thanks to the new potential for railroad commuting. Olmsted's naturalistic design attempted to provide the healthful, restful benefits of country living combined with the amenities of the city. The general plan for Riverside was brilliant and pioneering. It called for winding streets, prodigious greenery, ample open spaces, and use of native plantings. Today, these design elements are being used in many of our newest suburban developments, as more Americans seek the healing powers of nature in their home communities.

Olmsted and Vaux were commissioned to perform a feasibility study of the Riverside site, to assess its potential for development and recommend how a village might be built. The site, which Olmsted was asked to evaluate, was not an utter wilderness. In the 1860s, Riverside was the farm of David Gage, and included fifteen acres of intentionally planted trees, where a scant two decades prior, only a few old stands of oaks and hickories grew. In Gage's nursery, he had already planted a "dense growth of straight timber, varying in size from the young shoot of two or three years to tall stems, from sixteen to eighteen inches in diameter . . . The varieties are several sorts of Hickory, Elm, Ash and Cherry, with an undergrowth of Crab, Plum, Haw, Dogwood, etc."[75] Olmsted designed Riverside to capitalize on its surrounding natural assets—the gentle curves of the Des Plaines River with its banks of native trees and shrubs.

In explaining the proposed development to his wife, Mary, Olmsted wrote:

The motive is like this: [Chicago] is on a dead flat. The nearest point having the slightest natural attractions is one about 9 miles straight back—West. It is a river (Aux Plaines) or creek 200 ft wide, flowing slowly on limestone bottom, banks generally sandy with sandy slopes and under water a little limestone debris. As a river not very attractive, but clean water 2 or 3 ft deep, banks & slopes rather ruggard [*sic*] & forlorn in minor detail but bearing tolerable trees—some very nice elms but generally oaks mostly dwarf. The sandy, tree-bearing land extends back irregularly, so there is a good deal of rough grove land—very beautiful in contrast with the prairie and attractive. 1600 acres of land including a fair amount of this grove but yet mainly rich flat prairie have been secured, & the proprietors are now secretly securing land in a strip all the way to Chicago—for a continuous street approach—park-way. I propose to make the groves and riverbank mainly public ground, by carrying a road with walks along it & to plan village streets with "parks" & little openings to include the few scattered motes on the open ground. An excellent R.R. passes through it & a street R.R. parallel with the park way is projected.

The place has been somewhat resorted to by picnics & is now occupied by a famous stud farm with a training race track & is known as Riverside farm. They had proposed calling it Riverside Park but I objected & they told me to suggest a better. This I refer to you . . . The river should be the important circumstance as the centre of improvements. I am willing to make it a river park but not a park.

Please write me a name or some names. All the Old English river names are out of my head. River Groves is the best that has occurred to me but a more bosky word than grove would perhaps be better.[76]

Seemingly, Mary was unable to come up with a more bosky word, and Riverside remained the name of the village. Excerpts from Olmsted and Vaux's 1868 preliminary report show how the designers capitalized on Riverside's natural features, which are still key landscape elements today.

The Des Plaines River (or Aux Plaines River, as it was then called) was and is critical to the growth of Riverside. A natural portage near Riverside drew early Indian tribes, explorers, and missionaries, including Louis Joliet and Father Marquette. Olmsted's vision included recreational uses of the river, as well as strolling paths for quiet enjoyment of nature. Throughout the late nineteenth and early twentieth century, Riverside became a destination for sightseers seeking the wildflowers and native trees that bloomed along the riverbanks. Nature groups such as the Prairie Club of Chicago planned hikes here, the railroad promoted wildflower routes near Riverside, and social workers urged Chicago's tenement dwellers to board a bus or trolley to escape crowded city conditions and stroll along Riverside's namesake. Two dams were installed in the early 1900s, and residents enjoyed ice skating and ice sailing during the winter months.

Today, the river continues to attract outdoor enthusiasts and nature lovers. Rustic strolling paths border the riverbanks, traversing village parkland on one side and forest preserves on the other. Wildlife abounds near the river, thanks to recent efforts of this community and others along the Des Plaines watershed to reduce pollution. Blue and green herons, egrets, and kingfishers are common sights, and fishing reports include walleye, northern pike, large mouth bass, and sauger fish.

Greenery is one of the most obvious and valued characteristics of Riverside. Triangles of green parkway planted with massed groups of shrubbery and mature trees delineate the intersections of Riverside streets. Olmsted required that each homeowner plant two trees in the parkway in front of his house, a tradition that continues today. The 1871 village prospectus stated that 47,000 shrubs, 7,000 evergreens, and 32,000 deciduous trees had been recently planted.

Villagers treasure their trees and plants. Former residents such as Virginia Cross West reminisced about the natural beauty: "Children could walk in the forest preserves without fear. We always walked in Indian Gardens along the river. We picked wild flowers, searched for Indian arrowheads and got polliwogs out of the river."[77] The profusion of trees offered cooling shade in pre-air-conditioning days, as resident Mary Frick recalled. "I worked at Swift and Company summers when I was away at school, and that was a hot drive home from the stockyards. I can remember coming into Riverside, the temperature would drop. It was just amazing how cool it was here."[78]

Olmsted, the maker of New York's Central Park and Boston Commons, was intent on providing ample public green space in Riverside. This took the form of small green islands where streets converged, larger neighborhood parks where children could play and residents gather, and long stretches of green space, the "commons" bordering the village and major streets. Longcommon Road, for example, the main diagonal street running southwest to northeast, parallels a continuous ribbon of green parkland.

Preservation and appropriate treatment of these natural charms has been a long-standing battle for the village. Olmsted himself had to dissuade the first village president from building his home in the middle of the public Longcommon. Today, Riverside has 134 acres devoted to parks, including forty-one small parks and five large parks. It is a continual challenge to balance the needs of passive with active recreation in the parks—soccer fields versus unbroken greenery are a matter of constant compromise.

Virtually all of Riverside's streets wind and twist through the landscape. Olmsted's report and writings suggest that this is to capture natural vistas at turns in the road, and imitate the curving lines of the Des Plaines River. Other scholars point out that the winding streets also precluded commercial or industrial traffic and buildings—an artful urban-planning device in the days before zoning laws were common. Today, commercial traffic is still rare, because even dedicated cabbies or pizza delivery drivers get lost in Riverside's circuitous routes. Not as readily apparent today is the fact that Riverside's roads were initially depressed lower than normal into the ground. This, Olmsted felt, helped roads disappear into the landscape. From the right vantage points, you can still look across the Longcommon, for example, and see only the top portions of cars, not the roads or wheels touching them.

Olmstedian design principles suggest that the "hand of man" is not seen, that is, that the landscape appears to have been entirely natural. Contrived gardening practices such as geometric floral beds and exotic plantings were not favored. Of course, Olmsted was realistic in assessing that real estate values were contingent on proper amenities, typically available only in cities. "Artificial improvements" like water, transportation, and street lights were built into Riverside's plan to attract prospective residents. The village's historic water tower designed by Olmsted's contemporary William Le Baron Jenney was restored in 2005. The village retains its gaslit streetlamps, now considered quaint, but then a sign of modern convenience.

Riverside became a showplace for architects. Although Olmsted, like many landscape architects, had an uneasy relationship with architects, the latter were drawn to the sylvan surroundings in Riverside. Frank Lloyd Wright, Louis Sullivan, Jenney, and many others created masterpieces in the village. Nonetheless, Olmsted recommended that homeowners plant trees in the parkway lest a home's architecture offend the eye. Riverside gardens tended to make great use of these "borrowed views" from the parkway and neighboring yards, or visually extended the landscape to the plentiful green spaces. Some landscapes capitalized on river views, such as that of Owen L. Rea, whose 1920s prizewinning riverfront garden was framed by a rustic arch covered with 'Dorothy Perkins' climbing roses.[79]

Lot sizes were ample in Riverside's early days—the smallest lots were one hundred feet wide by two hundred feet deep.[80] Intended for wealthy citizens, the grounds of a typical early villa might easily extend around the block. Early engravings and descriptions in the suburb's 1871 prospectus show curving driveways and both natural and cultivated plantings. The Jennison residence on Bartram Road, for example, included "grounds [that are] tastefully laid out, and ornamented with rustic grape arbor and fence, while a fountain of rare beauty attracts the attention of all."[81] Other homes on Scottswood Road (in the wooded division of the village where the river makes a bend) were described as being "well shaded by a natural growth of fine forest trees." The prospectus describes the L. W. Murray landscaping: "The grounds are broad and open, and the absence of the usual, and one might add, unnecessary fence, gives an air of good taste and broad hospitality well worthy of imitation."[82] Murray, an early village official, had commissioned Jenney, who worked on many of Chicago's West Parks. Olmsted's firm collaborated in Riverside with Jenney, who had his own home just down the street from the Murrays. Most of the Riverside homes Jenney designed included one or more verandas to offer shade and outdoor views. Some of Jenney's homes included conservatories, such as that of Riverside Road resident W. T. Allen, "to render the house cheerful in weather too cold or unpleasant to permit the enjoyment of the verandas or lawns."[83] The Riverside Garden Club, one of the charter members of the Garden Club of Illinois, offered novel tours of homeowners' cultivated gardens.

Early plots were large enough to support the home, coach houses, outbuildings, drives, walkways, and gardens associated with the upscale lifestyle of the late nineteenth century. As the Riverside 1871 prospectus claimed, "The fortunate dweller at Riverside . . . has plenty of elbow room, and can dig to his heart's content, raise his own fruits and vegetables, keep his own cow, and even make his own butter."[84] Of course, not every Riversider exercised his green thumb. Hired gardeners, who either lived in the coach houses or offsite, tended to the very large estates. Other homeowners dabbled in the gardens themselves. C. B. Beach, for example, hybridized iris and planted a lush garden surrounding his riverfront home on Bloomingbank Road. Catherine Mitchell, on Fairbank Road, was a wildflower enthusiast who led nature walks around Riverside and the Chicago area. (The Catherine Mitchell Lagoon is located in the nearby forest preserve on Harlem Avenue.) Two families (the Coonley and Babson estates) commissioned the designs of Jens Jensen, but most other homeowners created do-it-yourself gardens or employed local nurserymen and landscapers to lay out the grounds. In the 1880s, Riversider James D. Raynolds turned his rose-growing hobby into a profession and, according to historian A. T. Andreas, had the largest rose garden in the west, producing 150,000 roses per year. Charles Reissig built five greenhouses in Riverside, and the grounds around his home became known as Reissig's Exotic Gardens.[85] Other florists and gardeners established businesses in the village, and residents also had ready access to Vaughan's nurseries in nearby Western Springs.

Landscape Integrity

Riverside suffered financial setbacks almost immediately after incorporation. Olmsted's design for the western part of the village, across the river, was never fully realized. In the northern portion of the village, streets were mapped out, but few homes were built there until the twentieth century. Riversiders have fought hard to retain the original layout and ambience as conceived by Olmsted and Vaux. Although lots have been subdivided, the village has refused state funding for road repairs, when contingent on "straightening out" the curved roads. Green spaces have been preserved, and the village follows a tree maintenance plan. Today, the village encourages tree planting and has published a list of native trees suitable for the landscape. Riverside's Olmsted Society, and other landscape preservationists, encourage the use of native plantings on public parkways and riverbanks. While more heavily populated, and although modern elements appear here and there, it is still possible to enjoy the "umbrageousness, and facilities for out-of-door recreation" provided through the vision of Olmsted and Vaux.

Pullman, Illinois

Bounded by 103rd Street on the north, 115th Street on the south, Cottage Grove Avenue on the west, and the railroad tracks on the east.
http://www.pullmanil.org/

Built as a self-sufficient company town for railroad-car magnate George Pullman, the town of Pullman used landscaping extensively in public areas. Landscape architect Nathan Barrett collaborated with architect Solon Beman to create elaborate green common areas and prototype scientific farms. Company grounds crews maintained workingmen's row house front yards, but the tiny backyards were privately planted.

"Pullman is a sermon in bricks and mortar on the humane and considerate treatment of employees; a dialectic statue to the efficacy of moral government," wrote historian A. T. Andreas in 1884.[86] "Pullman is un-American," countered author Richard Ely in 1885. "It is benevolent, well wishing feudalism, which desires the happiness of the people, but in such way as shall please the authorities."[87] The controversial town of Pullman fueled many news items from its inception and beyond the notorious labor uprising called the Pullman strike in 1894. In the design for Pullman's model workingman's town, it is interesting to see the role of landscaping and personal property rights.

George Pullman, inventor and industrialist, made his first impact on Chicago's landscape in the 1850s when he proposed raising the city's houses higher than street level to bring the city out of the swamp. His successful efforts with moving buildings earned him a sizable reputation and fortune. In the 1860s, recognizing the importance of the emerging railroad, he invested and built luxurious railroad cars, the Pullman sleepers, to

TOP Formal gardens such as these in Arcade Park were maintained by the Pullman Company. Workers' front yards were also maintained by the company and allowed little opportunity for personalization. Ca. 1885. Reproduced by permission of Pullman State Historic Site.

BOTTOM Workers flocked to available green space during the noon lunch break. Ca. 1895. Reproduced by permission of Pullman State Historic Site, Collection of Paul Petraitis, Pullman Research Group.

accommodate burgeoning leisure travel. A railroad strike of 1877 served as impetus for Pullman, as both a profitable and philanthropic exercise, to build his town for workers of the Pullman Car Company.[88]

In the late 1870s, Pullman began looking for a suitable location for his model town. It was rumored that many cities were viable candidates, including Kansas City and St. Louis. Despite favorable grants offered by the latter city, "it was [Pullman's] contention that during the summer a man could do at least ten per cent more work near Lake Michigan than in the Mississippi Valley in the latitude of St. Louis."[89] The search for land in the Chicago environs was handled with much secrecy. According to author Frederick Cook, Pullman initiated several red-herring visits to outlying areas to divert attention, and thereby avoid inflated real estate prices, from his real goal, the acreage near Lake Calumet that was to become Pullman. Then on the real estate beat of the *Chicago Times*, Cook dogged Pullman's efforts, and "finally I pinned down an actual sale of large dimensions, with Colonel 'Jim' Bowen as the ostensible purchaser."[90] The four-thousand-acre tract that Pullman purchased piecemeal through the Pullman Land Association via various agents was then said to be the largest land purchase in the history of the city.[91]

Pullman secured the services of Solon Beman and Nathan Barrett to develop the architecture and landscape architecture, respectively, for the fledgling town. New York native Barrett, had worked with George Pullman on landscaping his New Jersey oceanfront estate and later on his Prairie Avenue home in Chicago. One of the first members of the American Society of Landscape Architects (and president of this group in 1903), Barrett was a proponent of the formal style of landscaping. He claimed to be the earliest interpreter of the formal garden in America, perhaps as a reaction to Olmsted and other landscape artists gaining popularity in more naturalistic landscape styles.[92] His elaborate public gardens at Pullman, surrounded by the town's grand buildings, displayed sinuously shaped flower beds, planted in the carpet-bedding style of the day.

Pullman, who wished his town stripped of any unsavory elements including liquor or unsanitary conditions, must have deemed landscaping a wholesome endeavor. Public areas were extensively landscaped. Arcade Park, the broad grassy town center adorned with flower-bed cutouts, marked the entrance to the town and was an important backdrop for the busy Hotel Florence and Arcade Building, a marketplace now replaced by the visitor center. Mrs. Duane Doty, a resident and Pullman booster, waxed eloquent: "At intervals of thirty feet shade trees are planted along both sides of the streets, and on the main streets flowers are grown round the trees. Open spaces planted with shrubbery and flowers really constitute a large park, in the midst of which the homes of the people stand. The monumental buildings and vast shops in the long stretches of meadow, and the walks, lined with trees and shrubbery, emphasize the park features of Pullman."[93]

The streets were laid out primarily in a grid pattern, but as Richard Ely noted, "Skill has avoided the frightful monotony of New York, which must sometimes tempt a nervous person to scream for relief. A public square, arcade, hotel, market, or some large building is often set across a street so ingeniously as to break the regular line, yet without inconvenience to traffic. Then, at the termination of long streets a pleasing view greets and relieves the eye—a bit of water, a stretch of meadow, a clump of trees, or even one of the large but neat workshops." Ely further prophesied that the growth of shade trees, particularly the double rows of elms on the main boulevard, would break the "newness of things," which he characterized as the "mechanical regularity of the town . . . which suggests the epithet 'machine made.'"[94]

The environs around Pullman in the Calumet area were always known for their rich diversity in native plants. Lake Calumet itself, a lake bordering the east side of Pullman, supported many forms of aquatic plant and wildlife. Mrs. Doty noted that "500 native flowering plants have been found within a circle of three miles . . . 60 species of grasses and sedges, 100 of mosses, liverworts and fungi; and about 400 kinds of trees and shrubs."[95] Market gardens and farms in the immediate vicinity supplied Pullman residents with fresh produce, sold at the Arcade Building. A Pullman-owned greenhouse, including six acres of nursery plantings on the shores of Lake Calumet, annually raised 100,000 flowering plants, cultivated 70 species of shrubs and trees, and 125 species of plants used in the town parks and green spaces.[96]

Public utilities were widely hailed as engineering marvels of the day. Sewage was piped out to the nearby Pullman sewage farm, where the filtered water was used to grow crops for the town. A. T. Andreas reported that during the 1883 growing season, the farm produced "7,400 bushels of potatoes, 800 bushels of Onions, 36,000 ears of Sweet Corn, 400 bushels of Field Corn, 100 bushels of Carrots, 100 bushels of Beets, 250 bushels of Parsnips, 150,000 heads of Cabbage, 25 tons of Squash and 24,000 bunches of Celery."[97] Proper sanitation was important, and garbage and receptacles were provided each household, with daily removals of trash and ashes performed by town workers. Anyone wealthy enough to own a horse had it stabled at the Pullman barns, so there were no animal odors or waste.

While public grounds were elaborately landscaped, the individual personality of private gardens was absent. Since workers leased, rather than owned, their homes, landscaping was performed by the Pullman Company in a manner similar to today's condominium arrangements. Front yards of workers' row houses, twenty to thirty feet wide, were maintained by the company. There were limited opportunities for workers to plant favorite flowers, or perhaps a tree to commemorate a birth or anniversary. Pictures of row-house backyards, where workers could plant their own flowers or vegetables, are scarce, and we can only speculate. Backyards were enclosed by high wooden fences with gates leading to the sixteen-foot-wide alley. Each brick home had coal and wood sheds and a walkway to the door, with little yard left for planting.

Nonetheless, workers were occasionally rewarded for and with floral displays. Ely notes, "One warm-hearted [Pullman company] official . . . wrote a note of thanks to the occupant of a cottage which was particularly well kept and ornamented with growing flowers. In another case he was so well pleased with the appearance of a cottage that he ordered a couple of plants in pots sent from the greenery to the lady of the house, with his compliments." Ely posits that the company's example sparked a desire for home beautification and "even in a flat of two rooms . . . one sees prints and engravings . . . and growing plants in the windows."[98]

According to Ely, no one regarded Pullman as a permanent residence. "We call it camping out," said one of his interviewees. "The desire of the American to acquire a home is justly considered most commendable and hopeful," Ely opines. Although Pullman ultimately began to sell certain properties, the nature of the predominant landlord-tenant relationship hardly inspired setting down roots. The impact on landscaping was apparent—why plant a tree if you plan to spend only a few years in a leased home? Perhaps the legacy of Pullman lies in the early absence of trees on private grounds, and the absence of individualized gardens, which show how important landownership was to late-nineteenth-century Chicagoans, a city teeming with immigrants seeking homes of their own.

Landscape Integrity

In 1907, by court order from the State of Illinois, the town of Pullman was sold to its people. Two years later, the city of Chicago acquired Arcade Park, which was formally transferred to the Chicago Park District fifty years later. Significant changes have occurred in the environs surrounding Pullman. Once surrounded by farmland and prairie, the area is now heavily industrialized. The Calumet Expressway currently forms the eastern border of the town, heretofore bounded by the scenic and tranquil shores of Lake Calumet. Lake Vista, the manmade lake that offered a beautiful view north of Arcade Park, has been filled in.

But the overall spatial relationship among public and private green space has been preserved. Visitors can grasp the sense of disparity between small row-house plots and larger public gardens. Trees still line the streets, now grown into maturity. Ironically, community gardening has become a popular hobby in Pullman. Where once public gardens were maintained by the Pullman Company, now community gardeners proactively plant and maintain this green space.

The Arcade Park Garden Club of the Pullman historic district, working with the Chicago Park District, maintains the flower beds and assists in beautifying Pullman's public parks, including Arcade Park, Pullman, Hotel Florence Historic Rose Garden, Fulton Field, and the 115th Street interstate ramp. The club has purchased land and created more public green space. It tries to achieve a historic look in public areas like Arcade Park, although time and budget constraints often dictate lower-maintenance plant choices. The club encourages homeowners to plant their own gardens, and offers stipends toward plant purchases for this purpose. Recently, many backyards have become even smaller as some residents have added either garages or rear additions to their homes. Thus, as in the past, save for some novelty vegetable plots or postage-stamp backyard gardens, residents look to the public gardens of Pullman for landscape beauty. But now their landscapes are truly in the public domain.

Brookfield Zoo

3300 Golf Road
Brookfield, Illinois
http://www.brookfieldzoo.org

Brookfield Zoo might seem an odd choice as a landscape destination, but it is a prime example of how Chicago was (and is) at the forefront of the nation's environmental movement. Not only does the zoo protect endangered species, but since inception, it has employed landscaping to enhance the experience for both its inhabitants and visitors. Designed with both formal and informal elements, it retains some of the earliest woodland remnants in the area. The first zoo in America to house animals in a "barless," or unfenced, environment, Brookfield Zoo pioneered engineering feats to showcase animals in natural settings. Landscaping around the zoo was an interesting mix of formal and naturalistic styles, and greatly contributed to the public's enjoyment and interpretation of the experience.

While animals may be the star attraction at Brookfield Zoo, plants are not simply wallflowers. The zoo's landscaping is historically important as America's first attempt to display animals in a natural environment. Brookfield Zoo's siting of animal displays, elaborately constructed "natural" outdoor grottos, and sensitive plantings demonstrated to early-twentieth-century Americans an alternative to the prevailing practice of containing animals in small, cramped cages. As noted in the Chicago Zoological Society's 1927 yearbook, "No one will leave the Garden pitying the 'poor caged animals,' for the simple reason that the bears, lions, wolves and others will inhabit barless enclosures! They can be observed in an outdoor setting of rocks, trees and shrubs where they will have ample room for play."[99] Landscaping was integral not only for overall artistic effect, but for pragmatic elements such as the design of natural-looking moats to separate animals from patrons.

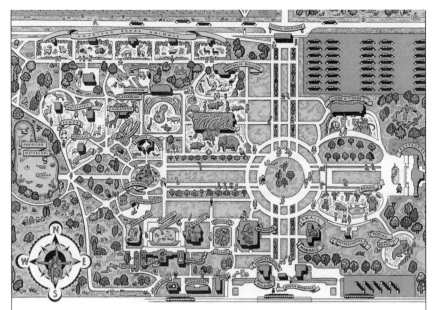

The formal symmetry of the allees leading to the grand fountain contrast with the naturalistic design on the west side of the Brookfield Zoo. The arrival of the automobile as a design consideration is evident in the provision for parking. From the author's collection.

Thanks to a major land donation from heiress Edith Rockefeller McCormick in 1919, the zoo was off to a promising start. The acreage was donated to the newly formed Cook County Forest Preserves, which would serve as the fiscal agent for the enterprise. The Forest Preserves had itself been recently created "to acquire … and hold lands … containing one or more natural forests or lands connecting such forests or parts thereof, for the purpose of protecting and preserving the flora, fauna and scenic beauties within such district."[100] Despite several false starts, the Forest Preserve District Act of 1913 was ruled constitutional in 1916 by the Illinois Supreme Court. As a new entity itself, the Forest Preserves appointed a small group of key individuals, including prairie architect Dwight Perkins, to validate the appropriateness of the Brookfield site. The Forest Preserves also had staunch allies for the zoo in its leadership. One month before his death, in his January 1921 annual public address, president of the Forest Preserves Peter Reinberg (a former Chicago-area florist and rosarian with one of the largest private greenhouses in the United States), predicted that the zoo would grow into "one of the most remarkable and attractive features of the Forest Preserve District."[101]

In February 1921, a blue-chip team was assembled to form the Chicago Zoological Society, whose goals included fund raising and obtaining legislative and tax support for the new institution. Many of these wealthy citizens were well versed in the art of landscaping, and placed a premium on tasteful landscape design in their own personal estates. Society member Charles L. Hutchinson, for example, was perhaps Chicago's leading patron of horticulture and the arts. Other familiar family names such as Armour, Brewster, Cudahy, Dawes, Donnelley, Insull, McCormick, Morton, Ryerson, and Wrigley were involved with the society, and had, for their own North Shore estates, commissioned landscape artists such as the Olmsted firm, Jens Jensen, and O. C. Simonds. The group did not lack in resources nor in appreciation for good landscaping. The zoo was a very high-profile undertaking, with the *Chicago Tribune* reporting, "The new zoological gardens are declared destined to be equal and probably the superior of anything of the kind in the world, taking rank with those of New York, London, Rome, Tokyo, and Bombay."[102]

Because of McCormick's large land donation, zoo organizers were able to develop a comprehensive plan, rather than piecemeal designs as often occurred in zoos where incremental donations were solicited. The landscape design itself was a collaboration among society members, the Forest Preserves, a German firm, Chicago-based architect Edwin Clark, and the Chicago landscape architecture firm of Simonds and West. Prior to designing the zoo, members of the society and the Forest Preserves visited a number of prominent European zoos and U.S. zoos in Cincinnati, Washington, Philadelphia, New York, and St. Louis for inspiration. Their consensus was that the barless animal exhibits in the Hagenbeck zoo in Hamburg, Germany, were worth consideration. Carl Hagenbeck's Thierpark, as it was then spelled, was begun in 1874 by an enterprising Hamburg fishmonger with a sideline wild-animal import business. His son, Carl Hagenbeck Jr., expanded the business to a park featuring fenceless animal exhibits separated from the public by moats. The Hagenbeck Tierpark (still operating today) opened to the public in 1907, and was much admired around the world.[103]

Clark's plan included a motor driveway encircling the park with small parking lots strategically placed for people to disembark and view the exhibits. (This drive was not implemented.) Landscaping complemented his buildings, which were constructed in an Italian Renaissance style, with clay-tile roofs and buff-colored buildings suggesting a country villa. The design was at once both formal and informal. The most striking landscape feature, both then and now, was the cruciform axes of two grassy malls with a large, circular fountain at the intersection.[104] The society report noted, "The Garden is divided approximately in half by an East and West mall, or avenue, one hundred and fifty feet in width and running from the refectory at the East end to the shore of the lake on the West. This wide street, for pedestrians only, has a formal arrangement of macadam walks, grass pots, trees, and pools, and has been planned along lines similar to the famous Tapis Vert at Versailles."[105] This formal mall arrangement was effective not only for pedestrian circulation, but also for providing many of the animals' exhibits with a sunny, southern exposure.

For five years between the founding of the society in 1921 and the ultimate passage of a referendum granting public tax dollars supporting the zoo in 1926, slow progress was made in shaping the grounds. Retaining the native trees of the landscape was a top priority, not only because the zoo was on forest preserve land, but also, no doubt, because of the influence of the Simonds and West firm. While landscape plans from this firm have yet to be uncovered for the zoo, their involvement is clear from correspondence in the zoo's archives. Roy West was the key contact with the fledgling zoo, and certainly shared Simonds's well-known love of naturalistic plantings.

The undeveloped site chosen for the zoo included prairie and oak savanna, with a few suburban cottages nearby. The Forest Preserves' chief engineer reviewed the McCormick tract before groundbreaking and observed, "At present the tract is, in part, covered with timber, there being much sturdy oak and linden, with a scattering of deep underbrush, which, coupled with its undulating condition, present many ideal features . . . The flat or present meadow areas may be easily converted into formal gardens, lending colorful beauty to the land, with its borders of majestic trees."[106] Reports submitted by the fore-man of the grounds crew in 1927 showed great efforts made to preserve native trees. In one two-week period in November, for example, the foreman wrote: "115 hawthornes, 35 cherry and 15 crabapple trees have been dug up and removed. Those found to be of suit-able size were transplanted in the wooded section of the property."[107] Hawthorns, cherry, and crab apples were the most frequent trees transplanted—not surprisingly, favorites of naturalistic landscapers like Simonds and West.

In total for 1927, Chairman Stanley Field reported that 376 cherry, crab apple, haw-thorn, and elm trees were transplanted, 452 "large dead forest trees and poplars have been fallen, eradicating all evidence of the former sub-division, 230 small trees were removed within the path of the malls, 155 dead trees of merchantable value for firewood were sawed to convenient handling lengths, and 275 stumps were blasted and removed."[108] Hundreds of cubic yards of clay were hauled to prepare the malls and roadways, and fifty cubic yards of cinders were collected for the walks and roadways. The grounds crew oc-casionally tried to save particularly large tree specimens, such as a sixty-foot elm trans-planted in 1928. The foreman reported, "Numerous difficulties had to be overcome, not only in its removal, but in securing equipment heavy enough to handle this enormous weight, but I feel we are to be amply repaid for our efforts."[109] The large transplanted trees not only added maturity to the new landscape, but also helped uphold the mission of the Forest Preserves to protect native growth.

Formal landscape elements were also planted during this time, such as the zoo's sig-nature double allee of maple trees that line the malls. Originally the planting consisted of a monoculture of Norway maples—with a first purchase in late 1928 of 246 trees at a cost of about $6,600.[110] Despite precautions of protective mulch and stay wires, of this initial planting only 26 survived due to poor drainage. By June 1932, all of the dead and doubtful trees had been replaced.[111]

Landscaping in this early period had to be sandwiched into the construction plan. With new buildings constantly under construction, planting schedules were juggled around finished exhibits, growing seasons, and availability of transplanted or bought stock. The grounds crew was initially limited to a small team working under the auspices of the Chicago Zoological Society. Nonetheless, sufficient progress was made that zoo director Edward H. Bean reported in March 1929, "A delegation from the Mississippi Chapter of the American Association of Landscape Architects [sic] visited the park and were much impressed . . . on the completed improvements and the landscape plan in gen-eral."[112]

With the devastating economics of the Great Depression, work progressed slowly between 1929 and 1933. Workers from the WPA assisted with landscaping and other tasks—laborers were paid $4.00 per day, and the grounds foreman was paid $175.00 monthly in 1932. But by 1933, work began again in earnest. The original McCormick do-nation was contingent on the zoo's opening by July 31, 1934, and the society also wanted to capitalize on the publicity and tourism generated by Chicago's Century of Progress Exposition in 1933 and 1934.

Inside and out, landscaping proceeded. As one of the northernmost zoos attempting to create barless, naturalistic exhibits, designers had to compensate for certain realities of Chicago's climate. Interior cages adjoining the outdoor exhibit area were built to pro-vide warm-climate animals with alternatives in the winter. Often, interiors of the exhibits were painted in thematic representations of the animals' environments; with desert scenes or tropical rainforests, for example, painted on the walls. But Roy West also was involved in planning interior landscapes. He developed plans and planting lists for the interior of the Primate House (then located just east of the present Reptile House and formal reflecting pool), which included bamboo and cane, tropical ferns, and other foliage plants. He proposed hanging baskets of *Asparagus sprengeri* ferns for a tropical effect. West also prepared sketches for the alligator and turtle exhibits—substituting tropical plants for look-alikes that would grow well in the light and temperature here. "Each one will have a quite dissimilar character, as the alligator exhibit is exposed to extreme heat and the plants that will thrive there will be for the most part rubber trees called *Ficus elastica*. The turtle room will be shadier and much cooler and there we can use a tropical evergreen that very much resembles a small pine, and various vines and other foliage plants."[113] With creativity and a bit of imagination, the interiors of the animal exhibits were made to look as natural as possible.

The zoo opened on Saturday, June 30, 1934, to much fanfare. Even after the opening, however, significant landscaping efforts continued as more exhibits were completed. In 1937, for example, over 5,000 shrubs, 140 trees and evergreens, and 1,200 balled and burlapped plants were installed. Plants were ordered from a number of local nurseries and were often delivered by rail to the nearby Riverside train depot. Today, the planting continues—even as the zoo roars forward into the twenty-first century.

Landscape Integrity

During the seventy-plus years that Brookfield Zoo has been open, many of the zoo's exhibits have been completely revamped to reflect advances in animal care. Landscaping around the buildings has changed accordingly. The zoo's signature axial views remain the same: maples still line the east/west and north/south malls. Many of the original Norway maples still exist, although they are coming to the end of their natural lifespan. As they get replaced, a more diverse mix of maples will be used to capitalize on new disease-resistant varieties and avoid the problems with monocultures.[114] (Most of the maples in the North Mall are not original as a severe storm caused major damage to these trees in the 1990s.)

In the so-called West Woods, the area between Ibex Island and Indian Lake, many original trees can be seen, including red oaks, hawthorns, cottonwoods, and some bur oaks that may be three hundred years old. This area reflects the naturalistic landscape that may have been present even before the zoo was laid out. The zoo has recently created other landscaped areas that, while not original, evoke the early prairie landscape. In the north entrance, an extensive prairie landscape was planted, and native-plant cultivars surround the mall's central fountain. Dragonfly marsh was created north of Indian Lake as a natural wetland. Indian Lake itself retains the "wild" feeling that serves as a great counterpoint to the formal pool at the east/west mall's opposite end. Throughout the zoo, many old trees can be identified: some big bur oaks in front of the "Be a Bird" exhibit and a giant elm near Café Ole on the zoo's north side. For the most part, the main pedestrian walkways, outdoor statuary, and uniquely constructed grottos remain true to the original plan.

Plantings near the inside and outside of the animal exhibits have undergone changes—primarily in response to better information about plant-animal nutrition needs and toxicity effects. It is a delicate balancing act to landscape exhibits in a naturalistic manner, yet be mindful of how the plant may affect the animal. There are river birch trees within the tiger's outdoor enclosure, for example, that would not be indigenous to Asia, but provide shade and preclude the animal from climbing. Some plants are not grown because their fruits, such as those of horse chestnuts, might be tossed by visitors into the animals' enclosures. Still, wherever possible, plant look-alikes are substituted for warm-climate plants that may be native to the animals' original habitats. Pinniped Point, an outdoor area for seals and walrus, with its many coniferous trees reminiscent of a northern climate, is a good example of landscaping that evokes the native habitat.

Cheney Mansion

220 North Euclid Street
Oak Park, Illinois
http://www.cheneymansion.com

Wealthy Chicago industrialists often gravitated to Chicago's North Shore to build their country estates, but nearby Oak Park afforded a suburban atmosphere with better proximity to downtown. While this near west suburb did not have the unique topography of the North Shore's ravines and lakefront, in the early 1900s it was a locus of independent thinking in architecture and the arts. Still close to the countryside, Oak Park offered homeowners a chance to develop gardens that took advantage of Chicago's ubiquitous level terrain.

Cheney Mansion, named after its last owner, Elizabeth Cheney, offers a charming view of an early-twentieth-century suburban garden of the well-to-do. While not as expansive as some North Shore landscapes, neither was it hidden behind long, winding driveways or secluded ravines. Like many of Chicago's early showcase properties, passersby could catch a glimpse of the gardens, and perhaps emulate them on their own smaller properties. The front and south side of the home contained major portions of the gardens, with a large, formal garden in back—consistent with the movement in the 1920s and 1930s to move showy gardens to the privacy of the backyard.

The provenance of this landscape is best picked up with the tenure of the Sharpe family. In 1913, real estate mogul Caswell Sharpe and his wife, Adeline Freer Sharpe, commissioned Charles A. White, a student of Frank Lloyd Wright, to design the Tudor revival home to replace their existing Victorian home. The estate comprised half a city block, and was one of the largest parcels in Oak Park. A complex of greenhouses once existed on the property, called the Freer Conservatory and Greenhouse, and was owned by Adeline's bachelor brother, Nathan Freer. Freer often opened up the greenhouses to the public to display many rare tropical plants. From its earliest days, the estate thus became known locally as a horticultural haven.[115]

A black wrought-iron fence, with brick piers matching the home, is original to this period.[116] The fence is largely ornamental, but may have kept out the occasional stray livestock that could well have inhabited Oak Park at the time. Generous spacing between the fence rails facilitated views of the garden within.

Andrew and Mary Hooker Dole purchased the home in 1922, after a brief period when

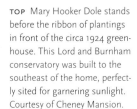

TOP Mary Hooker Dole stands before the ribbon of plantings in front of the circa 1924 greenhouse. This Lord and Burnham conservatory was built to the southeast of the home, perfectly sited for garnering sunlight. Courtesy of Cheney Mansion.

MIDDLE Perennials surround the water feature and sundial, both of which remain today. Courtesy of Cheney Mansion.

BOTTOM This wonderful formal garden is thought to have been sited directly east of the mansion during the Dole/Cheney period. Its perennial borders and exquisite art deco pergola are typical of the period. Courtesy of Cheney Mansion.

it was owned by a local developer. The Doles, another wealthy couple, were older newlyweds who enjoyed the estate for about twenty years, until they died in the 1940s. They must have been nature lovers, because in 1924 they added a Lord and Burnham greenhouse and in 1935 built a south-facing sunroom to better enjoy the garden views.[117] Mary Dole, a widow since Andrew's death in 1940, invited her niece, Elizabeth Cheney, to live with her in 1945. Upon Mary Dole's death in 1949, Cheney inherited the estate. Throughout the tenure of the Doles and Cheney, various gardeners were hired to maintain the estate.

Historic photos from about the 1930s show Mary Dole posed in her garden amid tall flowering perennials such as daylilies and delphiniums. Garden ornaments typical of the period, including a gazing ball atop a tall pedestal, were sited just outside the rim of the perennial bed in a sunny lawn. Taller canopy and understory trees ringed the outer edge of the property. The front of the house included pyramidal arborvitae mixed with what appear to be deciduous shrubs directly in front of the low brick wall that outlines the front entrance. A vertical wooden trellis leaned against the front of the house between the windows, but, in the photo, was not yet covered with vines.

A flagstone path led from the back of the house across the front of the greenhouse to its central entrance on the west side. Midsized, conical-shaped evergreens bracketed the entryway to the greenhouse, and a sandwich of ribbon plantings of shrubs fronted by annuals, edged by lawn, followed the curves of the flagstone path between it and the greenhouse. One photo, presumed taken in the backyard before the east half of the block was sold, shows an exquisite formal sunken garden, bordered by layers of perennials and shrubs, with a dramatic backdrop of an elaborate art deco trellis extending the entire width of the garden.

One of the most intriguing landscape elements is the water garden that still exists in the southeast portion of the garden. Enclosed with stratified stone, the pool is built into a hillock and is fed from a small piped waterfall. Early pictures with Mary Dole near the pool show it surrounded by deciduous trees arranged in an informal grouping. Trailing plants creep among the layers of stone, lily pads dot the surface of the water, and mounded groundcover and perennials soften the edges of the pool. Although the designer of the pool has not been positively identified, it certainly shows the sensibilities of a prairie landscape architect.

Landscape Integrity

When Elizabeth Cheney died in 1985, the Park District of Oak Park assumed management of the house and grounds, which were to function as a cultural center for the general public, according to her wishes. The two-acre landscape is a popular site for weddings, corporate events, and other public functions. Because of its conversion to a public property, many changes had to be made in plant material and layout, but the design intent is to keep the gardens period-appropriate and, through the use of plant materials such as grasses, native trees, and shrubs, to maintain a naturalistic landscape, where possible. Some portions of the garden are more formal, as appears to have been the case during the Dole period.

The sites of the buildings—the mansion itself, coach house, and greenhouse—are original. Front walkways to the main house and the ornamental fence have been preserved. The original greenhouse is still functional, and plants are grown in it for the Cheney grounds. The lily pool is fully operational, although large evergreen shrubs, possibly a 1950s or 1960s planting, have replaced the deciduous understory trees of the Dole period. Here and there on the grounds, trees such as the occasional ginkgo or an old mulberry growing through the east wooden fence, date to an earlier period. The original sundial and gazing ball's base pedestal also ornament the gardens. The overall effect, while not completely historically accurate, is of a gracious garden appropriate to a very prosperous family. The landscape, still visible to the public, continues to serve as an example of good taste in gardening.

SIX

FAIRS AND FLOWERS
Chicago Hosts a World of Fairs

Chicago may not be the Big Apple, but it has always been the core for fruit, flower, and vegetable fairs. With its enviable central location, Chicago has long played host to horticultural conventions and garden shows. At first accessible by plank road and then by boat, train, and plane, the city was a natural meeting place for national organizations such as the American Pomological Society, American Association of Nurserymen, Northwestern Fruit Growers Association, and more. The fairs not only drew professionals in the green industry, but were often open to the general public, greatly broadening their knowledge and familiarity with products of the field.

Chicago fairs began as humble displays of fruits and flowers hosted by local, then regional, gardening societies. The types of displays, premiums awarded, and rules for exhibitors evolved as Chicago's horticulture moved from subsistence agrarian to ornamental luxury. Ultimately, Chicago hosted two of the grandest world's fairs—where the horticultural displays drew hundreds of thousands of garden enthusiasts.

ANTEBELLUM FAIRS

In 1840s Chicago, groups of like-minded gardeners banded together in horticultural societies. These groups were the first to host fairs and flower shows where tips and techniques could be freely traded among exhibitors and participants. County organizations held the earliest fairs, which tended to showcase both horticulture and animal husbandry products. The Union Agricultural Society, of which Naperville nurseryman Lewis Ellsworth was an officer, held a fair in 1843, as did the Kane County Agricultural Society. In 1846, the Union Agricultural Society joined with the Mechanics' Institute to host the Cattle Show and Fair in Chicago. At this fair, horticulture played a small but visible role. Premiums were awarded for fruit (pears, peaches, plums, quinces, and grapes) and vegetables (potatoes, carrots, beets, onions, turnips, tomatoes, squashes, and pumpkins). The emphasis of this fair was decidedly utilitarian rather than ornamental horticulture: lectures were given on "grasses for western culture," indicating predominant interest in cattle-grazing needs. Silk and mulberry-tree culture were hot issues, attesting to some of the earlier horticultural experiments that proved unsuccessful.

By the next year, however, Chicagoans showed more interest in ornamental plants. On June 16, 1847, the CHS held its first Exhibition of Fruits and Flowers. This exhibition, and the four or five shows that followed under the auspices of the CHS, featured displays of prized flower and fruit specimens.

This first CHS exhibition exceeded members' expectations. It was not only a networking success for those in the trade, but also an educational opportunity for the general public. The fairs also reaffirmed the purpose of the fledgling horticultural societies. The Chicago Gardeners' Society held its first monthly exhibition in May 1859. Recent improvements in the growing railroad system now allowed even greater participation and expansive displays. The Rees and Ellsworth nursery hauled its plants for ten miles in a wagon and then brought them by the newly established railroad from Naperville, and, according to reports, the flowers still looked fresh. Still, this exhibition was a mixed success; although the displays were impressive, public attendance was low. The society blamed the local press, citing poor publicity.

Illinois state fairs, which included prominent horticultural displays, were traditionally held in Springfield; however, some early ones were held in Chicago. The first Illinois State Fair held in Springfield was not acclaimed by many Chicago horticulturists, who felt that the flower and vegetable exhibitions were lacking. Hoping to improve, the second Illinois State Fair was held in Chicago in 1854. John Kennicott urged exhibitors to focus more on natural specimens, or profitable or rare examples, rather than novelties. "The day of 'Monstrous Pippins' has gone by; and UTILITY, rather than SHOW, governs the present,"[1] he declared.

CHICAGO HORTICULTURAL SOCIETY'S FIRST SHOW

The first show sponsored by the CHS in 1847 featured then extraordinary specimens of ornamental flowers from the gardens of Chicago's elite and from professional nurserymen. Roses stole the show, with most displayed as cut-flower bouquets and others grown in pots. William B. Ogden showed a potted rose, 'Milledgeville', while John Kennicott, who exhibited eighty varieties, drew the most raves with his dark purple hybrid China rose, 'George the Fourth'. Four or five varieties of peonies were shown by various exhibitors, with the 'Whitleji', a Chinese white, and the scarlet 'Humei' getting the popular vote. Henry Thomas showcased his hothouse prowess with a display of cactus. Excepting some stalks of rhubarb, few vegetables were shown, but fruits, particularly Hovey's seedling strawberries and cherries, were in abundance. One of the few female exhibitors, a Miss Whiting, displayed a seven-foot-tall balm geranium.

The *Prairie Farmer* reported, "There was but a single Carnation in the room—a scarlet, shown by Samuel Brooks, Esq., one of the vice presidents of the society. The carnation is too difficult and capricious a flower for any but skillful and well prepared gardeners."[1] (Since carnations had but recently been introduced to America from England, even one entry of one carnation demonstrated the willingness of Chicagoans to experiment with new plants. Indeed, by the turn of the century, Chicago, and particularly the nearby Joliet area, became known for its carnation-growing industry. The Chicago Carnation Company of Joliet even wanted to name a light pink carnation, 'Sunbeam', after Chicago florist Edgar Sanders.)[2]

Local trends in floral arrangements were evidenced in the show. Flower-show rules dictated the shape and number of cut flowers, and the style of bouquets (round, pyramidal vs. flat). Special rules governed tulip displays, with a cup-shaped flower preferred. Exhibitors were cautioned that "all Tulips with lopping or dogseared petals [were] an abomination to a Florist."[3] Considering the crude preservation methods then available, that any flowers arrived in show condition seems prizeworthy in itself. CHS exhibition rules also provided insight to contemporary challenges for fruit growing. Flavor, size, rarity, and beauty were prized in fruits. A specimen's suitability to the soil and climate of northern Illinois was a key criterion, and an indication of early trial and error with local conditions. Early maturation for summer fruits, to accommodate Chicago's short growing season, and long keeping qualities for winter fruits were prized traits. Interestingly, foreign grapes were preferred and rewarded over natives for outdoor (vs. greenhouse) specimens.

1. *Prairie Farmer*, August 1847, 234.
2. James Nartshorne to Edgar Sanders, Chicago Carnation Company, September 22, 1900, Edgar Sanders Collection, Chicago History Museum.
3. *Prairie Farmer*, April 1849, 112.

Determined to show the region how to host a flower show, Chicagoans geared up for the 1855 Illinois State Fair, which was to be held in the Bridgeport neighborhood of Chicago. John Kennicott urged his colleagues to come and cited many inducements for exhibitors: excellent facilities on forty acres of "high, dry prairie and grove" on the south branch of the Chicago River—only two and one-half miles from the central business district. Railroads gave discounts on fares, and for those who couldn't find lodging, the City and Central Railroad provided for camping out, in a fashion that "will not be deemed a hardship by the hardy sons of the plow and shop." Kennicott reported, "To men of thought and observation, and the lovers of Nature's open volume, we offer a new feature at this Fair. Chiefly through the instrumentality of the Illinois Central Railroad Company; (you see 'corporations' *have* 'souls' in the West) the Society expects to make a show of the NATURAL PRODUCTS of Illinois, alone worth a journey of a thousand miles to see."[2]

In 1859, Chicago hosted its first national fair, the U.S. Agricultural Society's seventh annual exhibition. Kennicott was again a great booster for the fair, deeming it an educational opportunity and a sophisticated show. Although the timing coincided with the Illinois State Fair in Freeport, Kennicott urged readers to attend both: "The Chicago fair is to be no mere 'horse show,'—no great "buncombe display,'—as you seem to suspect."[3] He exhorted his colleagues to come to the U.S. fair: "It is scarce necessary to say that agricultural fairs are the only schools of demonstration open to farmers and country artisans of the prairie region."[4]

Fair visitors could learn quite a bit from these displays. According to the *Prairie Farmer*, cabbage "monsters"—one weighing forty pounds—were displayed by Pat Lannon of Bridgeport. Real estate developer S. H. Kerfoot showed some monster beets and a variety of white, red, and yellow onions. Critics agreed, however, that the show was stolen by the Tremont House exhibit of fruits and vegetables. The hotel's gardener, Levi Emery, was deemed a "scientific vegetable gardener." The *Prairie Farmer* reported,

> He exhibits between forty and fifty varieties of garden vegetables, grown near Chicago, on ground that is supposed abroad to be worthless.
>
> Mr. Emery said, part of the vegetables exhibited by him were grown on land where carrots, last year, would not grow; and, this year, he exhibits them two feet long. "But for the underdrains," said he, "we could not make this exhibition to-day" ... On this same land were grown Nansemond sweet potatoes, while, on the upland, none grew because of the drouth; cabbages, very fine ones; turnip beets as fine as we ever saw. We took off the first one we laid our hands on and wrapped our tape around it; it measured *two feet in circumference!*[5]

In this fair, the horticulture department had five subclasses: fruit and ornamental trees, pears and peaches, canned/preserved fruits, wines, and flowers. Prominent Chicago-area nurserymen and florists such as Arthur Bryant Sr., Samuel Brooks, Robert Douglas, and Lewis Ellsworth received awards in these categories.

Despite the beginning of the Civil War, the Illinois State Agricultural and Horticultural Society and the Chicago Mechanics' Institute jointly hosted the 1861 state fair in Chicago at Brighton, south of the city. This became a very con-

tentious fair, with accusations of mismanagement and political favoritism, and
with extremely unfavorable weather. Optimism initially prevailed for the fair's
success. The *Illinois Farmer* enthused, "Chicago is a good point for a great Fair; it
is not only convenient for our own State, as all the railroads lead there ... a city
like Chicago can furnish the accommodation and assistance."[6] As a state fair,
all forms of the region's products were to be showcased, including a separate
horticulture category with five subclasses: trees and fruit; canned and preserved
fruits; flowers and plants; native wines, cider, and vinegar; and floral designs and
bouquets. Judges included John Kennicott, Arthur Bryant Sr., Alice Kennicott,
and Charles E. Peck from the Chicago area.

Sadly, as the time of the fair drew near in September, preparations were not
complete and the weather turned frightful. "With the Fair came the rain. Two
weeks of unparalleled weather at this point, at this season of the year. The ground
was flooded—the exhibition was drowned." Horses and livestock escaped their
pens and ran amuck amidst the crowds on the eighty-acre fairgrounds. Buildings were flimsy with leaky roofs. Beer halls were plentiful, "flaunting their
prodigious signs in the most prominent places," and "frighten[ing] the moral
portion of the community, with the idea that this prided State Institution had
become a side show to a beer garden."[7]

The downstate *Illinois Farmer* concurred, deeming the fair a "bitter disappointment": "In ordinary times, the grounds at Brighton would not have been
so bad. With a small expense, these grounds can be made valuable. Our objec-
tion to them is, mainly the distance, want of facilities, the aroma from slaughter
houses, etc., at Bridgeport."[8] The editor of the *Republican and Telegraph* wrote
angrily: "We are surprised at the selection, by the Executive Committee of the
Society, of the 'Brighton Course.' The result will be perhaps but a slight injury to
the future interests of the State Society, but such an advertisement of the mud,
slime, and disagreeable effluvia of 'Brighton,' and the way to and from it, to all
the people of the State, that no man in his senses will ever desire to visit it hereafter ... The officers of the Society would be justified in not allowing another Fair
to be held at Chicago for all time to come."[9] "The Fair of 1861 is dead," eulogized
the *Prairie Farmer*. "Wrapt in its robes, not of a spotless white, but all bedraggled
and besmeared with mud and mire, it lies buried in the Slough of Brighton. No
jury was needed over its prostrate form. It died of a concatenation of diseases
arising from bad location and bad weather, terminating in the worst form of
anasarea."[10]

Learning from the mistakes of the 1861 Brighton state fair, the Illinois State
Horticultural Society held its first fair in 1862 independent of other state groups
in Chicago. The fair was located in Bryan Hall, conveniently located downtown,
and underwritten by horticulture patron Thomas B. Bryan. As society secretary
C. Thurston Chase noted in announcing the fair, "It is to be held at Chicago as

the point of easiest access to the greatest number—and, as the Fruit Emporium of the West, where producer, dealer and consumer may meet and confer."[11] In addition to this 1862 fair, despite the gloomy forecasts, Chicago hosted quite a few more fairs in the 1860s, having learned lessons from the 1861 Brighton fair. During the Civil War, Chicago was the site for two great sanitary fairs, run under the auspices of the U.S. Sanitary Commission (the forerunner of the American Red Cross). The 1863 fair, the so-called North-Western Soldiers' Fair, was only the second sanitary fair conducted in this country, after one in Lowell, Massachusetts. Organized and managed largely by women, the president of the commission was Ezra McCagg—not surprising, given that his friend Frederick Law Olmsted was general secretary of the national commission. The fair was held during the last week of October and first week of November in Bryan Hall, with opening addresses given by Thomas B. Bryan, the great benefactor of horticulture in Chicago.

A great variety of items were donated and displayed at the fair—needlework, fancywork, musical instruments, manufactured items, an original draft of the Emancipation Proclamation—all intended to raise monies for the war effort. Four display classes were related to agriculture and horticulture: evergreens for decorative purposes (wreaths, shields, stars, crosses), fruits (by the box, barrel, or basket), flowers and floral designs, and agricultural products. With great pomp and circumstance on the second day of the fair, "the great feature of this day was the Farmers' Procession of loaded wagons of vegetables . . . The procession of wagons paraded through the principal streets, thousands of men cheering it as it passed, and thousands of ladies waving their handkerchiefs at the welcome sight."[12] This regional fair drew exhibitors and visitors from all around the Midwest.

During 1863, the Second Annual Illinois State Horticultural Fair was hosted in nearby Rockford in September. This fair was extremely important to Chicagoans in that it showcased the horticultural prowess of John Blair, a Scottish-born landscape artist who would soon relocate and leave his mark on Chicago's parks and private estates. The Rockford fair was perhaps the first time in local history that landscape effects were used to display groupings of floral exhibits. Heretofore, most exhibitions consisted of cut flowers, bouquets, and potted plants set up in military rows on tables awaiting the verdict of show judges. In Rockford's Floral Hall, thanks to John Blair's design, the 100-by-50-foot building integrated evergreen pillars and evergreen boughs laced across the ceiling to give a forestlike effect. Swaths of green lawn on the ground were traversed with paths, further lending the impression of a garden. Edgar Sanders reported, "Directly on entering, a very large design struck the eye, composed of evergreens, flowers, birds [etc.], its shape being an imitation of a church spire, and four pinnacles, the former crowned with the American Eagle, bearing the stars and stripes . . .

Passing on to the centre was a noble fountain in full play, ensconced by many aquatic flowers, plants and animals, which, backed up, as it was by the flowering plants, made it a very attractive feature."[13] Creating an overall horticultural ambience presaged the trend not only in later regional garden exhibits and fairs but, to a greater extent, in world's fairs and expositions.

Chicago exhibitors at the Rockford fair were few; however, J. Asa Kennicott showed petunias, verbenas, pansies, Japan pinks, and Delaware grapes; and the firm of Williams and Wittbold displayed foliaged plants, verbenas, and "the only roses on exhibition." Jonathan Periam showed fifty varieties of apples, and Princeton's Arthur Bryant and Son displayed one hundred named varieties.[14] Chicagoans' somewhat low turnout is readily explained when the sheer number of fairs during September and October 1863 is considered. In addition to the

John Blair's Rockford estate designs attracted the attention of Chicago-area horticulture aficionados. Blair incorporated rockwork in many of his designs, including this 1880s grotto in Robert Tinker's conservatory. Tinker had married Blair's client, Mary Manny, whose grounds were connected by suspension bridge across the Rock River to the estate of her future husband. Blair's relationship with the Manny/Tinker family extended for decades beyond the 1860s. Courtesy of Tinker Swiss Cottage Museum, Rockford, Illinois.

Illinois State Horticultural Society's fair in Rockford, the Illinois State Fair took place in September and October, as did several nearby county fairs, including those of De Kalb, Kankakee, McHenry, Bureau, and Kane counties. Nurserymen and florists needed to be selective about which fairs were worth attending.

This 1863 fair also recognized two significant events: the death of John Kennicott and the land grant made by Congress to establish an agricultural and mechanical school. John Warder, renowned Cincinnati horticulturist and attendee at the society's meeting, eulogized Kennicott as "one of the truest friends of agriculture and the pioneer of horticulture in the West."[15] In recognizing the land grant, the society unanimously judged that a single independent scholastic institution be established, rather than multiple schools, an option being discussed in many circles. In terms of Chicago's landscaping, the 1863 fair was notable for introducing Rockford's landscape artist John Blair to Chicago's tastemakers.

SANITARY AND INTERSTATE SHOWS

Blair was chosen to landscape Chicago's 1865 sanitary fair, held downtown on Michigan Avenue. This marked the first time that exterior landscaping complemented a horticultural show in Chicago. Seedsman A. H. Hovey was chair of the horticultural committee and commissioned Blair to design a park that extended several blocks from the intersection of Washington Street and Michigan Avenue, where stood the 370-by-62-foot Horticultural Hall. Blair transformed the lakefront park into a fantasy garden filled with a Canary Island aviary, a fairy grotto, fountains, and long parterres of colorful flowers. Like other immigrant landscape artists of the day, Blair fashioned a garden reminiscent of his Scottish homeland. His manmade Lookout Mountain, the signature attraction of the park, was as out of place on the prairie as was the closely shaven lawn bounded by trim arborvitae hedges.

Nonetheless, Blair's design was well received by the public. The *Voice of the Fair*, a weekly newsletter published by fair management, enthused, "Imagine Central Park epitomized and intensified and you have the [Horticultural] Hall ... The entire circumference is fringed with cedars, deciduous trees are scattered here and there, and interspersed among them all varieties of evergreen found on the eastern continent. The remaining space is artistically laid out into meadows, lawns, ponds, flower plats, and broad gravelled walks."[16] Blair had already earned a reputation in town, having worked on such properties as Kerfoot's Lake View development. The 1865 fair brought him even more prominence, and helped stir the general agitation for better parks in the city. Nurseryman Edgar Sanders, in describing the sanitary fair's landscape, said, "It is a pity our city had not the benefit of such a man to improve its parks and cemeteries, if allowed the chance, they would not be the treeless, unkempt places they now are."[17]

In September 1865, Chicago also hosted the Illinois State Fair. Once again, the weather conspired to dampen the enthusiasm of the crowd. The *Prairie Farmer* reported: "Exhibitors denounced Chicago as the slop basin of creation, those from a distance enquiring if it always rained here, or remembering the 'big muddy,' declared that Providence was everything but favorable toward holding a State Fair at Chicago." Transportation to the fair was unreliable, with horse cars apparently getting stuck and running slow. Railroads like the Illinois Central gave reduced fares, but the tracks were too far from the grounds, "leaving passengers with a long distance to walk."[18]

As a comprehensive state fair (versus a horticultural society fair), the exhibits of course included far more than agricultural or horticultural products. But representation in these latter fields proved unsatisfactory. The *Prairie Farmer* called the exhibition of farm products and vegetables "more meager than we have ever before seen at a State Fair, indeed altogether unworthy a second rate county fair." Some exhibitors came from downstate and central Illinois, but from the Chicago region only A. H. Hovey, P. S. Meserole, and Charles Snoad of Joliet were in evidence. Neither was the *Prairie Farmer* impressed with the flower display: "The floral exhibition was pretty much like that in vegetables—very homeopathic—a disgrace to

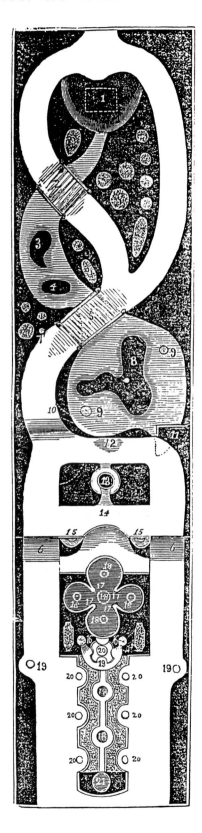

In developing the plan for one of Chicago's first landscaped regional fairs, John Blair created winding paths along the rectangular site. Much of the landscaping reflected the current era's fascination with fanciful flower creations, including Lookout Mountain (1), rocky water fountains (7), and rustic bridges (2, 5). Other features were a spring (3), an aviary (4), an island inhabited by eagles and cranes (8), fountains (9, 18), a rustic stand where bouquets were made (10), a bower where ladies could rest (12), a well (13), a terrace (14), a music stand (15), summerhouses (16, 21), statuary (17, 19), and vases (20). (Thanks to William Dale for identifying this design.) From *Prairie Farmer*, June 10, 1865, 462.

the florists of Chicago; and but for the show from abroad, would have been very meagre indeed." Of Chicago-area florists, only Sulzer and Brothers brought begonias and foliage plants. Edgar Sanders was notable for fuchsias, heliotropes, phloxes, dahlias, and verbenas, and others, like Lewis Ellsworth (Naperville), T. S. Shearman (Rockford), and J. T. Little (Dixon), displayed some plants. Female exhibitors included Miss W. S. Steele of Evanston (phloxes and pansies), Mrs. A. Bryant of Kankakee (floral designs and bouquets), and Miss Ada Kellogg and Mrs. Cargill, both of Chicago (other floral displays).[19]

After the Civil War years, local shows continued, with the *Prairie Farmer* sponsoring a series of strawberry exhibitions. The newspaper trumpeted the series of shows in its May 26, 1866, issue: "As this will be [the] first attempt at an exhibition of this kind in the Northwest, we trust that all interested in the success of small fruit culture in our midst, will unite with us in the endeavor to make it entirely worthy [of] the Garden State and the great fruit metropolis of the West."[20] These events, celebrating the newfound horticultural achievements in strawberry culture, drew good crowds in June 1866. Premium winners included Seth Wadhams of Cottage Hill[21] for best new seedling and Charles Reissig for best display of cut flowers. Ruth Hall, one of Chicago's early female garden writers, recognized the democratic potential of the show: "Never before were so many flowers brought into Chicago and sold as this season; and narrow back streets where the windows are garnished with blossoms, became bright and pleasant to the passerby."[22]

The Chicago Fire of 1871 disrupted countless lives and businesses. But the city did not take long to reaffirm its reputation as a central meeting place for the nation. A group of leading Chicago businessmen built the Inter-State Exposition Building, near the present site of Chicago's Art Institute, to host the first Inter-State Exposition from September through November 1873. The genesis of this exposition was said to have dated back to organizational efforts by the Northwestern Mechanical and Agricultural Association as early as 1869. Initially conceived to showcase Chicago as a convention city, the project took on even more importance as a symbol of rebirth after the fire. In a planning meeting in June 1872, civic leaders resolved to host a fair that would be "national and metropolitan in its scope . . . and that our friends from abroad be assured of a warm welcome."[23]

The fair was held near the "Burned District" in the heart of the city, during the two-year anniversary of the October fire. W. W. Boyington, Chicago's premier architect, designed the exposition building, a multitiered iron and glass structure with three prominent cupolas. A precursor to America's grand Centennial Exposition of 1876 in Philadelphia, the Inter-State Exposition was similarly organized into various departments reflecting the many fields of the blossoming Industrial Age. There were eight main departments, including categories for fine and liberal arts, minerals and manufactured products, raw materials, and instruments and machinery of the useful arts. Department F, "Products of the Farm, Orchard, Nursery, Garden and Greenhouse," included four subsections: fruits and vegetables; flowers, plants, ferneries, and aquaria; grains, seeds, vegetables, and dairy products; and ornamental pottery, rustic work, and birds in cages.

Setting the precedent for the format of many American world's fairs to follow, the railroad companies featured prominently in the displays. The entire fruits and vegetables section was organized by railroads, demonstrating produce raised in farmlands along each rail line. Fittingly for Chicago, the railroad hub of the nation, railroad barons saw the exhibition as a marketing tool to entice investors to buy land along their routes. Published reviews of the displays by the Atchison, Topeka, and Santa Fe Railroad, for instance, noted, "Altogether, this display presented an invitation to the agriculturalist full of promise in the yielding fullness of its soil, its genial climate and its cheap lands, that will attract an industrious population to the lands along the line of the road."[24] The Burlington and Missouri River Railroad displayed glass jars of corn, oats, rye, and other cereals. Written reviews noted, "The lands which produced these cereals ... are not exceeded in fertility, beauty, and any attractions or advantages of locality and soil."[25] Similar claims and displays of horticultural abundance were made by the Iowa Railroad Land Company, the Union Pacific Railroad Land Department, and a consortium of railroads represented by the State of Kansas.

Only seven exhibitors offered displays in section 2, "Flowers, Plants, & Aquaria." Most of these participants were local nurserymen and florists, perhaps attesting to the difficulties in the existing technology for refrigerated shipping. Live plants and cut flowers were still too difficult to transport long distances. The only out-of-town florist represented was James Vicks of Rochester, New York, who displayed gladiolus. Chicago-based florists and seedsmen such as William Desmond, Hovey and Company, and William T. Shepherd showed live and cut flowers, rustic work, wirework, and other fashionable garden ornament. Edgar Sanders was there, of course, with a large display of annuals including many new varieties—visible signs of horticultural advances. According to a contemporary account, "It would be curious to note the changes that have taken place in the florist business within seventeen years [presumably the interval between the exposition and when Sanders started his business in Chicago]. In that time all our present showy-leaved bedding plants, such as the coleus, achyranthus, centaureas, tri-color geraniums, etc., have come into existence."[26] Sanders was complimented on displaying the latest in coleus and other new varieties.

The Inter-State Exposition Building was used frequently over the next several years for multidisciplinary fairs. Although the building readily accommodated horticultural products, participation was nonetheless limited. In 1879, a writer who styled him/herself "Horticulture" noted, "Although costly green-houses are attached to our Exposition Building, and every convenience is at hand for the purpose of exhibiting choice collections of plants, flowers, and floral designs, this always-attractive department is sadly neglected, and hence a disgrace to our city." Horticulture theorized that the cost of displaying perishable items discouraged most florists and nurserymen, and that they should be enticed by meaningful premiums.[27] The Chicago Florists' Club, which held its own flower shows beginning in 1888, did award premiums, and boasted a grander display.

Twenty years later, in 1893, the Inter-State Exposition Building was torn down to make room for the Art Institute, built in conjunction with Chicago's 1893

TOP Broad-leaved and spiky-leaved plants arranged in a cluster were part of the horticulture display in the 1874 Inter-State Exposition. From the author's collection.

BOTTOM A multitiered water fountain represented some of the garden statuary and ornamentation at the Inter-State Exposition. Ca. 1874. From the author's collection.

Columbian Exposition. This world's fair, sometimes called the Great American Fair, began in discussions among principal directors of the Inter-State Exposition Company in 1885. The movement for a world's fair had been contemplated earlier, in other prominent venues, and would continue to build momentum until 1890 when, through U.S. congressional approval, Chicago won the right to hold the next world's fair. It was a defining moment in Chicago's history, and landscaping and horticulture played a critical role.

Also in 1890, the CHS was reinstituted, with its membership counting many of Chicago's leading businessmen, some of whom held important positions on world's fair committees. Representative industry captains in the CHS in 1892, for example, included W. C. Egan (son of horticulturist W. B. Egan), Philip D. Armour (meatpacking magnate), Charles L. Hutchinson (long-time patron of the arts and horticulture), William Rand and Andrew McNally (of mapmaking fame), and Edward G. Uihlein (brewer). Professional horticulture or landscaping members included J. A. Pettigrew and Frederick Kanst (Chicago parks) and J. C. Vaughan (seedsman). Unlike the earlier 1840s incarnation of this society, there were decidedly more patrons than practitioners.

The focus of the society shifted as well—from technical information exchanges and seminars to hosting flower shows. The shows were conducted in various venues, from the Armory Building at Michigan Avenue near Madison Street, to the Coliseum, to the Art Institute. Flower arrangements dominated these shows, and the list of judges for these entries showed the emerging importance of women in gardening, as well as the shift of the CHS membership to a more elite group. In 1892, for example, the judging committee was headed up by Mrs. Potter Palmer, and her fellow committee members included wives of industry titans: Mrs. S. W. Allerton, Mrs. T. B. Blackstone, Mrs. John Cudahy, Mrs. J. J. Glessner, Mrs. H. N. Higinbotham, Mrs. Charles L. Hutchinson, Mrs. Moses Wentworth, and Mrs. C. T. Yerkes. Horticulture had definitely arrived as a sign of culture in Chicago (plate 11).

COLUMBIAN EXPOSITION, 1893

Never in Chicago's history had the landscaping at a fair taken on such importance as at the 1893 Columbian Exposition. With the eyes of the entire world on them, Chicagoans wanted to project an image of a cosmopolitan, industrious city. Leading experts were appointed for the fair's horticulture and landscaping. Frederick Law Olmsted and his associates were commissioned to design the fair's overall landscaping. Olmsted, the nation's reigning king of landscape architecture, had created previous designs for the South Parks before the Chicago Fire: these would now be revisited with the intent to create a permanent park in Jackson Park, the main site of the fair. Complementing Olmsted's grounds plan,

Designed by William Le Baron Jenney, the Horticultural Palace offered seasonal displays of floral interest both inside and out. The formal plantings adjacent to the palace contrasted sharply with the naturalistic designs of the Wooded Island across the lagoon. 1893. From the author's collection.

William Le Baron Jenney designed the Horticulture Palace. Jenney and Olmsted had collaborated on other projects, notably the nearby Chicago suburb of Riverside. Although better known for his architectural works,[28] Jenney landscaped a number of public and private properties in Chicago, and was a leading landscaper in his own right. To provide indoor and outdoor displays of horticulture, nurserymen, seedsmen, and landscape artists from around the world were invited to participate in the fair.

In sheer size alone, the Columbian Exposition eclipsed previous world's fairs, both in America and abroad. More than six hundred acres of undeveloped swampland were rapidly transformed into a green park. The classical White City architecture of the fair, which often receives top billing over the landscape, influenced urban planning and design for years. In fact, landscaping at the fair was equally influential as a key component of the subsequent City Beautiful design movement that emphasized the use of green space. The landscape department of the fair was created on December 15, 1890, and work began in earnest a month later. Nurseries were established on the Wooded Island and Midway Plaisance to cultivate the fairground plantings. Lagoon islands were created, and twenty-seven carloads of semiaquatic plants were brought from nearby Lake Calumet for planting along the shoreline. Several hundred thousand trees, shrubs, perennials, and annuals were ultimately planted for the fair.

The fairgrounds occupied an irregular, triangular-shaped piece of shoreline. Although Olmsted developed the general landscaping plan, individual state and country buildings were often landscaped by private contractors, many from Illinois.[29] Choice areas around the Horticulture Palace and Wooded Island were in demand, and leading nurseries often won planting rights. This, among other competing demands from exhibitors, hampered Olmsted's ability to create a harmonious design for the fair in its entirety. The use of water as a thematic element—in the lagoons, canals, and Lake Michigan itself—was critical to Olmsted's design.

Formal elements of the fair, such as the Grand Basin and Court of Honor, received harmonious landscape treatment, with mass plantings of bulbs and shrubs. Wide expanses of lawn offered simple relief to the dazzling white brilliance of the Beaux-Arts buildings. But Olmsted's most treasured landscape element, and the most popular among fair visitors, was the naturalistic Wooded Island. The island (still extant south of today's Museum of Science and Industry) was accessed through a number of bridges from the main exposition areas. Olmsted's oasis at the epicenter of the fairgrounds was designed as a respite from the sensory overload of the other exhibition buildings. Surrounded by lagoons, the Wooded Island was fringed with native plantings of willows, sumac, wild roses, cottonwood, honey locusts, and silver maple. Judicious use of color in the outer border was achieved through minor plantings of wild verbena, sunflowers, coreopsis, goldenrod, and liatris. Winding paths traversed the island in counterpoint to the more formal architectural features of the rest of the fair.

Contrary to Olmsted's wishes, the interior of the island was consumed by a patchwork of floral displays showcasing exhibitors' wares. Formal plantings such as the rose garden and stiff exhibits of "show" flowers in other beds were inconsistent with Olmsted's ideal of naturalistic plantings. Only the outer borders and curved walkways remained true to his plans. The Japanese Ho-o-den temple and garden on the northern part of the island was somewhat sympathetic to Olmsted's design in its overall contemplative spirit. It was not, however,

2940. The Virgin Fountain, German Exhibit, Horticultural Hall, Columbian Exposition.

Inside the Horticulture Palace, visitors could view displays from countries around the world. Here the German exhibit included statuary and a fountain. 1893. From the author's collection.

FLOWERS AT THE FAIR: MONTH BY MONTH

The opening and closing dates of the 1893 World's Fair capitalized on Chicago's short growing season—from May to October.[1] During these months, the fairgrounds never looked exactly the same from day to day, as seasonal blooms varied and new shipments of fruits and flowers arrived daily. In June, pansies and rhododendrons took center stage, later replaced by anemones, larkspur, columbines, pyrethrums, Iceland poppies, and forget-me-nots. Most of these plants had been started from seed in the spring of 1892 and nurtured in greenhouses over the winter. Later in the month, the hybrid pyrethrums peaked, and Oriental poppies nodded in jaunty suits of red and orange.

The rose garden on Wooded Island came into its prime in July, but the first bloom was short-lived. Tender roses on the west side of the garden suffered the most from scorching sun. Several thousand gladiolus among the roses showed promise of blooming soon. Four beds of clematis in the center of the garden thrived in the full sunlight and outperformed all others on the fairgrounds. Hollyhocks, in delicate colors of peach pink, sulfur yellow, cerise, and pinkish mauve, made a fine show. Larkspur, foxgloves, and harebells graced the garden bed borders.

The cold-storage building suffered a fire on July 10, and thus displays of fresh and seasonable fruit were not impressive. That month, lemons and oranges in California's interior courtyard garden drew great interest in the Horticultural Palace.

German and English beds of phlox in mixed colors were stunning in August. Around the Japanese garden, Japanese iris that had been planted late began to thrive. Unusual varieties of morning glories, with pink, white, and reddish purple hues, drew attention. A good show of lilies, particularly *Lilium auratum* and *L. longiflorum*, decorated the island. In the Pomology Department, New York's showing of 175 varieties of gooseberries and 20 varieties of currants caused a stir. Illinois began showing apples from the crop of 1893, including Golden Sweet, Pennock, Summar Pearmain, and others.

Nine weeks of unrelenting dry weather took its toll on September's display. Trees and shrubs were scorched, and the dahlias suffered. By the third week of September, roses were in poor condition, many of them badly mildewed. Nonetheless, phloxes continued their fine display, and the gladiolus in the rose garden came into flower. Asters grown in the German bed were the best on the island. The cannas on the east lawn of the Horticultural Palace were very striking in September. Among those in front were 'Mme. Crozy', 'Alphonse Bouvier', 'Chas. Henderson', and 'Paul Marquant'.

October brought a harvest of fruit to the Horticultural Palace. Later in the month, preparations began for the Chrysanthemum Show, which noted horticulturist L. H. Bailey proclaimed "the best and biggest flower show yet held in the West." A horticultural world congress was held with participants from around the world, closing the fair with one of the best ever displays and exchanges of horticultural knowledge.

1. Sources for this feature were L. H. Bailey, *Annals of Horticulture in North America for the Year 1893* (New York: Rural Publishing Co., 1894), and *Gardening*, June–October 1893.

truly representative of that country's horticulture, and tended to use stereotypical landscape elements.

The Horticultural Palace also displayed floral and vegetable exhibits from nurserymen, states, and foreign countries. The promenade gallery on the second floor of the dome showcased herbaria, models of villa gardens, pressed plants,

FAIR OBSERVATIONS FROM A CHICAGOLAND NURSERYMAN

Most Chicago horticulturists visited the fair, but for those who couldn't make it, written accounts were provided in CHS meetings. The 1894 *Transactions of the Horticultural Society of Northern Illinois* included this essay by Princeton nurseryman L. R. Bryant.

> None attracted more attention, or were given more space, or did as much to beautify the White City as Horticulture in all its forms . . . The wooded island was, indeed, one of the most pleasing features of the fair and stood as a worthy exhibit of the landscape gardener's skill. In the heat of summer it furnished most delicious rest and coolness to the weary sightseers. The oaks and other forest trees stood here and there in groups . . . I doubt not that on the long settees, under their protecting shade, occurred many a romantic episode . . . The mass and frequent clumps of shrubbery, . . . lined the shores on all sides . . . Often, passing at a distance, I have stopped to admire the peaceful scene—such a contrast to the hurrying crowd that passed and repassed at my side . . . The old log cabin of Davy Crockett, scarcely visible through the foliage, seemed to add forcibly to the reality of the scene. In the center of the island was situated the famous rose garden, where fifty thousand rosebushes yielded their fragrance during the early morning hours. Everywhere, lining the walks on either hand, were beds of flowers of every kind and description. Even the well-known sunflower, towering high above its companions smiled sweetly on the passer-by. The wooded island was full of treasures for all . . . [W]e consider the most delightful place . . . the fruit section of Horticultural Hall . . . Here was elysium itself . . . Covering 1,600 square feet of space, the luscious fruits of Illinois were displayed on two long tables.[1]

1. L. R. Bryant, "Observations at the Fair," Transactions of the Horticultural Society of Northern Ilinois, 1894.

and photographs of showcase gardens. The northeast gallery, with more foreign exhibits, had a greater variety of plants. A middle island of staghorn and tree ferns (some nearly sixty feet tall) separated plantings on the east and west walls of the gallery. Japan's exhibit of dwarfed trees drew attention, and German roses and azalea, Mexican cacti, and Great Britain's cannas and begonias were crowd pleasers.

Pomological displays were west of the ornamental plant curtains, separated by interior courts (plate 12). A California orange orchard was in the north court, while a German wine building and lily tanks adorned the south court. Viticulture products from around the world were displayed in the south court, and seed displays and gardening implements were shown in the north pavilion. A masonry lily tank with forty species of water plants graced the front entryway of the Horticultural Palace. Outlined by extensive plantings of pansies in spring and cannas in fall, the lily tank made a graceful transition from the Horticulture Palace to the lagoon and Wooded Island to the east.

Foreign nations were well represented at the fair—a coup for a city recently considered a frontier outpost. In total, twenty-three foreign exhibitors brought fruit, grain, and plants from around the world. To simplify logistics, foreign nations tended to favor nonperishable exhibits. Still, several plucky countries brought fresh fruits or live plants. The apples from New South Wales, for example, were fifty-two days in transit by land and sea. Of the nine exhibitors who braved weather conditions on the Wooded Island, Germany had the most displays (ten), followed closely by France (nine). The Netherlands and Great Britain also had a good showing.

Even after the fair closed, achievements in horticultural technology and landscape design lived on. Over six hundred new plant introductions were made to the American trade in 1893. Technological advances in watering and irrigation and other gardening tools were displayed. The World's Horticultural Congress, a meeting of notable professionals from around the globe, improved information and knowledge sharing among the nations of the world. A few days after closing, by preordained plan, a match was set to the White City, and an all-consuming fire erased all traces of the Great American Fair. The Fine Arts Palace, now Chicago's Museum of Science and Industry, is the only restored building. Yet, like the tough, resilient plants that emerge victorious over mighty prairie fires, the Wooded Island remains unscathed.

GARDEN CLUB SHOWS

The Columbian Exposition set the standard for subsequent world's fairs in America and across the globe. As a fair city, Chicago continued to be popular for conventions and meetings among professionals in the green trades. Horticulture and agriculture fairs, however, tended to be regional for the next forty years. Flower shows, with displays of floral arrangements, gained prominence as women's garden clubs assumed greater importance. The CHS sponsored annual flower shows throughout the first decade of the twentieth century. Such shows were always dependent on funding from wealthy patrons. Occasionally, a show was cancelled; in 1903, according to the *Chicago Tribune*, "petty jealousies and intestinal strife among the members of the Horticultural Society" caused the show to be abandoned. Beginning in 1927, the Garden Club of Illinois hosted a number of flower shows in downtown Chicago. Entrants were typically garden enthusiasts from local suburban garden clubs, although professional nurserymen sponsored the shows through advertising and booth exhibits.

Garden clubs founded largely by Chicago-area women became prominent in the early 1900s. As an outgrowth of the Columbian Exposition's City Beautiful movement, the clubs often focused on civic improvement for their local communities. Flower shows, hosted in a public meeting space or member's private estate, soon became popular. Shows tended to display attractive arrangements of flowers and prized specimens, rather than new hybrids. As such, the shows

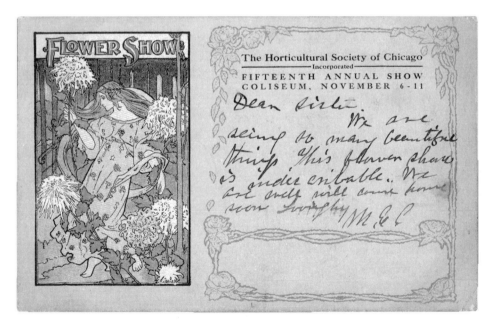

TOP LEFT The Coliseum was the site of many local and regional flower shows in the early 1900s. From the author's collection.

TOP RIGHT The CHS elevated horticulture to a cultural event by hosting many of their shows at the Art Institute along with musical programs featuring symphonic ensembles. 1913. From the author's collection.

BOTTOM The CHS shows prompted this visitor to write, "This flower show is indescribable." The CHS and garden clubs hosted shows in the early 1900s. Ca. 1905. From the author's collection.

were less for the scientific community than for those interested in floral aesthetics. Local club meetings were social gatherings, with many clubs requiring sponsors for membership.

The impetus for the centralized Chicago Garden and Flower Show beginning in 1927, and indeed for the formation of the Garden Club of Illinois, can perhaps be traced to the arrival in Chicago of John Servas, the promotion manager of the Hotel Sherman. Servas had previous success with the International Exposition Company, which operated, among other extravaganzas, the New York Flower Show. He wished to bring events of this scale to his new place of business, the Hotel Sherman.

According to Servas, in 1926 he invited Chicago-area garden clubs to discuss the possibility of a major show, and from this nucleus of nearly thirty clubs, the Garden Club of Illinois was formed. Beginning in 1927, the shows were increasingly successful, growing from thirty participating clubs in 1927 to fifty-three in 1928 and sixty-four in 1929. By 1930, the show had outgrown its space in the Hotel Sherman, and was moved to the newly opened Chicago Stadium, to the Merchandise Mart in 1931, and then to a large space at the Palmer House in 1932. The 1931 show was the first under the complete control of the Garden Club of Illinois. Servas noted, "By staging the show under its own control, the Garden Club of Illinois places itself in a dominant position in the matter of the flower show in Chicago. It has achieved the confidence, good will and support of the trade, press and the public."[30]

Giving further significance to the role of fairs, in 1924, the University of Chicago hosted the First Annual School in Fair Management, a pioneering curriculum. Fairs—local, county, and state—of many different themes had become a major business. There were an estimated 2,400 fairs in the United States at that time,[31] and expositions were recognized as an industry unto itself. Together with the International Association of Fairs and Expositions, the university sponsored this five-day workshop at its Hyde Park campus. As R. J. Pearse, a fair designer and speaker at the seminar, noted, "How many of us realize that we are standing on more or less hallowed ground, and that there should be just a little bit of sentiment connected with holding this first School of Fair Management on the grounds that at one time ... was held probably the largest, most magnificent and most attractive exposition that has ever been held in the history of the world. I refer to the World's Fair in 1893."[32]

The school offered lectures on fair accounting, advertising, public subscriptions for funding, building design, corporate organization, and more. Many of the lecturers noted that horticulture and agriculture were bedrock exhibits of most fairs. The school also recognized the major changes in fairs' audience—from producer to consumer. One speaker, J. C. Simpson, vice president of the Eastern States Exposition, proposed a shift in thinking from exhibitor to visitor. Old exposition models offered premiums for exhibitors, in the hopes of drawing more varieties of produce for the general edification of other growers. "This policy of the fair was sound at the time when the question of determining best varieties was paramount to quality and production," Simpson noted.[33] In the fairs of the 1920s, he opined, the visitor should be more engaged and edified—and tastefully displayed arrangements of fewer varieties might be more beneficial than of hundreds of novelty varieties. Simpson also suggested the heretofore heretical notion that displays of nonregional produce, such as oranges and grapefruits in the Midwest, would benefit local exhibitors and not their competition.

CENTURY OF PROGRESS, 1933

The growing professionalism of fair management was never more apparent than in Chicago's last world's fair, the 1933 Century of Progress (COP) Exposition. Celebrating one hundred years since the city's founding, the COP fair was among the few international American fairs to pay for themselves. Fair organizers began planning in 1928 and, despite the severe economic depression, again created one of the greatest American expositions. The fair was so popular that it continued over two growing seasons, during 1933 and 1934. The sophistication of the landscaping and horticulture was on the scale of the Columbian Exposition, and clearly highlighted changes in both fields through the intervening forty years.

The 1933 fair was alike in many ways to its 1893 predecessor. Both fairgrounds were created from scratch on the south lakefront under particularly aggressive time frames. Both suffered from surly weather and hostile growing conditions. The untimely death of a key landscape architect also affected each fair. (Ferruccio Vitale, the chief landscape architect in 1933, died in February 1933, and Henry Codman, Olmsted's right-hand man, died before the Columbian Exposition planning was complete.) Many of the fair exhibitors, like Vaughan's Seed Store, Lord and Burnham greenhouses, and Philadelphia's Henry A. Dreer nurseries, were represented in both fairs.

Despite these similarities, in the four decades between the two fairs, basic plant culture had been mastered. It was less important to showcase individual plant specimens and novelties. Fruit and vegetable products, so prominent in the earlier fair, were very minor elements in 1933. Instead, garden design theory and holistic garden compositions became the COP focus. By 1933, the modern "outdoor living room" eclipsed the kitchen garden so important to the Columbian Exposition.

The Beaux-Arts White City of 1893 gave way to sleek edifices awash in electric rainbow hues of color. The asymmetrical design of the 1933 fairgrounds encouraged eclecticism in buildings and landscaping. Difficult soil conditions were particularly taxing in this fair. Thousands of cubic yards of fill were brought in, and the fairgrounds were built on entirely new, manmade land. One writer exclaimed, "Lovely gardens growing in pure washed sand! Impossible you say? . . . Yet it was on washed sand that the beautiful exhibition gardens in connection with Horticultural Hall had to be grown. The early spring of 1933 found the site of the gardens a barren, sand expanse, anything but interesting. Here in a period of about two months beautiful lawns, flowers, shrubs and trees would have to be growing."[34]

The challenge fell to landscape architect Ferruccio Vitale, an Italian-born artist with an East Coast office. Vitale had been commissioned for a number of estates on the North Shore, and through those connections, plus his national reputation, was awarded the position. After Vitale's untimely death, his associate, Alfred Geiffert, assumed the post. Although it is somewhat disappointing that East Coast landscapers were hired for the top position when there were so many qualified landscape architects in Chicago's backyard, at least the advisory committee included local landscaping greats such as Jens Jensen, Franz Lipp, and Roy West. Significant input was also received for the displays in and around the Horticulture Building from local nurserymen and seedsmen, including Otto H. Amling, Leonard H. Vaughan, and Swain Nelson.

The north end of the fair site, anchored by well-established institutions such as the Field Museum and Adler Planetarium, was already landscaped, but the rest of the fair site was unimproved. The fairgrounds stretched for a narrow three miles, sandwiched between the lakefront on the east and the Illinois Central Railroad tracks on the west. Two lagoons on the north end of the grounds and, of course, Lake Michigan provided water features.

Over 1,600 trees of considerable size were planted in winter and spring of 1933 to provide a mature-looking landscape. About 25,000 shrubs, 24,000 lineal feet of hedging (e.g. privet, forsythia, and barberry), and 2,000 vines were planted. Flower beds, including beds of ageratum, marigold, petunia, salvia, heliotrope, begonia, and geranium, covered 75,000 square feet and complemented the vivid hues of the rainbow-colored buildings.[35]

While Vitale and Geiffert had responsibility for the overall design of the grounds, sponsors of individual buildings contracted separately for their own landscaping. Corporations, more so than individual state buildings, hosted many of the displays. Where once railroads had commandeered exhibit space (at early fairs such as the 1873 Inter-State Exposition), now the automobile had a major presence. General Motors' grounds had a wide plaza landscaped with hedges and lawn. The initials *GM* were sculpted in plant material in the center of the lawn. The Firestone building had a long reflecting pool in front of it with a rainbow-hued "singing" fountain. Ford's display included terraced lawns with formal appointments of evergreens and conifer hedges.

Some of the lessons and thinking expressed at the 1924 School in Fair Management were evident at this fair. School lecturers had observed that previous exhibition buildings used windows inappropriately—exhibits were often backlit and couldn't be seen well, and copious windows consumed valuable display space. With this thought in mind, and in deference to emerging modern styles of architecture, the 1933 fair buildings were windowless. Without windows, there was no need to consider garden views from within the buildings. Landscape designs could focus on the garden experience from the outside only. Landscapers also experimented with night-time lighting in the gardens, and achieved very dramatic effects with tree silhouettes painted in shadow against the stark, modern buildings.

One such modern building was the sleek, sculptural Horticulture Building, built by C. S. Colinan. This windowless structure on the east side of the south lagoon housed seasonal flower shows and dioramas. These garden dioramas were living compositions of plant material, realistically arranged in front of painted murals of natural scenery. Built into recessed, glass-fronted showcases, the dioramas brought the world to Chicago even as the Great Depression threatened to keep everyone home.

Each week, indoor flower shows were held in the Horticulture Building. The lineup for 1933 included June's peony shows, July's gladiolus show, August's water plant show, September's dahlia and autumn harvest shows, and October's chrysanthemum show. Shows in 1934 were similar, and included carnations in June, orchids in July, lilies in August, cut roses in September, and a grand finale in October. Many of the national specialty flower societies had recently come into existence, and this fair was a prime opportunity to show their wares. The American Gladioli Society had displays of over 375 varieties and 200,000 bulbs on the southeast and southwest sides of the north lagoon and along the avenue south of the Hall of Science. The Central States Dahlia Society hosted one of the largest dahlia shows near the Kohler Building.

Taking another page from the 1924 School in Fair Management, the horticulture exhibits of the COP fair were much more geared to visitors. Whereas the collections of plants featured in earlier fairs, including the Columbian Exposition, were arranged for a plantsman or botanist, the COP gardens were designed for the average homeowner. East of the Horticulture Building, over twenty gardens gave homeowners new ideas for their own backyards. The outdoor displays changed between 1933 and 1934, but all were built to the scale of a typical suburban or city backyard. Some were designed with a nostalgic theme, for example, the Victorian Garden and the Old Mill Garden (plate 11). Some captured the latest garden trends, such as the Water Garden and Alpine Garden. Others recalled the classics, such as the Formal Italian Garden and Rose Garden.

Other gardens showed how horticulture, too, benefited from the "Century of Progress." Two exhibitors interpreted the modern garden in 1933. The Men's Garden Club of Aurora (a western Chicago suburb) made extensive plantings in triangular, pizza-shaped beds with annuals such as verbenas, grandiflora, purple petunias, zinnias, and marigolds. Vaughan's Seed Store's version of the modern garden used a sharply defined "saw-toothed" bed of deciduous shrubs, sandwiched between a grass path and a bed of ribbon plantings. Designed by local landscape architect Kenneth W. Bangs, the central motif of this garden was a circular pool, enclosed by four apple trees. Turf walks, bordered by an arborvitae hedge with occasional plantings of *Syringa japonica*, led to the pool.

Reflecting the efforts of the conservationists and prairie landscape architects who attained prominence since the Columbian Exposition, there were two interpretations of an informal garden in 1933. Charles Fiore's second display, in counterpoint to his historical Italian garden, used a harmonic composition of *Viburnum carlesi*, maples, and white lilacs for a subtle look. Thomas Lynch of Winnetka used native plants around a naturalistic pool. Using stratified stone such as that popularized by Jens Jensen, Lynch's pool was surrounded by red dogwood, flowering crabs, and sumac. In 1934, there were no less than four gardens labeled "informal" by different exhibitors, including one by a consortium of groups from Joliet, Illinois.

Other indications of the role Chicago's conservationists played in the horticultural field were demonstrated by some of the dioramas and an exhibit by the Cook County Forest Preserves. The "Indiana Dunes Landscape" diorama showcased the newly preserved lakeshore ecology with natural plant material such as phlox, yellow lady's slippers, fireweed, and rare fringed gentians. "Ferns of the United States" had over seventy native species from all over the country. Chicago's forest preserves, amounting to over thirty-three thousand acres in 1933, were one of the greatest signs of horticultural progress since the Columbian Exposition. The foresight of such luminaries as Jens Jensen, who actively lobbied to protect the vanishing countryside, resulted in the permanent preservation of woodland and prairie around three sides of the city. This Chicago innovation in conservation was celebrated in "A Bit of Forest Preserve." Built atop two sloping hills with an intervening valley, the display included a rough-hewn footbridge over a rock-strewn stream. The scene was framed by native trees and shrubs including ash, elm, dogwood, hawthorn, and crab apple. Seasonal wildflowers provided colorful interest.

After a successful two years, as one of the few world's fairs that actually turned a profit, the COP closed in November 1934. The Horticultural Building, like all of the temporary buildings, was torn down. The landscape, although no longer containing the intimate gardens of the Horticulture Building, became a permanent addition to Chicago's lakefront. And the legacy of this fair, like that of many of Chicago's fairs, was to elevate the progress of horticulture into the next decade and beyond.

LEGACY LANDSCAPES: FAIR REMNANTS

Expositions and fairs are, by nature, temporary affairs and are often dismantled shortly after closing day. Larger expositions, however, often included a planned redevelopment as an important by-product of the fair. Although many of Chicago's lesser fairs exist only in photographs today, there are legacies from the two major world's fairs of 1893 and 1933. These fairs put Chicago in the international limelight, and made permanent enhancements to the city's lakeshore landscapes.

Wooded Island, Jackson Park

57th Street and Lake Shore Drive
Chicago, Illinois
http://www.chicagoparkdistrict.com/index.cfm/fuseaction/custom.natureOasis16

The centerpiece of Frederick Law Olmsted's design for the Columbian Exposition of 1893, the Wooded Island was intended as a natural oasis amid the glitter and glamour of the world's fair. Olmsted and his team carved the space out of a barren wasteland of sand dunes and stagnant ponds. While Olmsted had hoped that the island would conform to his naturalistic design principles, the interior of the island was used for floral displays by exhibiting nurserymen. Today, without the staged floral shows, the island perhaps resembles Olmsted's naturalistic views even more. The serene Osaka Japanese garden remains on the northern part of the island, approximately near the site of the original Hoo-o-den temple, part of the Columbian Exposition's Japanese exhibit.

It's hard to imagine the grand spectacle that was once Chicago's Columbian Exposition of 1893. Arguably the greatest of America's world's fairs, the exposition drew over twenty-seven million visitors from around the world. Its buildings, sculpture, and exhibits hailed from the preeminent architects, sculptors, and inventors of the day. The so-called White City, named for its glimmering white buildings, heralded the City Beautiful movement, which influenced urban planning efforts for decades thereafter.

Today, just one building of the many mighty edifices and only about 20 acres remain intact of the original 634 acres that once sheltered the White City. But those 20 acres are perhaps the most precious remnant of the fair: the exquisite Wooded Island, which was the crown jewel of Frederick Law Olmsted's landscaping plan for the fair. The island was central to Olmsted's vision of a naturalistic sanctuary, a place where fair visitors could relax from the plethora of overstimulating exhibits and attractions. He and his design team worked well in advance of the fair, testing plants that could survive the fluctuating water levels of the lagoons surrounding the island. His plan called for native shrubs, trees, and

8142.   Wooded Island from Horticultural Hall, Columbian Exposition.

Copyright 1894, by B. W. Kilburn.

Jackson Park, Chicago, Rose Garden on Wooded Island.

TOP The Wooded Island as seen from the Horticulture Palace offered a naturalistic oasis amid the glitter of the 1893 World's Fair. From the author's collection.

BOTTOM The formal rose garden in the Wooded Island, while not endorsed by Olmsted, was a crowd favorite and was maintained for many years after the fair ended. From the author's collection.

plants. Only hints of color were to be supplied by flowers interspersed here and there, a design approach that would mimic nature and avoid artificial contrivance.

Olmsted's design philosophy, while esteemed by many, ran counter to other interests represented at the fair. Nurserymen and seedsmen wanted elaborate flower displays to advertise their products. This Victorian practice of bedding out gardens was deemed "garish" by Olmsted. Yet John Thorpe, the fair's chief of the Bureau of Floriculture, wrote in 1892, "I want to see pansies, balsams, phloxes, asters and zinnias so exhibited as to have it said that they are the very best. Herbaceous plants of all kinds will be seen in great numbers."[36] Olmsted was forced to continually protect his island from private interests who saw its location as prime real estate to showcase their flowers, garden ornaments, and exhibits of any kind.

The Wooded Island was the largest of three islands Olmsted planned in the middle of an interconnecting series of lagoons and formal water basins. Today, you can still imagine the grandeur of the fair buildings that once surrounded it. A hint of the scale of the surrounding buildings can be inferred from the extant Museum of Science and Industry to the north, then the exposition's Fine Arts Palace. Every other building was destroyed after the fair, but with imagination, you can conjure up their images. Directly west, where streams of cars crowd present-day Cornell Avenue, the luminous Horticulture Palace with its glass dome reaching over one hundred feet, glowed in the sunset. On the east side, where now a minor strip of parkland exists to buffer Lake Shore Drive's multilane traffic, the immense Manufacturers' and Liberal Arts Building towered over the island. On the south and southwest, the Electricity, Mines, and Transportation buildings presided. Except for iconoclast Louis Sullivan's multicolored Transportation Building, all the surrounding structures glittered bright white in the sun.

As Princeton, Illinois, nurseryman L. R. Bryant wrote, "The wooded island was, indeed, one of the most pleasing features of the fair and stood as a worthy exhibit of the landscape gardener's skill. In the heat of summer it furnished most delicious rest and coolness to the weary sightseers … Often, passing at a distance, I have stopped to admire the peaceful scene—such a contrast to the hurrying crowd that passed and repassed at my side."[37] The esteemed horticulturist L. H. Bailey of Cornell University similarly praised the island: "The Wooded Island was the one free and native feature of the great Fair, an emerald jewel set in the midst of a most dazzling wilderness of architecture."[38]

The margins of the Wooded Island retained Olmsted's naturalistic style, with a perimeter planting belt ranging from a few feet to more than a hundred feet between an inner walkway and the lagoon waters. These outer edges included a profusion of plants such as bulrush, adlumia, Madeira vine, catbrier, virgin's bower, brambles, sweet peas, jimsonweed, and milkweed. Moisture-loving willows were planted in abundance around the island. The island's shoreline itself was deliberately irregular, providing interesting bays and coves to engage the eye and contribute to the sense of a natural wilderness.

Within the island, however, most of the land was forfeited to staged horticulture displays. The exhibits were very popular with the masses, but many horticulturists and those schooled in the emerging field of landscape architecture regretted the lack of cohesion among the displays. Nurserymen were assigned plots among crisscrossing walkways on the island's interior. There they exhibited specialty flowers and plants. Many plants were not especially suited for Chicago's soils, nor the weather conditions that season, and resulted in an uneven showing. The south end of the island in particular was carved up into a number of plots, including German exhibits of tea roses, dahlias, cannas, and Japanese climbing cucumber, said to be the only kitchen-garden vegetable growing outside at the fair. Belgian, Dutch, English, and French exhibitors also occupied plots on the south end of the island, as did local companies. New York exhibitors showed Japanese irises, azaleas, and rhododendrons. The latter are notoriously unreliable in Chicago's climate, but L. H. Bailey reported that the display was "a source of astonishment to most visitors," and indeed might boost sales of the plants throughout the United States.[39] Perhaps the sandy soils in Jackson Park, coupled with the short display period, contributed to favorable showings of these acid-loving plants.

A one-acre, rectangular plot on the south end of the garden displayed the very popular rose garden. Surrounded by a chain fence, the garden "was laid out in severe geometric fashion and comprised forty beds," according to Bailey, who likened the design to a "gigantic hot-air register."[40] Major exhibitors from California, Germany, the Netherlands, and Kentucky displayed standards, shrubs, moss, and other rose favorites. Chicago-area nurseryman J. C. Vaughan displayed in the rose garden, as did the venerable Dingee and Conard Company of Pennsylvania. Many varieties, whether old-fashioned favorites or novelties then, are available today, such as 'Clothilde Soupert', a light pink polyantha introduced in 1890; 'Perle des Jardins', a light yellow tea rose (1874); 'Grace Darling', a creamy pink tea rose (1884); lipstick pink hybrid perpetual 'American Beauty' (1875); and medium pink hybrid perpetual 'Mrs. John Laing' (1887).

Space was allocated to the Japanese garden on the north end of the island, which included one of the few structures allowed, the Hoo-o-den Temple. America was in the throes of its love affair with all things Japanesque, and many marveled at the temple and the garden. More informed horticulturists recognized that the garden was not truly representative of a Japanese style. There simply was too little time, and the soil conditions too poor, to do justice to a Japanese garden.

After the fair, the buildings of the fair were all razed except the Fine Arts Palace. As the original intent was to create a permanent park from the fairgrounds, adjustments had to be made for changes in traffic patterns and usage. In 1896, *Garden and Forest* magazine published a description of the revised plans for Jackson Park, as envisioned by Olmsted. Roads around the Fine Arts Palace (then renamed the Field Columbian Museum, and ultimately the Museum of Science and Industry) were to be resculpted for greater symmetry. The pond immediately south of the building was to be made more symmetrical to provide for ice skating and for sailing toy boats. Decorative plantings were to surround this whole vista. Initial plans called for the Wooded Island to remain fairly intact, as reported in *Garden and Forest*:

> The North Haven and Wooded Island will remain substantially as they were during the Fair, although the shore of the lagoon is to be more varied with islands and bays. The west shore of the west lagoon, where the water seemed too narrow, will be carried a little farther to the west. The south end of this lagoon will be widened and extended, leaving a long and narrow island where the Electricity Building stood. The sites of the Government Building and Manufacturers' Building will be turned into slightly undulating meadow, which will be used for tennis and other lawn games. The Grand Basin in the Court of Honor will have its outline changed into irregular forms, and it will become the northern end of the South Lagoon and the Middle Bay.[41]

Landscape Integrity

The Wooded Island is perhaps one of the few historic landscapes in this book that may be better today than it was originally. It does not contain any of the floral displays that were once there, so we cannot say that it represents an accurate historic reconstruction. With its naturalistic plantings and overall "wild" appearance, it is perhaps more true to Olmsted's original vision. The Osaka Japanese garden, placed approximately where the 1893 Japanese garden existed, at one time included structures relocated from Chicago's 1933 World's Fair. It is an exquisite garden, but does not attempt to recreate the 1893 Japanese garden as seen in pictures of that fair.

Minor interior walkways have been removed, but the outer perimeter still contains a paved strolling path. The south end of the island, approximately where the formal Rose Garden once lay, is now a nature preserve. The island is a favorite site for birdwatching and casual strolls. To the west, where the Horticultural Palace once gleamed in the sunset, Olmsted planned a men's and women's gymnasium and field. These areas today are planted green spaces between Cornell and Stony Island avenues. Where Olmsted planned a circular basin (between 59th and 60th streets at Stony Island) as the terminus to his long canal along the nearby Midway Plaisance, there is a beautiful circular perennial garden originally designed by one of Chicago's early female landscape architects, May E. McAdams.

International Friendship Gardens

601 Marquette Trail
Michigan City, Indiana
http://www.friendshipgardens.org

The International Friendship Gardens, a one-hundred-acre collection of ethnic gardens amid a rolling landscape of duneland, forest, creek, and ponds, resulted from the dream of the Stauffer brothers, nurserymen from Hammond, Indiana. At Chicago's 1933 World's Fair, the Stauffer brothers created an outdoor exhibit that attempted to bring the nations of the world together. Despite rumblings of war, the brothers tried to immortalize this dream by transplanting the gardens into a permanent site in nearby Michigan City. Today, the gardens are emerging from a period of neglect, and are a delightful mix of history and progress.

At the relocated International Friendship Gardens in Michigan City, the Italian Garden captured the sense of a formal Mediterranean garden with architectural columns, reflecting pool, and columnar trees. Ca. 1945. From the author's collection.

The International Friendship Gardens in Michigan City, Indiana, has been host to queens, kings, and dignitaries, yet the garden owes its very existence to the hard work and unshakable faith of a handful of volunteers. Like the threads of true friendship, stories about the garden overlap and interweave over time. Perhaps the best place to start is with the dream of three brothers, and with Chicago's great COP World's Fair in 1933. Despite the Great Depression, the horticultural exhibition at the world's fair was a magnificent sight, with outdoor gardens planted by the nation's leading landscapers. Clarence Stauffer and his brothers, Virgil and Joe, nurserymen from Hammond, Indiana, created one of the most memorable gardens of the fair—the International Friendship Garden.

The brothers dreamed of an idyllic garden where all the nations of the world could share the same backyard. The centerpiece of the exhibition featured the Old Mill Garden, symbolic of tranquility and peace. It was surrounded by display gardens from around the world. On any given day, according to a fair brochure, visitors could get autographs from famous dignitaries strolling through the gardens. The Stauffers surveyed famous world leaders for their flower favorites, and published the list in a promotional brochure:

> Chancellor Adolf Hitler, Berlin, Germany, *Edelweiss*
> General Italo Balbo, Rome, Italy, *Rose*
> Dr. Sigmund Freud, Vienna, Austria, *Gardenia*
> President Franklin D. Roosevelt, United States, *Mountain laurel*
> Mrs. Eleanor Roosevelt, United States, *Pansy and Rose*
> Imperial Family of Japan, Seoul, Japan, *Chrysanthemum*
> Benito Mussolini, Prime Minister of Italy, *Lover of all flowers*[42]

The brothers wanted to keep their dream of world peace alive even after the fair closed. Fortune smiled, and Michigan City philanthropist George Warren donated one hundred acres of land for two consecutive lease terms of ninety-nine years. This land, a beautiful mix of rolling duneland and towering trees located near Michigan City, is the present site of the International Friendship Gardens. The transplanted gardens, interspersed among the valleys formed by Trail Creek and woodland hills, were designed with international help. King Gustav of Sweden sent blueprints of his gardens. King George of England sent his personal gardener for consultation. Queen Wilhelmina of the Netherlands donated 200,000 tulips. Thus, despite the rumblings of World War II, the International Friendship Gardens opened in the spring of 1935.

The French garden displayed a boxwood maze of interlocking squares. An Italian garden boasted a long reflecting pool appointed with statuary and surrounded by pillars of "ancient" ruins. The Chinese garden featured a small pool spanned by an arched wooden bridge. The Netherlands had its tulip parade, while Persia flaunted a rose garden. Scotland's garden showcased hedges trimmed in amazing "cradle" shapes. In a matter of a few acres, you could literally tour the world.

All friendships have their ups and downs, and by the 1970s, supported only by private donations, the International Friendship Gardens began to decline. Weeds launched their inevitable march on the gardens, and trees fell victim to storms and disease. But thanks to recent grants and increased volunteer help, the gardens are being brought back to life. The African American garden is a relatively new addition to the garden, and is indicative of its rebirth. When natural disaster strikes, volunteer gardeners put reuse and recycling to work. For example, when a large beechnut tree was felled, the Norwegian gardeners carved the downed tree into the fanciful likeness of a Norse Viking. This frightful fellow now stands guard over the charming Norwegian garden.

The guest book for the International Friendship Gardens has always included comments and best wishes from visitors. Some of the more notable comments include

> "We need more gardens, art centers, inspirational centers, throughout the world like the wonderful International Friendship Gardens." Gandhi of India
> "I have attended the inspirational Friendship Gardens and have admired its beauty and activity over the last 40 years. I believe it will continue to grow and do more good the world over throughout its perpetuity." The late mayor Richard J. Daley of Chicago
> "It will help us all to build a better world." President Jimmy Carter

Landscape Integrity

The spatial relationships among the gardens and festival areas remain true to the original design. While much of the herbaceous plant material has deteriorated over time, many of the trees date to the 1930s, and remind us of the garden's architectural outline. Although erosion and invasive plants have claimed portions of the creek and lagoon respectively, both water features remain highlights of this special garden.

One of the garden's historic features, the Alaskan cabin transplanted from the 1933 World's Fair, recently fell victim to a fire. Although this treasure no longer remains, a new visitor center has been built at one of the garden's entrances. Plans call for the visitor center to interpret not only the ethnic gardens, but also early Native American use of the site, which was once an important camping and portage area.

A landscape plan was recently developed for the gardens. In the new plan, Trail Creek is a natural dividing line between the traditional festival area on the south and a regrouping of the ethnic gardens in the north. This accomplishes two key objectives: high-traffic festivals and events will be contained, and the effort to maintain the gardens can be more focused.

Nature trails will trace the different geographies between lowland river bottom and upland dunes. Ethnic gardens would be arranged in chronological order to show the progression from, for example, ancient Persian gardens to modern American. Plants will be chosen to approximate species native to the countries' cultures, but hardy in this environment.

SEVEN

PRAIRIE PASTIMES
Entertaining in the Garden

Throughout the decades, Chicagoans have turned to their gardens and to the outdoors for entertainment, food, medicine, and important traditions. Many garden customs were practiced across America, but some have Chicago's own imprimatur on them. As a mosaic of America's immigrant cultures, as the market basket of the nation, as a city built by pioneers in the industrial age, Chicagoans brought to and gathered more from the garden than simple fruits and flowers.

FOOD FOR HEALTH AND HARVEST

Regional differences were pronounced in how garden produce was used for food and medicine. Chicago may have been the meatpacking capital of the world, but fruit and vegetables were prominent, despite observations such as those of E. A. Popenoe, secretary of the American Horticultural Society. He wrote to Edgar Sanders in the 1890s, encouraging more emphasis on garden produce: "Chicago is noted for pork—but we must have the applesauce with the pork."[1] Even as early horticulture thrived in the city, Chicago's reputation, and typical diet, needed more fresh produce.

Fruits and vegetables became a staple in Chicago homes, with supply and demand often dictating what was on the menu. New technologies, including methods of heating and freezing foods, along with food fads also changed the way we ate. In the 1840s, fruits and vegetables were scarce. Debates ensued not over the niceties of elegant fruits and vegetable presentations, but rather whether they were even edible or not. Tomatoes, for example, had only recently been cleared of suspected poisons in popular opinion. Popular literature still implicated fresh vegetables in cholera outbreaks and other illnesses.

Along with validating the healthfulness of fresh fruits and vegetables, an equal priority was finding good coffee. With imported coffees difficult and expensive to obtain, Chicago settlers sought alternatives. An 1840s *Prairie Farmer* correspondent signed J. K. (John Kennicott?) suggested mixing dried, ground carrots with coffee grounds to make the coffee sweeter and last longer. In the 1860s, there was general excitement over the possibility of an "Illinois coffee," made from chickpea.[2] This was a short-lived hope, however, as the coffee substitute received lukewarm reviews: "From all the testimony . . . we think it fully settled that no large crops can be grown, and that it will never take the place of coffee."[3]

The changing importance of fresh produce in diets was clearly reflected in Chicago's cookbooks. Early cookbooks, such as Chicago-based Rand McNally's *Kitchen on Every-day Cookery with Directions for Carving* (1885), capitalized on Chicago's meatpacking expertise. It included specific instructions on cuts and types of meat, with over half of the book devoted to meat-based recipes. Only five pages contained vegetable recipes, and these focused on the basics—mashed and boiled potatoes, stewed corn and succotash, various squashes, cauliflower, and stewed tomatoes (since fresh tomatoes were still questionable). Salsify, a popular whitish root vegetable, was a staple, often prepared with white gravy. Vegetables were frequently served as relishes or as preserved products. Before household refrigerators, and before improvements in markets and transportation, fresh vegetables were eaten only in season. Preserved products became a big market, and Chicago became known for pickles, with major nurserymen such as the Budlongs on the North Side managing both cucumber growing and a thriving bottling business.

By the turn of the twentieth century, "scientific" approaches to nutrition started to outweigh local folklore. The Department of Household Science, organized within the Illinois Farmers' Institute, began in 1893. This group proposed bringing modern scientific theory to household management, including proper techniques for cooking and growing vegetables and fruits. Mary L. Wade, a featured speaker from Chicago at the 1913 annual meeting, observed that food prices, especially meat, had nearly doubled in the past decade. She proposed a number of meat-substitute dishes, many of which included vegetables. Wade maintained that the key to tasty vegetables lay in proper cooking techniques. "There is no vegetable better for us than the very plebian vegetable, the cabbage. We pay a high price for culture. And we haven't got anything more healthful or anything more delicious if we learn to cook the cabbage right. I think I have taught more people to cook cabbage than any one thing." A lively question-and-answer period followed Wade's presentations with such detailed information as whether to cook vegetables covered or uncovered, salted or unsalted, rapidly or not, with or without draining the water. It was important to preserve the "mineral salts" in cooking vegetables, she said. Spinach, for example, the "broom of the stomach," should be cooked in a minimal amount of water. Carrots were usually "cooked to death." With proper, "scientific" preparation or cooking techniques, Wade assured the audience that cabbage would not be indigestible, as was typically thought, and carrots could be "eaten raw if thoroughly masticated."[4]

In 1919, Vaughan's Seed Store published the fourth edition of the *Vegetable Cook Book*. Subtitled "How to Cook and Use Rarer Vegetables and Herbs, a Boon to Housewives," the book was a compendium of recipes from the *Chicago Daily-Herald* and other trusted sources. Vaughan's, with its huge seed and vegetable business, surely had a vested interest in helping homemakers learn to cook and enjoy vegetables. The general recommendations included steaming versus boiling vegetables, using soft water (by adding a half teaspoon of bicarbonate of soda per gallon), and using uncovered pots. Other handy tips included adding sugar to improve beets, turnips, peas, corn, squash, and tomatoes, a "pinch of pearl ash" to render old yellow peas tender and green, and a small piece of red pepper, charcoal, or bread crust to reduce unpleasant cooking odors. Vaughan's book described a smorgasbord of vegetables far greater than the measly potato and corn recipes found in Rand McNally's 1885 book. From artichoke to turnip salad, sixty-four pages were devoted to vegetables alone.

Ten years later, the *Modernistic Recipe-Menu Book* was published in Chicago by the DeBoth Homemakers' Cooking School. These schools, in Chicago and New York, were run by Jessie Marie DeBoth, A.B., a "consultant in Home Economics for Educational and Industrial Organizations." DeBoth opened her book with the observation that "it is only at the time of the present census-taking that the lawmakers have . . . consented to list every housewife as 'homemaker' instead of as 'unemployed.'"[5] DeBoth explained that her book would not cover technical cooking matters—she felt that every cookbook contained such information. Instead, she would provide workable, everyday menus.

Vegetables and fruits were an integral part of the DeBoth menus. There was even a whole chapter devoted to "Meatless Menus." A sign of the times, another chapter was entitled "Refrigeration Menus," offering homemakers the option of make-ahead cooking with the now readily available home refrigerator. Preserved vegetables and relishes, once the staple of every Chicagoan's larder, were relegated to DeBoth's chapter called "The 'Company Shelf.'" As she explained, "It is a great convenience to have a well-stocked emergency shelf against the times when the unexpected guest drops in."[6] Homemade pickles and preserves, once Chicago's signature dishes, now were afterthoughts, and served only "to make many leftover dishes more acceptable." Finally, bringing the culinary quest to full circle, the chapter on beverages began with coffee. But no longer were Chicagoans hunting for coffee in carrots: this book included no less than five recipes with store-bought coffee—from simple "Boiled Coffee," to café au lait, to "Perfection Coffee" with whipped cream and the modernistic miracle of cracked ice.

Nearly all household manuals or cookbooks had chapters on special diets for invalids. Most illnesses were treated at home, and debates ensued over restorative properties of certain foods. Menus for invalids, from today's vantage point, seem more likely to send a person back to the sickbed. Amid recipes for dubious mixtures of porridge, gruel, broth, and panada (stale bread bits soaked and mashed), the 1878 *Lakeside Cook Book No. 1* listed the plant-based ingredients for flaxseed tea, barley water, blackberry syrup, arrowroot custards, and Irish-moss blancmange. To take one's mind off one's ills (or perhaps one's menu), this book also offered to invalids a blackberry wine, wine whey, and blackberry cordial.[7]

Jonathan Periam's 1884 *Home and Farm Manual* offered definitive prescriptions for invalids' cookery. All fruits and vegetables were to be perfectly fresh, and vegetables were to be boiled in salted water to remove insects. Periam also provided a "Table of Foods and Time of Digestion," which suggested, by food type, the average time required for proper digestion. A sweet apple, for example, required the shortest time (1½ hours), whereas boiled turnips took a full 3½ hours. Individuals were warned to monitor the recommended digestion times, especially when ill. In addition to some gruels and jellies, Periam provided recipes for baked rice and apples, Iceland moss jellies, orangeade or lemonade, chamomile tea, and white wine whey.

The 1930s *Modernistic Recipe-Menu Book* downplayed vegetables and fruits in invalid diets. The book did encourage elegant meal presentations for the convalescent, including fine silverware and a fresh flower in a bud vase. It also offered a "constipation diet." In years past, people who were troubled with constipation took their nightly "pills" or herbs, and ate what they chose in the daytime. Between that type of treatment and the book's present one was a period of "roughage-eating," which, however, "did not give the results expected of it."[8] The cookbook prescribed two or three months of cooked fruits and vegetables and coarse grains. Then, raw fruits and vegetables were added for another two months. After four months of a predominantly fruit and vegetable diet, the sufferer was allowed some eggs and a little bit of meat.

Vegetarianism, as a lifestyle, had been practiced in the United States for some time, with the first American Vegetarian Society formed in New York in 1850.

FAVORITE EARLY RECIPES

Throughout the decades, households in the Chicago region created favorite dishes using produce from their garden. Of particular interest is how refrigeration and increased knowledge about food hygiene affected fruit and vegetable recipes. Here are just a few.

Carrots for Coffee

Prairie Farmer, January 1849

Wash and scrape the outside off; then cut them in pieces the size of about half an inch square; then dry on a stove. Parch and grind like coffee; or mix equal portions of carrot and coffee and grind and make your coffee as usual.

    If you know it to be mixed you may say that it tastes a little sweeter than coffee generally. We got our information from our neighbors who came from Germany a few years ago; and who say in their country there are large factories where it is packed in pound papers and sold.

Tomato Catsup

Prairie Farmer, 1846

Wash the tomatoes and press them through a fine sieve. To six quarts of the juice and pulp, add the same quantity of the best vinegar. Then set it over a slow fire to boil. When it begins to thicken, add pimento, cloves, and pepper, each half an ounce, cinnamon, quarter of an ounce, and two nutmegs— all very finely powdered. Boil to the consistency of thin mush; then put in four tablespoonfuls of salt and take it out of the vessel immediately. When entirely cold, bottle, cork and seal. It must be boiled in a brass or bell-metal kettle, and if the catsup is allowed to remain after the salt is added, it will taste of the kettle. If you have a tin-lined vessel it will be the best.

Salsify, or Vegetable Oysters

Lakeside Cook Book No. 1, 1878

Wash and scrape them thoroughly, and as you wash throw them into a bowl of cold water. Cut into pieces about half an inch long, boil three-fourths of an hour; when tender, pour off all of the water, season with pepper and salt, a small lump of butter, and enough cream to almost cover them; if no cream, use milk, with more butter, and thicken like gravy with a little flour. They are nice served on toast.

Brown Betty

Lakeside Cook Book No. 1, 1878

Grease a pudding dish, put into this a layer of cooking apples (sliced), then a layer of bread crumbs, with sugar sprinkled over, and small bits of butter. For three apples use one cup of bread crumbs, one-half cup sugar, and a piece of butter the size of an egg. Put a layer of bread crumbs on top; bake. It is nice either with or without cream.

Pie Plant Charlotte

Lakeside Cook Book No. 1, 1878

Wash and cut the pie plant into small pieces, cover the bottom of a pudding dish with a layer of pie plant and sugar, then a layer of bread crumbs and bits of butter, or thin slices of bread nicely buttered, and so on until the dish is full. Allow a pound of sugar to a pound of fruit. Bake three-quarters of an hour in a moderate oven. If preferred, turn over the charlotte a boiled custard when ready for the table.

Creamed Old Potatoes

Cooking Recipes Compiled by the Ladies' Social Union, Oak Park, Illinois, 1901

Old potatoes can be made like new. Peel and cut them in dice, about an hour before using; cook them in salt and water. When nearly done drain and add milk almost to cover, with a little butter and a few pieces of mint, pepper and salt. Let it come to its own cream.

Orange Baskets

Cooking Recipes Compiled by the Ladies' Social Union, Oak Park, Illinois, 1901

Cut with a sharp knife the orange skin in shape of a basket (leaving a strip for handle); remove skin or either side and take out pulp without breaking the handle. Cut pulp into small pieces and add to these slices of banana, pieces of pineapple, and a few candied cherries; fill orange baskets with the fruit, sweeten and moisten with maraschino. Tie the handle of the baskets with narrow ribbon.

Inexpensive Fruit Cake

Orange Judd Cook Book, 1914

Cream 2 cups brown sugar and 1 cup white sugar with scant ¾ cup shortening, add 2 eggs, (one will do), 1½ cups buttermilk, in which dissolve 1 teasp. Soda (or 1½ cups black coffee and 3 teasp baking powder), add flour to make the dough stiff enough so it will hardly drop from spoon. Sift with the flour 1 teasp each cinnamon and ginger and a little less of cloves. Lastly add 1 cup each raisins and currants well dredged with flour. Bake two hours in moderate oven. I use milk pans. When the cake is cool cover with the following icing: Boil scant ¾ cup white sugar and scant ½ cup sweet milk until it hardens when dropped from end of spoon. Then remove from fire and add 1 teasp vanilla and ½ cup each chopped seeded raisins and walnuts, or any other nut meat. Beat until it begins to harden and then spread quickly over the cake. This cake will keep for months and is better when two or three weeks old, than when fresh.

Homemade Cereal Coffee

Orange Judd Cook Book, 1914

To 1 qt wheat bran add 1 pt corn meal, ½ cup molasses, and 1 egg. Mix well together, then spread in a pie pan about ½ inch thick, and brown slowly and evenly in the oven. Watch carefully, as it burns easily, and if burned it would spoil the flavor. Some people like to add a very little butter or salt to the mixture. When browned and dry, store in covered tin cans. To make coffee use 1 cup of this cereal mixture for 1 qt coffee, and let boil 1 hour or more—it requires long boiling to bring out the flavor. If desired, 1 tablesp ground coffee may be added during the last 5 minutes of boiling. Serve with cream and sugar.

Lemon or Orange Ice

Orange Judd Cook Book, 1914

For a lemon ice boil together 20 minutes 2 cups sugar with 4 cups water, or until it syrups, then add ¾ cup strained lemon juice, let get cold, and then freeze. Proceed the same for orange ice, only use less lemon—about ¼ cup lemon juice—and 2 cups orange juice, with the grated rind of 2 oranges.

Dandelions

Vaughan's Vegetable Cook Book, 1919

Use the dandelions in the early spring when they are young and tender. They take the place of spinach and are treated the same. Dandelions may be used as a salad with a French dressing.

Flower Sandwiches

Vaughan's Vegetable Cook Book, 1919

Make a filling of two-thirds nasturtium blossoms, one third leaves, lay on buttered bread, with buttered bread on top, sandwich style.

Preserved Rose Leaves

Vaughan's Vegetable Cook Book, 1919

Put a layer of rose leaves in a jar and sprinkle sugar over them, add layers sprinkled with sugar as the leaves are gathered until the jar is full. They will turn dark brown and will keep for two or three years. Used in small quantities they add a delightful flavor to fruit cake and mince pies.

"Victory" Spinach

Vaughan's Vegetable Cook Book, 1919

Carefully wash the spinach, scald it in boiling salted water, then pour cold water over it, drain and chop fine. Stew an onion in butter until it is soft, add the spinach, sprinkle flour over it and cook for ten minutes stirring constantly, add salt, pepper, a little grated nutmeg, and cover with meat stock or gravy. Boil a few minutes and when done, add a little sour cream.

Frozen Tomato Salad

Vaughan's Vegetable Cook Book, 1919

Peel and chop fine a half dozen solid tomatoes, season with a teaspoonful of salt, a saltspoonful of pepper and a teaspoonful of lemon juice. Freeze the pulp until solid in an ice cream freezer, when frozen, mold it into fancy shapes and serve on lettuce with a tablespoonful of mayonnaise over each mold.

Many of these early movements were interwoven with philosophies of medicine, religion, and the temperance movement. In 1889, the Chicago Vegetarian Society was formed by Mrs. Le Favre, with about twenty members—most of them women, although a few male doctors also subscribed. Shortly thereafter, in 1893, Chicago hosted the Third International Vegetarian Congress at the world's fair. Participants of the congress, while appreciative of the content and presentations, could not help but notice the fumes from the stockyards nearby.

Some enterprising and committed individuals took vegetarian precepts to another level and made a business out of it. Henry and Anna Lindlahr developed the "Nature Cure," a vegetable-based diet that, along with exercise and other programmatic elements, promised health and well-being. In 1915, the Lindlahrs operated a three-year College of Nature Cure and Osteopathy (later converted to a sanitarium) on South Ashland Boulevard in Chicago, and a health resort in Elmhurst, Illinois. The first edition of their *Nature Cure Cook Book* was published in 1915, and was evidently so popular that a sixteenth edition, renamed the *Lindlahr Vegetarian Cook Book* was produced in 1918. The Lindlahr Nature Cure proposed a system of classifying vegetables and fruits into starches, minerals, and so forth, and assembling recipes that balanced the "ideal" proportion of these food types. Henry Lindlahr had himself overcome diabetes and obesity with his own diet, and his son, Victor, also became a nutritionist doctor with a very popular radio program broadcasting nationwide for over twenty years. Victor Lindlahr's 1948 book, *201 Tasty Dishes for Reducers*, also emphasized vegetables not only for health, but for losing weight.

HOLIDAYS IN THE HEARTLAND

Flowers, fruits, and greenery were traditional elements of many holidays and special celebrations in Chicago. With a majority of Chicago's pre-1900 immigrants hailing from Germany or Ireland, both Christian countries, Christmas was always a major holiday. In the late 1800s, Christmas trees harvested from upper Michigan and Wisconsin were brought to Chicago via so-called Christmas Ships. The most famous of these, *Rouse Simmons*, is the subject of books, documentaries, paintings, and a play. This legendary Christmas Ship, a wooden schooner owned by Captain Herman Schuenemann, made an annual pilgrimage from the Michigan woods hauling thousands of fresh Christmas trees to eager families waiting at the Clark Street Bridge in Chicago. Schuenemann and his crew steadfastly tamed the wild waves of Lake Michigan from the late 1800s through his tragic last sail in 1912, when the *Rouse Simmons* and all hands on board succumbed to the furious gales of a winter snowstorm. Schuenemann's family kept the tradition alive by selling fresh greenery and trees from the dock until the early 1930s.

This classic painting by the late Charles Vickery depicts the excitement when Christmas trees were unloaded from famous schooners such as the *Rouse Simmons* near the Chicago River. *Christmas Tree Schooner* by Charles Vickery. Courtesy of the Clipper Ship Gallery, www.charlesvickery.com.

Christmas trees were decorated in all manner of styles, often reflecting the ethnic heritage of the family. While treasured heirloom glass ornaments were sometimes available, more often Christmas tree decorations were handmade from natural materials. Apples were hung from trees—and then quickly eaten so as not to waste the precious fruit. Walnut shells were silvered, and filled with tiny surprises. Popcorn was strung on a chain, or individually clipped to the end of each tree sprig to resemble a snowflake. Oranges, and other out-of-season or nonnative fruits, were often given as presents.

Decorating with holly and mistletoe was a new phenomenon in the late 1800s. Writing in 1888, Edgar Sanders noted, "It is but a very few years ago since the first pieces of holly for decoration was seen in Chicago." Holly was used earlier in the East Coast because its climate was more favorable to the plant. Chicago's mistletoe, although grown with some success in the American South, was largely imported from England. Sanders noted, "Unlike the holly, there is nothing much that is charming in its appearance. Its use, therefore, is from association [with traditions from native lands], and that mostly among young people in their merrymakings."[9] Ferns, grown nationwide and distributed regularly to florists, were popular decorative greens during the holiday season in the late 1880s.

In the 1890s, Chicago-based *Gardening* described Christmas plant decorations from around the country. For Christmas trees, white and black spruces were recommended for their pyramidal form, whereas hemlocks were deemed too slender, and Norway spruce too coarse. For potted plants and cut-flower decorations, East Coast cities such as Boston and New York favored Otaheite dwarf orange, tulips, *Lilium harrisi* (a form of white trumpet lily), marguerite daisies, and poinsettias. Demand for potted plants in Chicago lagged behind that of eastern states. Potted poinsettias were also scarce in Chicago; however, cut bracts sold well. Roses, particularly the Beauties and Meteors, were favored in Chicago, as were carnations and violets.

In the early 1900s, traditional indoor decorations might have included a table adorned with holly or evergreen boughs and mistletoe sprigs tied to place-card holders. Walnut shells were still a favorite ornament—colored or stained. Christmas trees, no longer as precious a commodity as in the Christmas Ship days, were sometimes cut to size, adorned with candles, for a tabletop decoration instead of candelabra.[10] Yule logs were traditionally burned, or a so-called Devonshire yule fagot: a bundle of ash sticks encircled with a lucky number of nine bands. One peculiar custom involving fruit was the snapdragon game. In this Christmas sport, raisins were put in a shallow dish, set aflame with brandy, and from this fiery centerpiece, guests would try to pluck out the raisins.[11]

By the 1920s and 1930s, living Christmas decorations were no longer a luxury, but instead a big business. In his book, *Commercial Floriculture*, Highland Park florist Fritz Bahr urged his colleagues to heavily promote Christmas floral arrangements, and locate and order stock weeks ahead of time. "Push and keep on pushing a little harder each year, and it can be made one of the most important flower days of the whole year," Bahr encouraged.[12] He exhorted florists to make unusual basket arrangements with high-quality flowers to compete against the "inferior arrangements" sold in grocery stores. His book featured a Christmas basket made from birch bark, filled with poinsettias, oranges, pussy willow, and winterberries. Holly wreaths could be made in varying sizes in prices ranging from seventy-five cents up to five dollars. Boxwood was coming into fashion as centerpieces or as topiaries created with moss-filled wire balls.

The arrival of spring was in and of itself cause for celebration to winter-weary Chicagoans, and numerous holidays in April and May accentuated the festive air. The *New Idea Entertainer*, a 1903 compilation of "entertainments" from various Chicago sources, listed a wide variety of spring flower–related fetes, including a flower social, a forest social, an apple blossom luncheon, and children's May party. May Day, the first day of May, waxed and waned in popularity as a formally recognized holiday. The *Chicago Record-Herald*, among other papers, encouraged its revival in the early 1900s. May Day included floral-themed traditions such as hanging baskets of flowers on friends' doors, or hosting luncheons, profusely decorated with flowers.[13]

Thanksgiving was always a major American holiday where harvest bounty was celebrated. Turkey was the mainstay, but vegetables and fruits were indispensable side dishes. A typical menu from 1907 included cream of tomato soup, mashed potatoes, boiled onions, sweet potato croquettes, celery, fruit salad, individual pumpkin pie, and cranberry sherbet. The *Modernistic Recipe-Menu Book* featured slightly more elegant treatments of the same produce: scalloped pumpkin, cranberry frappe, grapefruit salad, fluffy potatoes, celery au gratin, tomatoes stuffed with mushrooms, fruit cup, and radish roses. Although the grapefruit and frappe evidenced improved market distribution and refrigeration technologies, the menu stayed fairly constant over the years.

Along with its religious connotations, Easter allowed Chicagoans to celebrate the start of spring. Many festivities took place outdoors, including egg-rolling contests, egg hunts, and so forth. *Dame Curtsey's Book of Novel Entertainments* describes decorations for an Easter luncheon as including window and doorway decorations of garlands of trailing smilax and feathery asparagus vines. Floral decorations included yellow and white tulips and lilies. On Easter Monday, the celebrations continued with a "Floral Card Party," wherein guests brought a potted plant, fern, or flower bouquet. Whoever brought the plant with the most blossoms won, with all proceeds and the plants themselves going to charity and local hospitals.[14]

Mother's Day, a major florist holiday today, did not become a national holiday until 1914. Many turn-of-the-twentieth-century Chicago books on social customs do not focus on this holiday, but by the early 1920s, florist Fritz Bahr wrote, "It is hardly necessary to call attention to the importance of Mother's Day. Everyone knows that this is the latest addition to the list of great flower days, yet in some instances it has already outstripped Christmas or Easter."[15] Bahr noted that most retailers could grow their own stock for Mother's Day, including potted-up bedding plants or cut flowers such as snapdragons, sweet peas, annual larkspur, lupines, tulips, and forget-me-nots. Memorial Day, formerly Decoration Day, was also celebrated with flowers, providing great revenues for florists located near cemeteries. Cut flowers, wreaths, and sprays of blossoms were used to decorate the graves of fallen soldiers.

Summer was a slow season for florists, with many customers out of town on vacation or using flowers from their own backyard. In the 1920s and 1930s, roadside markets competed with the florist for the cut-flower business. But the social season was rife with opportunities for garden parties, lawn fetes, and other venues to display a bounteous garden. Those Chicagoans who did remain in the area enjoyed the outdoors from swinging hammocks on the front porch to tea parties on the back lawn. As autumn approached, harvest parties were plentiful, with the bounty of the season prominently displayed. *The New Idea Entertainer* recommended an autumn harvest church fundraiser with banquet and display tables arranged under a grove of trees. Tables were to be decorated with "garlands and sprays of autumn leaves and flowers, the beautiful roadside asters being especially adapted for table decorations."[16]

RITES OF PASSAGE—WITH FLOWERS

Major life events in early Chicago were marked by floral traditions. Much of the symbolism carried over from European and colonial traditions, updated with the flowers available in the Chicago region. Whether in season or not, flowers or artificial substitutes were used to accentuate the ephemeral nature of joy and sorrow.

Victorian mourning rituals were quite elaborate. Floral arrangements played an important part of these traditions. As previously noted, Chicago's cemeteries were early civic improvements, and many nurseries and florists (e.g. Peterson's Nursery, Willis N. Rudd, Conrad Sulzer) obtained a substantial portion of their income by supplying arrangements to grieving families. In the 1870s, Edgar Sanders recommended "immortal flowers," or everlastings, which could be woven into emblems such as wreaths, crosses, and anchors, as suitable memorials. Everlastings were to be combined with sphagnum moss, dyed green, obtained either from nearby woods or imported from France. Sanders's recommended everlastings included acroclinum (of a deep rose hue), globe amaranth (*Gomphrena globosa*; white, orange, or crimson), helichrysum (scarlet, yellow, purple, white), and neranthemum (purple, white, and yellow).[17] While white flowers were preferred, Sanders noted that this color was scarce among everlastings, so multihued flowers were deemed acceptable.

Jonathan Periam's etiquette for funerals in the 1880s specified that flowers decorating the coffin and viewing room should be white. If the deceased were young or unmarried, a wreath should adorn the coffin; if elderly or married, a cross was the appropriate ornament. Many of the wirework designs such as harps, crosses, and lodge emblems persisted in the florists' trade in the 1930s, but Fritz Bahr opined that a spray of loosely arranged flowers was more becoming. Bahr urged florists to maintain needed wirework in stock so that a suitable memorial could be made at a moment's notice. However, he cautioned that such symbols of death should not be displayed in the open.[18] Apparently, the Victorian view of death as commonplace gave way to modern sensibilities where death should be kept at a distance. Flowers no longer needed to be white, according to Alex Laurie, a writer for Chicago's Florists' Publishing Company. "Colored flowers and colored plants are general at funerals, the dead white being gradually

TOP Floral arrangements in many designs—gates, wreaths, wheels—were common for departed family members. Elaborate decorations were left at the cemetery, particularly on Decoration Day. From the author's collection.

LEFT Studio photos were typical for confirmations and first communions, and often included a candle and flowers. Ca. 1909. From the author's collection.

relegated to the realm of antiquity," he noted.[19] Door baskets of flowers were also later fashionable, according to Laurie, "replacing the wreath and the door spray as the badge of death."

While flowers were intricately woven into funeral ceremonies, they were also key elements of many happier occasions, such as first communions and confirmations. Boys and girls alike posed in photographer studios with the requisite Bible and bouquet or boutonniere. Baptisms and christenings, particularly in the early days when infant mortality was so high, were great causes for celebration, and baby baskets were covered with floral tributes.

Weddings, then as today, were infused with flowers—as much as the families could afford. Victorian studio wedding photographs were often taken before the actual event because of the elaborate procedures needed in early photography. The photos invariably showed brides with lush bouquets or at least sprigs of orange blossom, the traditional bride's good-luck bloom. Weddings continued to be big business for florists throughout the decades, with improved refrigeration techniques offering more than seasonal flowers for bouquets. Bridal clothes fashions, with bridesmaids' dress colors ranging across the rainbow, offered further diversity in color and shape of bouquets and corsages. In the 1930s, for example, with America's colonial revival period, florist Alex Laurie suggested that bridesmaids' bouquets could have a "colonial effect," with tightly and symmetrically arranged flowers. Sweet peas, roses, and spring flowers matching the bridesmaids' dresses were common, with white or pale pink roses the norm for brides of the 1930s.[20]

Bridal showers and wedding parties often featured flowers. For a floral shower, guests were each asked to bring one or two stems of flowers, each symbolizing the sentiment wished for the bride. These were then assembled into a bouquet, which was dried and saved. In another variant of the floral shower, guests' bouquets were gathered into a floral ball suspended over the doorway. As the bride-to-be left, petals from the bouquet were allowed to fall, to symbolize good luck. Another themed shower, the "glass and jar" shower, asked guests to bring a jar of fruit, pickles, marmalade, or other homemade gift to stock the bride-to-be's pantry. Recipes for same were included as part of the gift. Table decorations at showers and wedding receptions were elaborate. For a 1907 "pre-nuptial luncheon," "The table was a dream. In the centre, to simulate a lake, there was an oblong mirror, surrounded with smilax and trailing vines ... A very pretty feature was the crowning of the bride-to-be with a wreath of myrtle, which signified good luck."[21]

MEDICAL USES

Plants figured prominently in Chicago's medical history—not only as popular home remedies, but also in the development of the medical profession. More so than their East Coast counterparts, Chicago physicians were intimately associated with the natural world. A physician's visit to patients on the prairie involved much more than a short trip across the civilized cobblestone streets in New England. Instead, Midwestern doctors might have traveled for days to make their house calls. They could not help but come in close contact with local vegetation. As one of the few practicing physicians in the early days, John Kennicott, for example, was a familiar figure on the prairies. He traveled a wide circuit on his brown Indian pony, Potawatomie, making house calls as far north as Waukegan and Lincolnshire and as far west as Elgin. In his travels, he kept his eyes out for interesting or unusual plant specimens. One early historian wrote of this odd sight: an upright tree moving along the prairie on top of a small pony. On closer inspection, the writer realized the "moving" tree was actually the Old Doctor on horseback bringing a prize botanical specimen home.

Like most early settlers in the Chicago-area, physicians also sought a living from the land, as the practice of medicine in those days was not sufficiently remunerative. In his book, *Medicine in Chicago*, Thomas Bonner quotes an early practitioner: "Anterior to eighteen hundred and forty, nine-tenths of all the physicians who had located themselves in this region, had done so with reference to pursuing agriculture, and with the avowed intention of abandoning medical practice; most of whom . . . divided their attention between farming and medicine."[22] Thus individuals such as W. B. Egan dabbled not only in real estate, but also in marketing his own plant-based medicine, sarsaparilla. Early physicians who also studied the natural sciences were strong lobbyists for many of Chicago's green spaces: J. H. Rauch advocated more city parks to alleviate "miasmatic" conditions, and J. V. Z. Blaney helped establish Rosehill Cemetery to improve unsanitary burial customs. George Vasey greatly contributed to the botanical explorations and collections of native flora.

Chicago, with its wildly rumored and occasionally true tales of unsanitary and unhealthy conditions, drew many newly minted, and some self-ordained, doctors. In the mid- and late 1900s, medical licensing requirements were not stringent, and many forms of medicine were practiced. Bonner notes that in 1850, seventeen different types of medical doctors practiced in Chicago, including those who relied heavily on plant-based medicine such as botanic, eclectic, and homeopathic physicians. Homeopathy, while not strictly plant-based, often used herbal remedies and was endorsed by leading Chicagoans and horticulture patrons, including William B. Ogden, J. Y. Scammon, and John Wentworth. By 1905, Chicago trained more homeopathic students than any other city in the nation.[23]

Physicians were not the only source of plant-based medicine. Druggists and chemists liberally dispensed medical advice and treatments. The Murray and Nickell Company of Chicago described itself as a leading importer and miller of

American and foreign botanical drugs. In 1880, J. M. Nickell compiled a handy reference guide of common and botanical names for use by druggists and physicians alike. From *Abies communis* (common spruce) to *Zizia aurea* (meadow parsnip), the J. M. Nickell's *Botanical Ready Reference* listed 2,465 plants and their medicinal uses. With its German translation, the book was very useful to druggists who had Chicago's large German population to serve. Nickell commented in the preface, "I am well aware that druggists are, almost daily, perplexed by the use of common names, as applied to botanical drugs, and even botanical names (as many have two or more); and in the absence of a book to which you can readily refer, a sale is often missed, when you have the goods wanted."

With frontier doctoring largely do-it-yourself, medical misinformation often spread as rapidly as disease. Controversy over plants as poison or panaceas raged in Victorian Chicago. Typical prairie homesteaders, miles away from city doctors, self-diagnosed and treated their own illnesses with herbs from their garden. Early autumn brought the onslaught of cholera, and with it the persis-

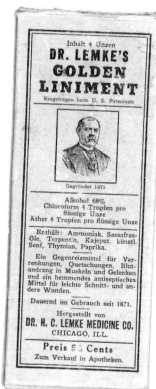

Medicines included a number of herbal remedies—some still popular today. From the author's collection.

tent rumor that fresh fruits and vegetables were the prime culprits. (More likely it was the tainted water supply and generally unsanitary conditions.)

By the 1880s, fruits and vegetables were no longer suspect as sources of disease. In fact, many legumes were highly regarded for their purported medicinal properties. John Neltnor's horticulture magazine recommended these plant-based remedies:

> Celery is highly recommended as a cure for nervous diseases.
>
> Carrots are used by the peasants of Savoy in the shape of an infusion as a specific for jaundice.
>
> Lettuce has a slight narcotic action, and when properly cooked is a salad really very easy of digestion.
>
> Asparagus is a strong diuretic and forms part of the cure for rheumatic patients at some of the leading health resorts.
>
> Lemons eaten raw are excellent for rheumatism and are recommended for this troublesome ailment by many of our best physicians.

HERBS FOR HEALTH

Herbs were popular home remedies, as recommended by books such as Jonathan Periam's *Home and Farm Manual*. His time-tested prescriptions included chamomile flowers for an emetic tonic, dandelion decoctions for "torpid conditions of the liver," sarsaparilla for chronic rheumatism, and infusion of roses for fevers, among others. Two of his recipes follow.

Home-Made Bitters

Home and Farm Manual, 1884

Take half an ounce of the yolk of fresh eggs carefully separated from the whites; half an ounce of gentian-root; one and a half drachms of orange-peel; and one pint of boiling water. Pour the water hot upon the ingredients mentioned, and let them steep in it for two hours; then strain, and bottle for use.

Blackberry Syrup

Home and Farm Manual, 1884

Periam noted that "syrups are used principally for their medicinal qualities. Blackberry and elderberry are generally used, but all fruits have more or less cooling, anti-febrile qualities. The directions here given will apply as well to any other berries of which the extract can be gotten, as to blackberries." A key ingredient in his recipe is the brandy, and he recommends a wineglassful as the proper dose for adults.

Make a simple syrup of a pound of sugar to each pint of water, and boil it until it is rich and thick. Then add to it as many pints of the expressed juice of ripe blackberries as there are pounds of sugar; put half a nutmeg, grated, to each quart of syrup; let it boil fifteen or twenty minutes; then add to it half a gill of fourth-proof brandy for each quart of syrup; set it by to become cold, then bottle it for use.

Figs are an excellent food for invalids. They are best if boiled about five minutes and eaten hot about fifteen minutes before breakfast.

Sorrel is cooling and forms the staple of that "soupe aux herbes" which a French lady will order for herself after a long and tiring journey.

Onions, if slowly stewed in weak broth and eaten with a little Nepaul pepper, are an admirable article of diet for patients of studious and sedentary habits.[24]

While many herbal remedies were based in some science, many more were outright snake oils. One popular medicine man, John Austin Hamlin, brought his powerful Wizard Oil to antebellum Chicago. Hamlin's Wizard Oil Liniment (65 percent alcohol) contained an interesting array of herbal ingredients including oils of thyme, sassafras, and fir, along with a worrisome tincture of turpentine and ammonia. Hamlin elevated the medicine man's science to an art. His traveling road shows—covered wagons with troupes of musicians, pump organs, and hefty supplies of Wizard Oil—crisscrossed the country.

As Chicago grew in the latter half of the nineteenth century, so did its immigrant population. Along with the newcomers came their favorite herbal recipes from the Old Country. Chicago became an ethnic smorgasbord of do-it-yourself dosing. The mass appeal of these folk remedies was evident in the multilingual advertising of some of Chicago's early herbalists. H. C. Lemke began to bottle his concoctions in Chicago in 1871. Instructions and dosages on his early bottles of Herb Pectoral were written in both English and German. Reflecting the later shift in Chicago's demographics, a circa 1940 box of Dr. Lemke's Herb Tea was printed in five languages: English, German, Spanish, Czech, and Polish. Lemke's "stomachic drops," at fifty cents an ounce, were "vegetable medicine containing extracts of Cassia Bark, Ginger Root, Catechu, rhubarb root, capsicum, gum camphor, and oils of peppermint, cassia, anise and cloves." The drops were 45 percent alcohol. Lacking liquor, but perhaps a better bargain at thirty-five cents for two and a half ounces, Dr. Lemke's Herb Tea contained a mix of eighteen herbs and flowers, including yarrow and lavender flowers, althea, and licorice roots.

Chicagoans were not the only people targeted by suspect cure-alls. Recognizing the need for better self-regulation, Chicago physician Nathan S. Davis and others formed the American Medical Association as early as 1847. The AMA made its permanent headquarters in Chicago in 1902. Shortly after this, in 1906, the National Food and Drugs Act severely curtailed the blatantly fraudulent claims of patent medicine marketers. Beginning in 1907, the AMA spotlighted questionable cure-alls through its journal and its 1911 volume, *Nostrums and Quackery and Pseudo-Medicine*. In 1921 and 1936, attesting to continued fraudulent claims by questionable practitioners, the AMA issued new editions, exposing various patent-medicine schemes perpetrated throughout the country.

Quackery crossed all gender and race barriers. A Chicago woman implicated in 1932 promoted a diabetes cure by mail order—an elixir containing only sassafras bark, marshmallow root, couch grass, juniper berries, and round dock—"another diuretic," according to the AMA.[25] The AMA's book cited the C. H. Johnson Medicine Company of Chicago, "owned by Orville G. Johnson, a negro of Lima, Ohio," who sold Luculent, a fraudulent consumption cure made of sage leaves, senna leaves, thyme, bachelor's button, rhubarb root, and other herbs.[26] "Mother Helen," aka Helen Schy-Man-Ski, along with her son, sold Mother Helen's S-M-S Herb-Nu remedies for a variety of ills ranging from tuberculosis to kidney trouble. The AMA reported that Herb-Nu was essentially a bitter-tasting mix of twenty-five ingredients, most of them common herbs with no therapeutic value.[27]

PICNICS ON THE PRAIRIE

Along with nature hikes for science or exercise, Chicagoans flocked to the countryside in May and June. "The season of picnics is upon us, and right heartily is the overcrowded population of our city enjoying it," reported Chicago's 1865 *Voice of the Fair*. Countryside picnics were a favorite way to enjoy the outdoors particularly in May and June, either by buggy or exploring the newfangled "half hour's wisk [*sic*] on a crowded train." After lunching from a picnic basket, nature walks and dancing were popular, even though "a twig may transform a swan-like step into a waddle . . . and the bare earth itself, though level as a lake and firm as the sea-beach, forbids all the gliding, and the sliding."[28]

In the days before Cook County created its forest preserves and parks, city dwellers were hard pressed to find suitable picnic spots. This may seem incongruous given all the vacant countryside, until you remember all the appurtenances the typical Victorian family required in assembling a proper picnic. Amenities, such as comfort rooms for ladies to rest in the heat, tables and chairs to protect suits and dresses from stains and mud, water for horses or to wash up, were all desirable for an enjoyable picnic. Lacking thermal coolers, most picnickers did not bring their own foods, but sought them on site. As the *Prairie Farmer* noted in the 1860s, "Compared with eastern cities, the environs of Chicago afford but few places adapted for summer resorts. For several years our excursion parties have been weary journeys of forty to sixty miles to reach desirable points, and then from the necessities of the case, having but a few hours in the heat of the day for recreation and enjoyment."[29] The *Voice of the Fair* concurred, noting that, in the 1860s, "the picnic is the only sort of excursion that many may indulge in. Our city has no park worthy of the name, and until it has, these brief and unfrequent glances at green things, and gregarious sniffs of fresh air, will be popular and deservedly so."[30] Enterprising farmers were quick to spot an opportunity. In the 1860s, gardener and farmer George Davis looked for other ways to earn an income from the land. The *Prairie Farmer* noted,

> George is doing a great deal to develop the unoccupied land about the city. His example is to be duplicated by dozens of men who own property which brings them in nothing but taxes. A great many are going to abandon cabbages for small fruits, and we look to see our city become in reality a Garden City. And Mr. Davis insists that there need be no money sent out of the city and its vicinity for early fruit of this character ... he proposes to compete with Cincinnati and other southern cities in supplying the market early in the season. To this end he is going to add to his greenhouses, and grow fruit *the year round*.[31]

In addition to diversifying his crops to include such small fruits as strawberries, the entrepreneurial Davis converted the wooded portion of his forty acres some ten miles west of Chicago (in the town of Harlem) to a public picnic ground. The strawberry social was widely celebrated on the Davis park grounds. Accessible by the new Galena and Chicago railroad, excursion trains left at 10 a.m. and returned at 5:30 p.m. The Davis picnic grounds included "places for refreshment, where delicious fruits, fresh from the garden, with all the accompaniments, will be dispensed ... There will be seats under large trees, in shady nooks and grottoes—long and pleasant walks through the oaks and the hazels, swings, cricket, and quoit grounds, and a splendid lawn for dancing."[32]

Even after the Chicago park system was established, city dwellers still sought adventure in the outlying countryside. Railroads and private landowners continued to supply the need. The Suburban Railroad advertised several picnic groves along its route circa 1905 that could be secured "for picnic purposes, by churches, lodges, societies or family picnics."[33] Sunday school groups, including the Swedish Methodist, First Bohemian, Douglas Park Evangelical, and German Baptist groups, were counted among those picnickers along the Suburban Railroad route. (Charles) Becker's Grove, (Fred) Bergman's Grove, and (August) Schurz Park, all offered shaded trees and shelter near Berwyn, Riverside, and Grossdale (now Brookfield). Riverside's Mrs. Reissig, of Reissig's Florist, offered a large pavilion on her shaded grounds.

A special variant of the picnic grounds was the beer garden, particularly popular among Chicago's large German immigrant population. Beer gardens afforded opportunities to socialize outdoors in venues that might include dance floors, bowling alleys, vaudeville acts, and other entertainments. In his Web site, historian Scott Newman lists twelve beer gardens that flourished between the 1890s and 1930s, including the famous Bismarck Gardens and Rainbo Gardens.[34] Many of these beer gardens evolved from outdoor cafes into full-fledged amusement parks such as Sans Souci and Riverview Park.

TOP Mrs. Reissig's Grove, outside of Riverside, Illinois, included a pavilion for the comfort of picnickers, who flocked to the former rose-grower's grounds. From the author's collection.

BOTTOM This photo of a picnic in the woods on the North Side shows that outdoor air was considered healthy—even for those on crutches. Reproduced by permission of Rogers Park/West Ridge Historical Society.

SUMMER RUSTICATING

If a simple picnic outing was not sufficient to escape city conditions and enjoy the outdoors, more elaborate summer "rusticating" was in order. Ironically, even though Chicagoans had such a relatively short growing season, summer was the time to escape the city. The city's oppressive heat, and later, fears of cholera and other illnesses, sent people away for the summer. As early as the 1850s, horticulturist J. Ambrose Wright noted, "The people who claim to live at home cannot resist the invitation of railroads and cheap fares to go everywhere and see everything. Multitudes are half their time abroad; as our own city empty of all but laborers and strangers from June to October, can testify."[35]

Except for the very poor, nearly everyone sought sanctuary from the heat in the summer, at least for some period of time. Railroads brought inexpensive travel within most budgets. Chicago, the railroad capital of the nation, was seen only as a stopping-over point, however, not a place to linger for the summer. John Bachelder's 1875 book, *Popular Resorts*, listed the "Principal Summer Retreats in the United States." Most of his recommended resorts featured ocean or mountaintop hideaways—amenities clearly unavailable in Chicago. Chicago, he acknowledged, was a railroad center and business emporium, but not a popular resort. Everett Chamberlin, a major Chicago booster, declared in his own book that Chicago wished to attract a "better class" than tourists. "Chicago does not, however, aspire to notoriety as a summer resort. She has suburbs along the high bluffs of the lake shore at the north which have such aspirations, and the reader can easily see that those aspirations are likely to be gratified."[36] Indeed, Bachelder cited (now suburban) Waukegan with its "fifty-six beautiful lakes within its boundaries" as being unrivalled in all the western states as a summer resort. McHenry and Lake Zurich, Illinois, now considered northwestern suburbs of Chicago, were also praised for their resortlike atmosphere.

Bachelder suggested, for those readers who were "worn out by constant visits [to the] tourists' resorts of the East . . . we would invite them to look to the great North-west."[37] In Wisconsin, Milwaukee, Lake Dells, Sheboygan, Devil's Lake, Green Lake, and Geneva Lake were also recommended. Many of these resort towns had already been discovered by Chicagoans, and boasted thriving summer communities. The outdoor life was celebrated, often centering on restorative waters and healthful foods. In Perry Springs, Illinois, newspapers reported, "proprietors have extensive gardens where fresh vegetables and fruits are produced in abundance enabling guests with wholesome diet as well as delicacies found in large cities and towns."[38] The number of summer resorts grew in the latter half of the nineteenth century, appealing to various lifestyles and budgets. The travel page of the *Chicago Tribune* in 1907, for example, listed summer vacations ranging from $60 round-trip train rides to Seattle to weekly stays averaging $10–$12

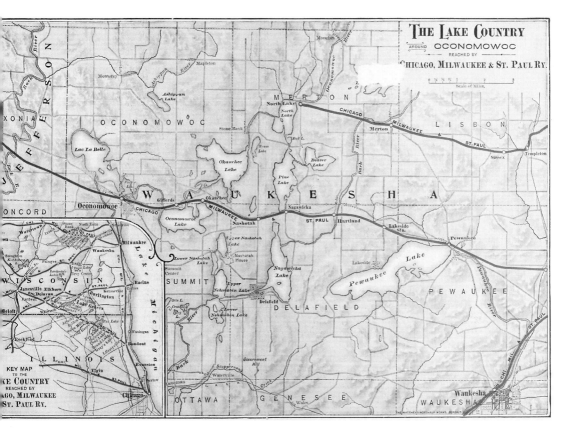

TOP Like many railroads, the Chicago, Milwaukee, and St. Paul published routes that catered to the summer resorts. Here, favorite resorts in northern Illinois and Wisconsin are highlighted. From the author's collection.

BOTTOM LEFT & RIGHT These photos show the contrast between a young boy's Chicago backyard (*left*), and visiting at his grandmother's summer home in Lake Villa (*right*), where gardens and outdoor fun were an important part of the experience. Ca. 1910. From the author's collection.

per week at nearby resorts in Fox Lake, Illinois, or inns in the northern suburbs (the Maplewood in Winnetka, Kenilworth in Kenilworth, and the Netherlands Inn in Evanston).[39]

The early 1900s brought about the country-estate era in America, where a back-to-nature movement countered encroaching industrialization. For the family of modest means, camps or cottages at nearby lake towns such as Fox Lake, Lake Villa, and Crystal Lake, offered a week or two of outdoor fun. Often, erstwhile summer resorts and camping sites evolved into towns of elaborate summer homes for wealthy Chicagoans. The Wisconsin towns of Lake Geneva and Oconomowoc are examples where summer homes were built for industry titans such as the Swifts, the Allertons, Montgomery Ward, the John M. Smyths, the Pinkertons, Sears, Levy Leiter, and the Wrigleys. Marshall Field's department store catered to the summer-home owners with ads declaring the "greatest assortment of summer home furniture ever assembled under one roof." This collection included natural willow furniture, china, Old Hickory furniture, "cottage style" beds, willow birdcages, and flower bowls, "complete with bulbs" for fifty cents.[40] Landscaping around these mansions often rivaled that of their counterpart homes in Chicago.

Frances Kinsley Hutchinson, wife of philanthropist and horticulture patron Charles L. Hutchinson, wrote of the extensive landscaping by the Olmsted firm of her Lake Geneva home, Wychwood, in her book, *Our Country Home*. This couple, one of the most benevolent and influential in Chicago's horticulture milieu, embarked on creating a "small home" in the woods. According to Frances

This formal rose garden was a small part of the overall naturalistic landscape at the Hutchinsons' summer home in Lake Geneva. From the author's collection.

Hutchinson, they wanted the home to blend in seamlessly with the outdoors, and used natural materials in its design. Yet what was small to the Hutchinsons would not represent a modest cottage to others: the multistory stucco and timber mansion included elegant terraces, porches, pergola, greenhouse, and formal garden. The kitchen garden alone was one acre. While the Hutchinsons extolled the value of native plants and were loathe to cut any trees, Frances was drawn to the woods because it was "a real American forest, one might almost say a New England forest."[41] This comment is disappointing for defenders of Midwest landscapes. For if Chicago's undisputed tastemakers at the turn of the nineteenth century, albeit lovers of naturalistic gardens, still pined for New England landscapes, how would rank-and-file Chicagoans ever embrace their own soil?

Dreams of country homes were not limited to the uberwealthy. Many upper-middle-class businessmen fancied a piece of country land in the early 1900s for recreation and income. But farming is a very specialized business, and most would-be country gentlemen bought more land than they need. Five to twenty acres within twenty miles of one's principal residence was recommended by many writers, so that frequent visits were possible.

Fitting this summer-home category was the South Haven, Michigan, hobby farm of Bertha and William Jaques.[42] In 1914, Bertha Jaques, of 4316 Greenwood Avenue, wrote a self-published book, *Country Quest*, about their five-acre farm and orchard. Their reasons for having a second home were typical: improved health, escape from urban congestion, a love of the outdoors, and the possibility of some profit from the orchard. Bertha, a member of the Wildflower Preservation Society, described, month by month, the joys of nature from May through October. She habitually took the five-hour train or boat ride to South Haven, and set up her Rose Cottage while her husband, a doctor, worked back in the city, and traveled to the farm whenever he could. They both relished simple pleasures. "Anyone who has a summer home with a single unnecessary thing in it does not know how to get the most out of a vacation," Bertha wrote. "Live as near like the birds as possible; be active outdoors; eat natural and simple food; be happy and sing. Isn't that a good recipe for freedom?"[43]

Nonetheless, Bertha knew her own limits, noting, "Agriculture is an excellent thing if one does not get too much of it."[44] She enjoyed walks in the woods and the scholarly study of flowers, but not the heavy lifting involved in raising crops. Her husband, an honorary member of the Illinois State Horticultural Society, offered both sides of the producer/consumer questions in marketing fruits. Addressing the society, he noted, "Now, there are two Jaques. In the summer time he is Farmer Jaques over in Michigan on a little farm largely for his health and for the supplying of vegetables and fruit to himself and any surplus for his friends, and in the winter time he is Dr. Jaques, and I want to tell you that Dr. Jaques has about all he can do to keep Farmer Jaques alive."[45] Jaques considered a breakeven position on the farm a good return for its healthful and joyful benefits.

More geographically accessible than the Jaques's South Haven cottage, and more financially achievable for average homeowners than the Hutchinson estate, was the country home of W. C. Egan in Highland Park. While Egan's property certainly rivaled that of many estates on the North Shore, his was more of a do-it-yourself experience, which he willingly and generously shared with others. His article "How I Built My Country Home" charmingly and candidly described the pitfalls and mistakes of a beginning gardener. "I am garden-bred," Egan began his story, "for in the early fifties my father's garden [W. B. Egan] was one of the show-places in Chicago."[46] Egan bought his country property after retiring from a thirty-five-year business career. "Flowers, shrubs and trees did not grow among my business affairs. Nevertheless, I was determined that the place should be of my own creation, and so I resolved to go ahead and make my own mistakes in my own way," he said.

Egan's tale is a lighthearted and unassuming story that rings true with every veteran gardener. He learned design theory through reading and trial and error. He planted exotic trees helter-skelter until his yardman "got dizzy dodging them with the lawnmower." He built a rockery in the center of the lawn that was "fearfully and wonderfully made," and grew flowers on top "but neglected to furnish a step-ladder that they might be seen." After a few years of experimentation, Egan redesigned his property to include more thoughtful plant groupings that provided views and better composition. He relocated the rockery, in a more naturalistic construction, to the top of a ravine where it more resembled a natural outcrop. He found delight in the native wildflowers and trees available. All of these lessons he willingly shared in his writings in books and in magazines such as *Garden and Forest*.

The Egan story is a Chicago gardening success, closing the circle from one generation to another. Where W. B. Egan's garden estate was once a very visible example of current taste in gardening, his son's Egandale, while removed to the North Shore, nonetheless was democratically available to those interested in horticulture. "My garden is open to all who love and appreciate its contents, be they from the mansion in the park or from the cottager's home," W. C. Egan wrote. "My flowers bloom not for me alone, but for those of kindred tastes, and it is a pleasure for me to show them."[47]

OTHER OUTDOOR PURSUITS

As often as possible, Chicagoans sought amusement in the outdoors. Earliest forms of fashionable entertainment included croquet—a game made possible for the leisure class through the careful cultivation of manicured lawns. Croquet,

Croquet was a fashionable game in the late 1860s through the late 1890s. It afforded an opportunity for families to enjoy the outdoors and display their precious lawns. Courtesy of Tinker Swiss Cottage Museum, Rockford, Illinois.

suitable for women and children alike in those genteel times, remained popular throughout the last quarter of the nineteenth century. Bicycling and golf became the next sport crazes to arrive in Chicago—both equally suitable for women, although significantly more strenuous than croquet. Chicago "wheelmen" were often accused of reckless driving on their bikes through the city's parks. Golf courses, in Wheaton, Lake Forest, Riverside, and other upscale suburbs, offered the 1890s outdoorsman and woman a fashionable venue for fresh air and spirited competition.

Hearty hikes were a favorite pastime of nature lovers at the turn of the nineteenth century and beyond. In addition to the Prairie Club, the city of Chicago, in conjunction with railroads, mapped out routes that could be enjoyed by individuals and families. Bird study became a popular hobby in the early 1900s, and various groups formed to participate in bird walks. *Birds of Lakeside and Prairie* by Edward B. Clark was published in 1901 as a popular guide to help identify local species. Residents began to notice the city parks as stopover points on bird migration paths. Alice Hall Walter published *Wild Birds in City Parks*, which offered descriptions and a bird count of over two hundred species as observed between 1898 and 1903 in Lincoln Park.

As the automobile gained ground in Chicago, the city began to promote itself as a center for beautiful scenery. Day trips could be had to outlying nature areas

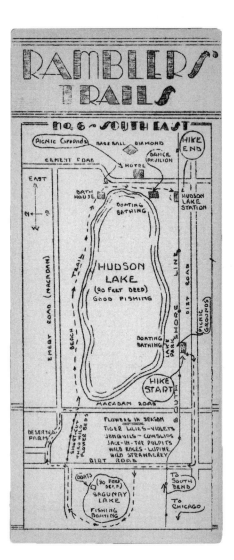

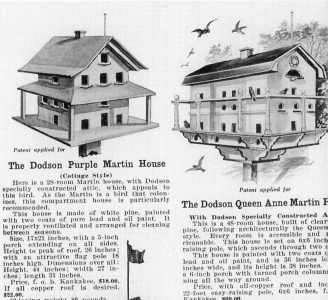

The Dodson Purple Martin House

(Cottage Style)

Here is a 28-room Martin house, with Dodson specially constructed attic, which appeals to this bird. As the Martin is a bird that colonizes, this compartment house is particularly recommended.

This house is made of white pine, painted with two coats of pure lead and oil paint. It is properly ventilated and arranged for cleaning between seasons.

Size, 17x21 inches, with a 5-inch porch extending on all sides. Height to peak of roof, 26 inches; with an attractive flag pole 18 inches high. Dimensions over all: Height, 44 inches; width 27 inches; length 31 inches.

Price, f. o. b. Kankakee, **$18.00.** If all copper roof is desired, $22.00.

Shipping weight 80 pounds.

Patent applied for

The Dodson Queen Anne Martin H

With Dodson Specially Constructed A

This is a 48-room house, built of clear pine, following architecturally the Queen style. Every room is accessible and cleanable. This house is set on 6x6 inch raising pole, which ascends through two s

This house is painted with two coats o lead and oil paint, and is 36 inches lo inches wide, and its height is 38 inches. a 6-inch porch with turned porch column ning all the way around.

Price, with all-copper roof and inc 22-foot easy-raising pole, 6x6 inches, f. Kankakee, $60.00.

Shipping weight 150 pounds.

The Purple Martin is one of the most valua we have. It is so preeminently aerial that its fe essarily consists of flying insects, principal of the mosquito. Every Purple Martin will catch c sume on an average of two thousand mosquitoe and with a colony of these wonderful birds you have a mosquito in your vicinity.

The Dodson Colonial Martin House

(With Dodson Specially Constructed Attic)

Here is a 56-apartment house, colonial style of architecture, built of clear white pine, with copper roof. This house has turned porch columns and paneled pilasters.

This house is set on 6x6-inch easy-raising pole. Pole enters base of house and ascends through two stories. This is a magnificent Martin house and, erected on your place, will cause a colony of Martins to take up their abode as your neighbors.

The dimensions of this house, are: 38 inches long, 26 inches wide, 44 inches high with a projecting 6-inch porch all around. Painted with two coats of pure lead and oil.

Price, all-copper roof and including 22-foot easy-raising pole, f. o. b. Kankakee, **$85.00.** Shipping weight 170 pounds.

Dodson Easy-Raising Pole for Bird Houses

Here is a device that I have perfected which makes an ideal support for all the different kinds of bird houses. It is made of selected yellow pine, thoroughly painted, and is set on a self-locking appliance for raising and lowering.

Size 6x6 inches, 22 feet long, for Martin house. Price, f. o. b. Kankakee, **$11.00.** Shipping weight 115 pounds.

Size 4x4 inches, 20 feet long, for Martin house. Price, f. o. b. Kankakee, **$8.00.** Shipping weight 70 pounds.

Size 3x3 inches, 14 feet long, for Bluebird, Fly Catcher and Swallow houses. Price, **$7.00.** Shipping weight 40 pounds.

Patent applied

Page 3

LEFT Rambles brochures were produced by the Outing and Recreation Bureau of Chicago, encouraging use of public transit to natural wonders such as the Indiana Dunes and Hudson Lake. From the author's collection.

RIGHT Kankakee resident Joseph H. Dodson, director of the American Audubon Society, produced birdhouses of his own design for bird-watchers and gardeners. From the author's collection.

that once took weeks to reach. Picnics, ever popular, were even more accessible as families packed up in the old Model T and toured the countryside, stopping at roadside farm stands on the way. Recognizing the new transportation form, landscape architects such as Jens Jensen strove to design parkways that retained their naturalistic views. The Lincoln Highway near Dyer, Indiana, included an "ideal mile" landscaped by Jensen and funded by Edsel Ford that was to be a prototype for beautiful roads everywhere. Like the railroad barons before them, automakers now looked to Chicago to help spread the beauty of the Garden City outward.

In the 1920s, Chicago positioned itself as a destination city for motorists who enjoyed natural scenery. From the author's collection.

EIGHT

THE NEXT CENTURY

The Century of Progress Exposition in 1933–34 was a fitting finale to one hundred years of horticulture in Chicago. The city had emerged victorious from a swampy lakeshore outpost to a cosmopolitan center, with international gardens elegantly displayed against the breathtaking backdrop of Lake Michigan. Hundreds of thousands of visitors who strolled through world's fair gardens designed by internationally acclaimed landscape architects surely concurred that this was indeed the ultimate Garden City.

Progress continues relentlessly, despite artificial milestones such as centennials or world's fairs. But this century in Chicago's history is particularly important in that it encompasses both Chicago's formative years and the golden era of American horticulture. Between 1833 and 1933, Chicago's achievements figured prominently in the nation's overall horticultural progress. Chicago's noteworthy accomplishments included a market gardening exchange that supplied the nation, world-class landscapes for public parks on the lakefront and in the neighborhoods, a unique prairie style of landscape architecture of international acclaim, the birthplace of American ecology through the work of H. C. Cowles and others, elevated roles for women in horticulture through their strong women's and garden clubs, and innumerable discoveries and improvements in plant culture through mastery of prairie growing conditions.

After the 1933 World's Fair, what beckoned on the horticultural horizon? Many of the formative factors that led to Chicago's horticultural prominence had irrevocably changed. Chicago's role as a railroad hub, for example, and its related importance in distribution and marketing of fresh produce, was transformed with the nascent airline industry. In 1928, Graham W. Dible, a San Diego florist, flew to Chicago with a $1,000 consignment of orchids. "With the airplane," said Dible, "we in California can ship to the middle west flowers as fresh as they can be delivered in California. This trip is only a forerunner of what we intend to do."[1] Anticipating the change, Chicago actively prepared for the potential of air travel. In 1911, the city hosted the International Air Meet in Grant Park. From its small beginnings in the early 1920s as the Chicago Air Park, Midway Airport was declared the "world's busiest" in 1932. In 1946, the city purchased Orchard Field, whose agrarian roots would be forever commemorated in O'Hare airport's call letters, ORD. Planners of the 1933 World's Fair anticipated that Northerly Island would ultimately become an airport. In December 1948, despite the intervening Great Depression and World War II, Meigs Field opened—with seventy-five of the first hundred arriving aircraft operated by Flying Farmers, a nod to Chicago's horticultural roots.[2]

Although Chicago had prepared for the aviation world, the economics of market gardening were forever changed. While Chicago's acres of glasshouses were less relevant in growing flowers for retail, the city's central location still made it a distribution hub. Thus Kennicott Brothers, descendents of John Kennicott, became the first wholesale florist in the Midwest in 1915.[3] Modest truck farms could not compete in scale or price with vegetables grown in warmer climates, and shipped in a matter of hours instead of days. Land for market farming became scarce—post–World War II housing booms gobbled up the remaining small farms and rural homes in the immediate vicinity of Chicago. Corner grocery markets, key customers for local truck farmers, succumbed to retail chains. Locally owned Jewel Tea Company, for example, had one hundred retail stores by 1936 and by the end of World War II was among the ten largest grocery chains in the United States.[4] With such size came significant buying power and volume purchases—more than a local grower could supply.

Chicago became less notable as a grower of the nation's produce than as a distributor and processor of fresh foodstuffs. Still, the city's preeminence as a major center of horticultural activity continued. In 1935, the national headquarters of the Society of American Florists and Ornamental Horticulturists moved to Chicago from its previous location in New York City, where it had been for nearly fifty years. H. Mortimer Brockway, executive secretary of the society, explained, "For the last ten years a change in the center of activities in the florist industry has been under way. For that reason it was considered advisable to move our headquarters from New York to Chicago, which is now the largest flower market in the world."[5]

With the luxury of large retail grocers, the need for kitchen gardens lessened, except in periods of economic strife. Physic gardens for homemade herbal cures were less important as modern medicine became more accessible. Major pharmaceutical companies in the Chicago region (e.g., Abbott Labs ca. 1900, G. D. Searle ca. 1908, and Baxter Labs ca. 1931) advanced scientific drug formulas and rendered homespun tinctures less popular. Gardens for sustenance thus became oddities, an indulgence for hobbyists rather than essential elements of the backyard.

The Plant Patent Act of 1930 drastically altered the world of plant innovation, knowledge exchange, and garden design. Previously, professional nurserymen or amateurs who created new cultivars were rewarded only with the fleeting fame of the plant's introduction. There were no guaranteed financial rewards. Plant breeders hoped to capitalize on their inventions by getting as much public exposure to the new plant as possible—often through flower shows and world fairs. Now, with plant patents, despite the loopholes in early legislation, plant breeders could attain fair business profits from their hard labor.

Chicago-area plantsmen were among the first to secure plant patents. Among the big names, such as Luther Burbank and major businesses such as Jackson and Perkins and Stark Brothers, in the first twenty plant patents issued were two Chicago-area plant inventors. Patent 11 was issued in 1932 to Christian W. Hjermind of Maywood and Paul E. Weiss of River Forest from Premier Rose Gardens of Maywood. Their new hybrid tea rose offered a novel pinkish red color with semidouble flowers and excellent fragrance. That same year, Harold L. Ickes, attorney, reform activist, and politician from Hubbard Woods (who later became secretary of the interior under Franklin D. Roosevelt), obtained patent 19 for his new light coral red dahlia, which sported an especially sturdy stem.

The nature of flower shows and fairs changed in response to these advances in patent possibilities and in transportation. Chicago would not host another world's fair in the twentieth century. In 1936, however, the city hosted its first International Horticultural Exposition in the Amphitheater at 43rd and Halsted streets. Ironically, the Amphitheater was located in the midst of the stockyards, thus perpetuating Chicago's historical marriage of meat and potatoes. More than $16,000 in prizes was offered in premiums—a substantial lure for exhibitors.[6] Exhibits and visitors arrived primarily from Canada and the United States. A sign of the times, exhibitor Charles Garrety of California arrived by airplane to display his new dahlia. With new varieties no longer the great novelties they once were, Chicago-area flower shows of the 1930s and 1940s, particularly those hosted by local garden clubs, and the annual show hosted by the Garden Club of

Illinois, began to emphasize the aesthetics of floral arrangements as opposed to plant culture or new plant introductions.

As networking groups for serious horticultural knowledge exchange, many of the area's societies shifted focus. After the fanfare of the world's fair, the CHS foundered and was dissolved in 1937 after its charter was canceled by the State of Illinois. In 1943, in large measure to help inform and inspire the victory garden and vacant-lot community gardening, Fred G. Heuchling of the Chicago Park District and R. Milton Carleton, research director of Vaughan Seed Company, begin a movement to revive the society.[7] Two years later, the society reorganized as the Chicago Horticultural Society and Garden Institute.[8] The temporary president was Carl Cropp Sr. of Hinsdale, and the secretary was Edwin A. Kanst. The society operated a garden center, more an information center than the retail outlets we think of today, inside the Chicago Public Library (now the Chicago Cultural Center) in the Randolph Street annex. Early trustees of the reconstituted society included wealthy amateurs, such as Laurence H. Armour, Mrs. Joseph Cudahy, and Lester J. Norris. President of the society was C. Eugene Pfister, a rosarian from Mundelein, who was also president of the National Rose Gardens Association. The society exists today, and manages the world-famous Chicago Botanic Garden in Glencoe.

In the vacuum left by the CHS, in 1931 Frank K. Balthis of the Garfield Park Conservatory proposed a new horticultural society. His comments are telling: "This society, if formed, would not interfere with the present garden clubs, but would have as a prerequisite for membership a serious interest in horticulture."[9] While the implication may have been that many garden clubs now met for social purposes rather than knowledge exchange, Balthis bore no ill will toward the clubs. When the Midwest Horticultural Society did formally organize, in 1934, garden club members and professional horticulturists were included among the three hundred charter members. Lectures were geared to amateurs and professionals alike, and programs were more broad-based than the ones that individual garden clubs could offer. Founding members, in addition to Balthis, included Paul Battey of Glencoe, president of the National Iris and Peony Growers' Association; August Koch of Garfield Park; John Servas, director of the Century of Progress horticultural exhibitions; and various garden club presidents.

The greatest increase in horticulture networking groups was in garden clubs. In 1927, the Garden Club of Illinois included 17 women's garden groups. By 1950, the club coordinated activities of 175 local groups with more than 8,000 individual members. Men's garden clubs grew from the first established by Leo W. Nack in 1927, to groups in many Chicago-area suburbs with total membership of about 1,000 members. Still, as of 1950, distinctions were made among men's and women's clubs. "Women form clubs to study horticulture to make gardens. Men tend to create a laboratory," said Mrs. Raymond Knotts, one time president of the Garden Club of Illinois.[10] Despite these stereotypes, women horticulturists made strides during the war years. The floral equivalent of Rosie the Riveter, women gardeners secured jobs with the Chicago Park District during labor shortages in the 1940s. Unlike the hapless Annetta McCrea who was fired after a short tenure with the district, these women thrived. A 1943 *Chicago Tribune* article summarized an observation from Lincoln Park Conservatory's head horticulturist, C. Edwin Hewitt, that the "women gardeners are displaying a patience and care rarely shown by their male coworkers."[11] Following the park district's example, commercial florists began hiring women throughout the year, not just in busy seasons.

In the 1920s and 1930s, many leaders in Chicago's horticultural world either moved from Chicago or passed away. Of the notable wealthy amateurs, Charles L. Hutchinson died in 1924, W. C. Egan in 1930, Joy Morton in 1934, Cyrus H. McCormick in 1936, and Martin A. Ryerson in 1932. Although the fabled "country-estate" era had also declined, wealthy Chicagoans continued to explore emerging trends in landscaping. New generations of horticulture patrons carried the torch for innovative landscapes, many descendents of early Chicagoans, some newly arrived. One difference in the symbiotic relationship among patrons and practitioners of horticulture resulted from the smaller property sizes—even among North Shore estates. With the Depression and war years, many estates were subdivided. Newly built estates rarely included the outbuildings of 1900 gentlemen farms, and very few had sufficient space for gardener's cottages. The automobile further reduced the need for a resident coachman, who had often doubled as the groundsman cum gardener.

Thus, the gardener was no longer a part of the family household, no longer a collaborator, despite class differences, in the excitement of raising new plants. Landscape firms recognized the profit opportunities, and design/maintenance became a standard service offering. With maintenance outsourced to a third party, many patrons became less personally involved with the specifics of horticulture. The days of William B. Ogden hauling plants back from Calumet were long gone, and even casual, chance exchanges between gardener and estate owner would be rare with an impersonal weekly visit from the landscape crew. With less horticulturally informed homeowners, landscape designs bore the imprint of the designer, rather than of the client. Predictably, a certain similarity grew among estate landscapes, as clients imposed less of their personality in the design.[12] Nonetheless, Chicago's private/public partnership in horticultural endeavors has sustained an enduring successful relationship—as evidenced most recently by the extraordinary synergy in building the city's spectacular Millennium Park.

Just as the faces of Chicago's wealthy society members changed, so did those of its new immigrants and the gardens they created. The so-called Great Migration of black Americans from the South to Chicago and other northern cities during the 1920s brought new traditions. Like other newcomers before them, the African American community used horticulture to beautify their surroundings. In 1941, the *Chicago Tribune* reported of a South Side group, "The Community Garden club, whose object is to improve the appearance of the district, recently was organized by Negro housewives, who plan to convert empty lots into flower and vegetable gardens."[13] The club had 350 members and offered lectures on flower growing, canning, and pragmatic job skills such as typing and shorthand.

Even established immigrant groups continued their traditions after the 1933 World's Fair. World War II brought new waves of immigrants, some of whom assumed the traditions of their forebears, others who sublimated their identities due to wartime tensions. Those whose native language was not English continued to publish gardening material in their own language, and had stores catering to specific customs. Germans, once Chicago's major immigrant group

The *American Landscape Architect* magazine was among Chicago's twentieth-century efforts to promulgate good landscape design. 1930. From the author's collection.

and leaders in Chicago's farming and hothouse industries, often felt the brunt of suspicion during the wars. (Ironically, Ernest Gries, known as the kaiser's war gardener who kept German troops supplied during World War I, lived on Chicago's South Wallace Avenue during World War II—and was chairman of the Kiwanis' junior victory garden committee of Roseland.[14]) In another show of changing tastes due to world events, after Pearl Harbor, the fad for Japanese gardens waned.

Of the professional nurserymen, scientists, landscape architects, and florists, many early talents were also lost in the 1920s and 1930s. David Hill passed away in 1929, O. C. Simonds in 1931, Willis N. Rudd in 1925, J. C. Vaughan in 1924, H. C. Cowles in 1939, and Frederick Kanst in 1937. In 1935, Jens Jensen moved to his summer home in Ellison Bay, Wisconsin, to focus on his "school of the soil." Who then took up the torch for landscape design and for the discovery and exploration of nature? An educated eye is needed to recognize the undeniable grace in a neglected landscape from the last eighty years, or a lasting contribution from an inspired teacher. Some names of designers and teachers who took up the torch are readily recalled: Alfred Caldwell, Gertrude Kuh, Franz Lipp, and naturalists May Watts and Donald Culross Peattie. Others await rediscovery.

Landscape design trends naturally evolved. The modernist movement, as reflected in some of the gardens at the COP, enjoyed its heyday with smooth lawns, sculptured evergreens, and minimal floral fanfare. The automobile caused design changes—Gertrude Kuh, for example, embedded gracious motor courts in front of her clients' estates. Roadside plantings of thickets to block auto headlight glare were planned by the Illinois highway department landscape architects.[15] Predictably, some garden styles waxed and waned as fads reemerged from the past. Lawns, for example, were the Victorians' status symbol. During the Depression and war years, many lawns were converted to productive vegetable gardens. A "Save the Lawn" campaign was initiated in the 1940s. "We don't want the hysteria of 1917–18 to be repeated," Men's Garden Club president Hoyt F. Paxton of Elmhurst was quoted as saying.[16] Paxton recalled that there "wasn't a golf course, or flower garden, or a parkway safe from committees organized to plant fresh vegetables." While Paxton and others' main concern was cannibalization of public parks, the urge to have a glorious lawn was renewed—and encouraged through the marketing of new chemical fertilizers.

The integration of landscape architecture and architecture, emphasized during the 1920s and 1930s, intensified during the midcentury. Beginning with the Lake Forest summer school of architecture and landscape architecture that opened in the 1920s, the idea of a comprehensive house and garden plan radiated from the North Shore to other communities, both affluent and middle class. Like the suburban developers in the 1870s, modern-day developers emphasized green space, and often employed a landscape architect as an integral member of the planning team. In 1936, industrialist Kenneth G. Smith and "retired capitalist" R. A. Crandall bought 1,150 acres in Barrington to build their own estates and other upscale homes of not less than ten acres.[17] Landscape architect John Bell was commissioned to design the roads, artificial lakes, bridle paths, and plantings in the hilly development known as Knollwood Farms. In the same year, even the city's largest federally subsidized housing projects, the Julia C. Lathrop Homes on Diversey and the Chicago River, emphasized landscaping. Over 66 percent of the total thirty-five acres was devoted to playgrounds, lawns, and other open space, and was designed by Jens Jensen.[18] Lower-income housing now had gardens that earlier Hull-House clients could only imagine.

Between these housing extremes, in 1939, 326 acres of the venerable Swain Nelson's nursery in Glenview was developed into Swainwood by the Swain Nelson Realty Company. Each of the homes sold were to include landscaping—a new idea in suburban development. Swainwood was designed by Gerald F. Nelson, landscape architect, and, according to Seymour G. Nelson, the eighty-one-year-old president of Swain Nelson and Sons Company, there were more than a "million pieces of landscape material on the property."[19] More modest suburbs, such as Hometown, developed in 1949 between 87th and 91st streets on Pulaski (now Crawford) Avenue, featured smaller lots of forty feet. Yet William A. Dean, the master planner and landscape architect for the subdivision, included staggered setbacks, curvilinear streets, and cul-de-sacs to avoid Chicago's grid patterns. Dean became known as a designer for modest homes. In 1950, responding to post–Word War II building booms, a "budget house" was built at a Navy Pier Greater Chicago Home Fair. Even while the two-bedroom frame exhibit home was touted as the most economical, Dean was hired to landscape the grounds—an indication that gardens were now considered as integral as kitchens and baths. Indeed, gardens were synonymous with home—and Chicagoans were loathe to leave them.

It would be presumptuous and woefully inadequate to try to summarize the intervening decades between 1933 and the present day in a few paragraphs such as these. Chicago's second century of horticulture is still under way, and needs the objective lens of historical perspective for an accurate portrayal. Both the beauty and tragedy of a garden is its ephemeral nature. One can mourn the seeming disregard for the landscape art of yesterday, the heritage of a talented garden maker. But a more hopeful position takes account of the many lost gardens that have been reborn. Egandale, W. B. Egan's cherished sylvan retreat, was subdivided a few short years after his death, only to be reincarnated as Egandale in Highland Park, the garden glory of his son. O. C. Simonds's beautiful home and garden in Chicago were destroyed to make room for new developments, yet his retreat in Pier Cove, Michigan, remains. In only twenty short years, the outdoor theater crafted in Jensen's original Columbus Park masterpiece had been

Chicago's gardening tradition continues in Millennium Park. A newly planted prairie garden near the lakefront frames the city's skyscrapers, where once Fort Dearborn's garden flourished. 2006. Photo by the author.

largely forgotten. Through a case of mistaken identity, Jensen himself, while frail at eighty-nine years old, was reported as having died. Fame and floral tributes are fleeting, yet Jensen himself reassured us, "Let the garden disappear in the bosom of nature of which it is a part, and although the hand of man is not visible, his spirit remains as long as the plants he planted grow and scatter their seed."[20] Indeed, Columbus Park has been wonderfully restored in recent years.

Change is the promise of the future, and Chicago continues to push the gardening frontier while safekeeping its horticultural tradition. Mayor William B. Ogden christened the fledgling community the Garden City; today's Mayor Richard M. Daley's city beautification and environmentalism earns his moniker as the "green mayor." Where root vegetables were once barely grubbed out of Fort Dearborn, an exquisitely designed prairie garden flourishes in today's Millennium Park. Northerly Island, once home of the 1933 World's Fair gardens, became a busy municipal airport, and now is a serene wildflower garden and bird sanctuary. From the lessons of our horticultural heritage, Chicago will build the gardens of the future.

PLANT LIST BY TIME PERIOD

Plants in this chart are representative of those popular during the period as based on writings from influential horticulturists, nurserymen, and growers in the Chicago area. While not an exhaustive list by any means, it captures recommendations from leading tastemakers of the period. *Note that the plant names are printed as they were used in the actual source, and may not reflect today's nomenclature.*

PERIOD	VARIETIES
	Apples
1830–59	*Summer*: Carolina June, Sweet June, Kirkbridge White, Summer Pearmain, Early Harvest, Early Strawberry, Large Yellow Bough, Kenrick's Codlin, Red Astrachan, Autumn Pearmain, Maiden's Blush, Gravenstein, Wine Apple, Holland Pippin, Lyman's Pumpkin Sweet, Porter Apple, Rambo, Ross' Nonpareil
	Winter: Rhode Island Greening, Baldwin Apple, Belle Fleur, Gloria Mundi, Lady Apple, Newtown Pippin, Blue Pearmain, Swaar, Ladies' Sweeting, Esopus Spitzenburg, Rowle's Janet or Jeanneting[1]
1860–89	*For profit (market)*: Carolina Red June, Red Astrachan, Early Pennock, Pomme de Neige, Domine, Jonathan, Winesap, Rawles' Janet, Willow Twig, Gilpin
	For taste: Early Harvest, Carolina Red June, Benoni, Rambo, Pomme de Neige, Yellow Belleflower, Domine, Jonathan, Winesap, Rawles' Janet[2]
1890–1919	*Summer*: Astrachan, Duchess, Transparent, Bough, Benoni
	Fall: Wealthy, McMahon, Fameuse, Ramsdell, McIntosh, Milwaukee, Porter
	Early winter: Jonathan, Grimes, Dominie, Canada, Bailey
	Late winter: Minkler, Salome, Willow, Tolman, Northwestern, Staymans, Black Ben Davis, Black Twig
1920–39	Duchess, Yellow Transparent, Wealthy, Fameuse, McIntosh, Jonathan, Grimes, Delicious, Northwestern Greening, Minkler, Salome, Willow, Staymans[3]
	Ornamental Trees
1830–59	Honey locust, black locust, catalpa, Kentucky coffee-tree[4]
	Sugar maple, white (silver) maple, red maple, American white elm, red (slippery) elm, burr oak (*Quercus Macrocarpa*), laurel oak (*Quercus Imbricaria?*), beech, white ash, blue ash, cucumber tree, tulip tree (*Liriodendron*), buttonwood (*Platanus*), linden, honey locust, common locust (*Robinia*), hickory, coffee tree (*Gymnocladus*), buckeye, Osage orange, dogwood (*Cornus Florida*), American mountain ash[5]
1860–89	Sugar maple, ash-leaf maple, white ash, white elm, honey-locust, black-walnut, butternut[6]
1890–1919	Native sugar or rock maple, cut-leaved weeping birch, May day tree (*Prunus Maackii*), wild olive (*Elaegnus angustifolia*), dwarf maple (*Acer ginnala*) as a substitute for Japanese maples popular in the East, Colorado blue spruce[7]
1920–39	Red, scarlet and black oaks, red and sugar maples, flowering dogwoods[8]
	Street Trees/Shade Trees
1830–59	Soft maple, elm, ailanthus, beech[9]
1860–89	American white birch (*Betula populifola*), American elm (*Ulmus americana*), European larch (*Larix europoca*), American linden (*Tilia americana*), honey locust (*Gleditschia triacanthus*), ash leaved maple (*Negundo aceroides*), white walnut (*Juglans cinerea*)[10]

1890–1919   American elm (*Ulmus Americana*), sugar maple (*Acer saccharum*), silver maple (*Acer saccharinum*), Norway maple (*Acer platanoides*), white ash (*Fraxinus Americana*), American and Oriental sycamores (*Platanus occidentalis and orientalis*), American linden (*Tilia Americana*), European linden (*Tilia platyphyllos*), pin oak (*Quercus palustris*), red oak (*Quercus rubra*), honey locust (*Gleditschia triacanthos*), ailanthus (*Ailanthus glandulosa*), hackberry (*Celtis occidentalis*), western catalpa (*Catalpa speciosa*)[11]

1920–39   Ash (*Fraxinus Americana*), beech (*Fagus*), birch (*Betula*), catalpa (*Catalpa speciosa*), elm (*Ulmus*), honey locust (*Gleditschia triacanthos*), horse chestnut (*Aesculus Hippocastanum*), European linden (*Tilia platyphyllos*), mountain ash (*Sorbus*), Norway maple (*Acer platanoides*), oak (*Quercus*), Oriental plane (*Platanus orientalis*), poplar (*Populus*), purple leaf maple (*Acer platanoides Schwedleri*), sugar maple (*Acer saccarum*)[12]

Deciduous Shrubs

1830–59   Dwarf double flowering almond, rose of Sharon, syringa (mock orange), rose acacia, spirea, snow ball, high cranberry, lilac, barberry, white Tartarian, English fly, pink azalea[13]

Lilac, upright honeysuckle, snow ball viburnum, spirea, deutzia, weigela, Japan quince, dwarf flowering almond, philadelphus[14]

1860–89   Rose of Sharon (*Althea*), deutzia, almond, flowering currant, forsythia viridissima, Japan quince, lilac[15]

Serviceberry (*Amelanchier vulgaris* and *A. Botryapium*), aralia Japonica, Japan quince (*Prunus Japonica flore albo pleno* and *P. Japonica flore rubro pleno*), witch hazel (*Hamamelis Virginica*), hydrangea (*Hydrangea paniculata grandiflora*), honeysuckle (*Lonicera*), mock orange syringa (*Philadelphus coronaries, P. coronaries flore pleno, P. foliis aureis, P. Gordonianus,* and *P. grandiflorus*), currant (*Ribes sureum, R. gordonianum,* and *R. sanguineum*), sassafras, spirea, snowberry (*Symphoricarpus racemosus* and *S. vulgaris*), lilac (*Syringa Josekoea, S. Persica, S. rothomagensis, S. Siberica, S. Sinensis,* and *S. vulgaris*), arrow-wood (*Viburnum*), wild barberry (*Berberis Canadensis*), common barberry (*B. vulgaris* and *B. vulgaris purpurea*), wild Carolina allspice (*Calycanthus floridus*), button-bush (*Cephalanthus occidentalis*), dogwood (*Cornus Siberica foliis albo-marginatis*), American hazelnut/filbert (*Corylus Americana*), cotoneaster mumularia, American strawberry bush (*Euonymus Americanus*), St. John's wort (*Hypericum*)[16]

1890–1919   Spirea van Houttei, Persian lilac, Japan barberry, forsythia, green and red barberry, philadelphus, privets, Japan quince, spirea prunifolia, hydrangea, forsythia, lilacs, highbush cranberry, snowballs, Cornelian cherry, Tartarian honeysuckle, snowberry, Indian currant, weigela, rosa rugosa, sweet briar, elder, red and white dogwood[17]

1920–39   Japanese barberry, Spirea Vanhouttei, bush honeysuckle, philadelphus, hydran-
           gea, weigela, forsythia, red-twigged dogwood, snowberry, lilac, rosa rugosa,
           Persian lilac, golden dogwood, althaea, highbush cranberry, snowball, kerria,
           Japanese quince, golden elder, cutleaf elder[18]

           Japanese barberry (*Berberis Thunbergii*), red-branched dogwood (*Cornus San-
           guinca*), forsythia, sea buckthorn (*Hippophaea Rhamnoides*), hydrangea (*Aborescens
           Grandiflora* 'Snowball'), Chinese privet (*Ligustrum Ibota*), bridal wreath (*Spiraea
           Van Houttei*), lilacs[19]

           Bridal wreath, Japanese barberry, bush honeysuckle, pink flowering almond,
           mock orange.[20]

Evergreen Trees/Shrubs

1830–59   Balsam, white pine, arborvitae[21]

           Norway and native spruces, Scotch, Austrian and native pines, balsam fir, Sibe-
           rian and American arborvitae[22]

           Holly (*Ilex opaca*)[23]

1860–89   Norway spruce, Scotch pine, red cedar, common juniper, American arborvitae,
           hemlock spruce[24]

           Daphne (*Dephne Cneorum*), holly barberry (*Mahonia Aquifolium*)[25]

1890–1919 Norway spruce, balsam fir, white pine, Douglas spruce[26]

1920–39   Norway spruce, arborvitae, red cedar (*Juniperus virginiana*), Colorado blue spruce,
           balsam fir (*Abies balsamea*), hemlock (*Tsuga canadensis*), Sabine juniper, dwarf
           mountain pine (*Pinus Mugho*), Scotch pine, white pine, American yew, concolor
           fir (*Abies concolor*)[27]

Annuals

1830–59   Nasturtium (*Tropeolum majus*), hollyhock (*Althea Rosea*, white, deep scarlet, pur-
           ple, black, light cinnamon colored), China aster (*Aster Chinensis*)[28]

           Sensitive plant (*Mimosa perdica*), globe amaranth (*Gomphrena Globosa*), cocks-
           comb (*Celosia Cristata*), lady's slipper (*Balsamine hortensis*), Indian shot plant
           (*Canna indica*), African marigold (*Tagetes erecta*)[29] portulacca, petunia chryseis,
           phlox Drumondii, balsam, candytuft (red and white), German stock, rocket lark-
           spur, crimson cockcomb, German aster, cypress vine, morning glory, sweet pea[30]

1860–89   Asters, love lies bleeding (*Amaranthus Ruber*), balsams, Venus' paint brush (*Caca-
           lia*), cockscombs, calliopsis, collinsia, convolvulus, pinks (*Dianthus*), *Gomphrena
           Globosa*, sweet pea (*Lathyrus Odorata*), *Malope Grandiflora*, mathiola, nemophilas,
           Phlox drummondii, petunia, portulaca, rosea odorata, marigold (*Tagetus*), zinnia
           elegans

           Bedding plants: verbena, geraniums, heliotropes, petunias, fuschias[31]

1890–1919 China asters (Queen of the Market, Victoria), candytuft, single dahlias, China or
           Indian pinks (*Dianthus Chinensis*, *Diadematus*, and *Heddewigii*), Drummond phlox,
           marigolds (Eldorado orange, pale yellow African marigolds, striped French vari-
           eties), mignonette, nasturtiums (dwarf sorts for massing, Lobbianum for pick-
           ing flowers), pansies (English, Trimardeau, Cassiers, Bugnots, Odiers), Scarlet
           Sage (*Salvia splendens*), ten-week stocks, sunflowers (*Helianthus cucumerifolius* or
           'New Miniature')[32]

1920–39   Ageratum, alyssum, snapdragon (*Antirrhinum*), aster, calendula (pot marigold), dusty miller (*Cineraria maritime*), cornflower (*Centaurea cyanus*), cosmos, heliotrope, annual larkspur, marigold, love-in-a-mist (*Nigella*), nasturtium, pansy, petunia, *Phlox drummondi*, poppy, portulaca, mourning bride (*Scabiosa*), sweet pea, stocks, zinnia[33]

Sweet peas, larkspur, colliopsis, calendula, candytuft, pansies, poppies (California, Shirley, Flanders), gaillardia, alyssum, snow on the mountain, Euphorbia Variegata, forget-me-not[34]

1. "A List of Apples," *Prairie Farmer*, March 1846, 80–81.

2. *Transactions of the Illinois State Horticultural Society*, 1863, 74.

3. "List of Fruits for Northern Illinois," *Transactions of the Illinois Horticultural Society*, 1933, 39.

4. M. L. Knapp, "Arboriculture," *Prairie Farmer*, 1843, 221–22.

5. *Prairie Farmer*, April 1853, 159–60.

6. J. H. Garrison, "Ornamental Trees, Shrubs, Hedges, etc.," *Transactions of the Horticultural Society of Northern Illinois*, 1876, 366–71.

7. Edgar Sanders, "Heavy Trees for Western Prairies," *Prairie Farmer*, 1893, 7.

8. Frank Ridgeway, *Flower Gardening by Farm and Garden Bureau* (Chicago: Chicago Tribune, 1929), 40.

9. "What Is the Best Shade Tree?" *Prairie Farmer*, May 1847, 161.

10. "Tables of Fruits, Etc.," *Transactions of the Illinois State Horticultural Society*, 1868, 14–19.

11. J. H. Prost, "What Chicago Is Doing for Its Trees," *Country Life in America*, August 1910, 19–20.

12. Fritz Bahr, *Commercial Floriculture* (New York: A. T. De La Mare Co., 1922), 227–33.

13. "Orchard and Garden," *Prairie Farmer*, May 1846, 146–47.

14. John A. Kennicott, "Theory and Practice of Horticulture on a Farm," *Prairie Farmer*, 1859, 246.

15. Edgar Sanders, "A Chapter on Flower Gardening," in *The Prairie Farmer Annual* (Chicago: Prairie Farmer Co., 1868), 25.

16. O. C. Simonds, "Mr. Simonds' List of Shrubs," *Prairie Farmer*, 1887.

17. E. Bolinger, "Arrangement of Shrubs and Ornamental Plants in the Home Grounds," *Transactions of the Illinois State Horticultural Society*, n.s., 45 (1911): 62.

18. Bahr, *Commerical Floriculture*, ranked list of "Twenty Popular Shrubs," 141.

19. James H. Burdett, "Landscaping the Small Home Grounds," Garden Talks by Radio, broadcast by Vaughan's Seed Store and WDAP, April 22, 1924.

20. Ridgeway, *Flower Gardening*, 39.

21. "What Is the Best Shade Tree?" 161.

22. Kennicott, "Theory and Practice of Horticulture," 230.

23. Knapp, "Arboriculture," 221–22.

24. Garrison, "Ornamental Trees," 366–71.

25. Simonds, "Mr. Simonds' List of Shrubs."

26. Bolinger, "Arrangement of Shrubs," 61.

27. Bahr, *Commerical Floriculture*, 205–6.

28. "Orchard and Garden," 146–47.

29. "Annual Flowers," *Prairie Farmer*, 1850, 179.

30. "Annual Flowers," *Prairie Farmer*, May 1853, 168–69.

31. Sanders, " Chapter on Flower Gardening," 17–24.

32. "Fourteen Good Annuals," *Gardening*, February 15, 1894, 164–65.

33. Kate Brewster, *The Little Garden for Little Money* (Boston: Atlantic Monthly Press, 1924), 96–99.

34. Ridgeway, *Flower Gardening*, 17–18.

CHICAGO PLANTS

Here are some plants named after notable Chicago figures and places.
The names were gleaned from various flower and seed catalogs in the
author's collection.

NAME	PLANT	DESCRIPTION
Bridgeport Chicago Drumhead	Cabbage	Large drumhead cabbage as grown in Chicago's Bridgeport neighborhood.
Chicago (Westerfield) pickle	Cucumber	Particularly suited for pickling.
Chicago Parks lawn seed	Grass	Mixture of seeds from several states and foreign countries, named after the display lawns in Chicago's parks.
Mrs. O. W. Dynes	Water lily	A tropical water lily created in Garfield Park in 1938 through a cross between *Nymphaea* 'Mrs. Edward Whitaker' and *Nymphaea* 'Colonel Lindbergh'. Light blue with golden stamens. Named for a Garden Club of Hinsdale member.
Mrs. W. C. Egan	Chrysanthemum	Light pink chrysanthemum that opens gradually from the center outward to light creamy yellow. Semicompact, flattened hemispherical, 6–8 inches across. Broad, incurved petals and fairly good form as ideal chrysanthemum. Offered by Fred Dorner and Sons, Lafayette, IN.
William C. Egan	Rose	Climbing rose. Double flesh-pink flowers borne in large clusters, scent moderate. Parentage: *Rosa wichurana* × 'Général Jacqueminot'. Introduced by Dawson/Hoopes in 1900.
Egandale	Canna	Bronze-leaved, 3½ to 4 feet bedding canna. Strong, compact flower spikes, currant red color.
Mrs. John J. Glessner	Chrysanthemum	Bronze, irregularly shaped, incurved Japanese form petal
Alice Higinbotham	Carnation	Pure white carnation, created at Higinbotham's conservatories at Harlowarden.
Florence Higinbotham	Carnation	Deep pink carnation.
Mrs. H. N. Higinbotham	Chrysanthemum	Violet pink, curved, broad petaled. Named after wife of president of 1893 World's Fair.
Mrs. E. G. Hill	Chrysanthemum	Bright salmon, upper petals longer and narrower than bottom. Deeper color in center with tiny white base on lower three petals.
Mrs. Charles L. Hutchinson	Chrysanthemum	Large, white, Japanese form.
Mrs. Francis King	Gladiolus	Separate flowers measure 4 to 5 inches, stalk is 5 feet high, 5 to 6 flowers open on a spike. Winner of SAF certificate of merit.

NAME	PLANT	DESCRIPTION
Lake Forest Collection	Roses	15 everblooming roses, including 'Countess of Gosford', 'Etoile de France', 'General McArthur', 'Gruss an Teplitz', 'Kaiserin Augusta Victoria', 'Killarney', 'Lady Ashtown', 'La France', 'Madam Abel Chatenay', 'Madam Jules Grolez', 'Madame Jenny Guillemot', 'Rhea Reid', 'Richmond', 'Striped La France', 'W. R. Smith'.
New Chicago	Canna	Foliage rich green, spikes tall, scarlet flower.
Mrs. Potter Palmer	Carnation	Deep red carnation.
Mrs. Potter Palmer	Chrysanthemum	Yellow, Japanese form chrysanthemum.
Mrs. George Pullman	Chrysanthemum	Large, full flowers, of deep yellow.
Mrs. Julius Rosenwald	Gladiolus	Deep wine-red.
Mrs. M. A. Ryerson	Chrysanthemum	White seedling offered by Nathan Smith and Sons.
Mrs. John G. Shedd	Carnation	Variegated pink and white.
Selma Swenson	Sweet pea	Pink, grown by Gustav Swenson of Elmhurst ca. 1910.
J. C. Vaughan	Canna	Purple foliage, large flower clear, deep vermillion orange.
Leonard Vaughan	Canna	4½ feet high, rich deep bronze foliage, strong grower that flowers freely. Bright scarlet flower.
Mrs. J. C. Vaughan	Carnation	White flowers 3½ inches across. Vigorous growth and branching freely. Stems long, stiff, and wiry. Excellent keeping qualities.
Vaughan's Columbian lawn grass seed	Grass	Fine grasses that thrive under shade or in light sandy soil. This mixture was extensively used on the world's fair grounds.
Florence Vaughan	Canna	Green-leaved, 5-foot bedding canna. Flowers are rich golden yellow, thickly dotted with bright golden red.
Frances Willard	Peony	White/near white.
Winnetka Collection	Roses	12 everblooming roses marketed by Vaughan's, including 'Captain Christy', 'Caroline Testout', 'Duchess of Albany', 'Etoile de France', 'Gloire Lyonnaise', 'Gruss an Teplitz', 'Kaiserin Augusta Victoria', 'Killarney', 'La France', 'Madam Abel Chatenay', 'Mildred Grant', 'Red Kaiserin'.

KEY NAMES AND GROUPS IN THE GARDEN CITY, 1833–1933

ATWATER, ELIZABETH (1812–78) Born in Norwich, Vermont, Elizabeth Emerson Atwater was a consummate collector. Her plant herbarium collected during 1856–76 (now in the Peggy Notebaert Nature Museum) contains some of the earliest specimens from the Chicago area.

BABCOCK, HENRY H. (1832–81) Many of the earliest specimens available today in the herbaria of Chicago's prominent museums were collected by Dr. Henry H. Babcock. Raised in an educated family, Babcock attended Dartmouth for two years. The death of his father forced him to leave college and assume teaching positions in New England. He moved to Chicago in 1867. There he founded the Chicago Academy, deemed "one of the most successful private schools in the city."[1] He was president of both the Chicago Academy of Sciences and the Illinois State Microscopical Society.

BALL, GEORGE (1874–1949) George J. Ball began his growing operations in 1905, moving to the western suburbs of Chicago in 1909. His seed business first emphasized asters, sweet peas, and candulas. In 1927, Ball constructed several greenhouses in West Chicago. Ball Seed became an extended family enterprise with Ball's sons, with his grandchildren ultimately joining the company. Ball Seed Company, still based in West Chicago, flourishes today with international and national sales.

BLAIR, JOHN (1820–1906) Born in Callander, Scotland, John Blair moved to Rockford, Illinois, from Canada in 1854. Real estate developer and manufacturer H. W. Austin hired Blair as a landscape gardener in the 1860s. Blair landscaped Chicago's Sanitary Fair of 1865, which brought accolades to him and the city. He is credited with working on the early landscapes for Garfield, Humboldt, and Douglas parks. He later moved to Colorado.

BREWSTER, KATE (1879–1947) Lake Forest author of *The Little Garden for Little Money*, Kate Brewster was influential in the garden clubs. She and her husband, Walter, both art patrons, established Covin Tree, a Lake Forest estate designed by Olmsted Brothers.

BROOKS, SAMUEL (1793–1875) English-born Samuel Brooks is generally acknowledged as Chicago's first florist. He arrived in Chicago the same year the city was incorporated. He was instrumental in identifying the flora of the early region and is sometimes known as the Father of the Chrysanthemum for his work with these plants.

BRYAN, THOMAS B. (1828–1906) A leading Chicago businessman, Thomas B. Bryan was a devoted patron of horticulture. His home in Cottage Hill (now Elmhurst) was the meeting place for many visiting dignitaries.

BRYANT, ARTHUR SR. (1808–83) One of the Chicago region's earliest and most prominent nurserymen, Arthur Bryant and his sons were leaders in many state and local horticultural societies. Born in Cummington, Massachusetts, Bryant moved to Princeton, Illinois, in 1833. He started a tree-growing business in 1845, which eventually expanded into the Princeton Nurseries. In 1871, he wrote *Forest Trees for Shelter, Ornament, and Profit: A Practical Manual for Their Culture and Propagation*.

BRYANT, ARTHUR JR. (1834–1907) Arthur Bryant Jr. became a partner in his father's business in 1857. In 1869, he started Princeton Nurseries. Arthur Jr. was either president or vice president of the Horticultural Society of Northern Illinois from 1882 to 1893. He was one of three people in charge of the Illinois horticultural exhibit at the Columbian Exposition.

BUDLONG NURSERY Early greenhouse and nursery business famous for bottled pickles on the North Side. Lyman A. Budlong came to Chicago in 1857 and started a nursery in 1861 on the north branch of the Chicago River. His brother Joseph joined him a few years later, and the two became truck farmers and started a pickling operation. Joseph later established a nursery at Bowmanville. The pickles were sold directly from wagons in Chicago's market to retailers, and to wholesalers across the country.

CALDWELL, ALFRED (1903–98) Alfred Caldwell carried the torch for prairie landscape architecture beyond the death of his mentor, Jens Jensen. Caldwell designed a number of landscapes in the Chicago Park District, including Promontory Point in Hyde Park and the Lily Pool in Lincoln Park.

CALVERT, FRANK Frank Calvert was landscape gardener for developer S. H. Kerfoot in the late 1850s, and for H. M. Thompson of Lake Forest. Calvert was a Scottish gardener who opened a nursery near the Lake Forest train depot, and designed many of that suburb's early estates.[2]

CHICAGO FLORISTS' CLUB The Chicago Florists' Club was organized November 8, 1886, as an association for florists only. In 1887, the club held one of its earliest exhibitions, with prizes for floral designs for weddings, baskets, funerals, and table decorations.

CHICAGO GARDENERS' SOCIETY (1858–ca. 1862) The Chicago Gardeners' Society was an association of professional gardeners and nurserymen, including Edgar Sanders, John Ure, George Wittbold, J. Asa Kennicott, Charles Bragdon, Harman Khlare, Frank Ludlow, A. T. Williams, and F. Sulzer. The society had over one hundred members, including amateurs, as of 1860.

CHICAGO HORTICULTURAL SOCIETY The CHS was a prestigious society with its origins tracing back to 1847. Its membership consisted of professional gardeners and members of Chicago's captains of industry. The CHS became quiescent from the 1860s to the 1880s, but was reenergized during planning for the 1893 World's Fair. When reincorporated in 1890, the society's most visible activities included garden shows. The CHS today manages the Chicago Botanic Garden in Glencoe.

CHICAGO OUTDOOR ART LEAGUE An outgrowth of the American Park and Outdoor Art Association, the Chicago Outdoor Art League was organized in 1901. Prominent members included O. C. Simonds, Mrs. Charles L. Hutchinson, Jens Jensen, Dwight Perkins, Lorado Taft, Mrs. P. S. Peterson, Mrs. Coonley Ward, and Mrs. J. C. Vaughan. The league was dedicated to improving the beauty of outdoor surroundings at home, school, and settlement grounds. It also promoted Arbor Day events, tree planting, and school and community gardens. Working with railway corporations, the league was instrumental in beautifying the rights-of-way inside the city limits. The league sold penny packets of flower and vegetable seeds to children, and erected fountains throughout the city, including the relocated Women's Christian Temperance Union fountain at North Avenue and Lake Shore Drive.

CLEVELAND, H. W. S. (1814–1900) Colleague of Frederick Law Olmsted, Horace William Shaler Cleveland designed landscapes for the Chicago parks and private residences between 1869 and 1886, after which he relocated to Minneapolis. While in Chicago, Cleveland often partnered with William M. R. French, who later became director of Chicago's Art Institute. Cleveland wrote *Landscape Architecture as Applied to the Needs of the West* (1873) and a pamphlet, *The Public Grounds of Chicago: How to Give them Character and Expression*.

COOK COUNTY AGRICULTURAL AND HORTICULTURAL SOCIETY Short-lived group formed ca. 1856, whose membership included Edgar Sanders, John Kennicott, and Jonathan Periam.

COWLES, H. C. (1869–1939) Mentored by John Coulter, Connecticut-born Henry Chandler Cowles is often dubbed the Father of North American Ecology. His Ph.D. thesis on the vegetation of the sand dunes of Lake Michigan highlighted the delicate balance between plants and their environments. While at the University of Chicago, Cowles and his students conducted field experiments and study throughout the world.

DOUGLAS, ROBERT (1813–97) Robert Douglas (sometimes spelled Douglass) began one of Illinois's earliest nurseries in Waukegan. He was a very active member of the Horticultural Society of Northern Illinois. He also was a founder of the Waukegan Horticultural Society, which ultimately became the Lake County Agricultural Society, organized in 1851. Douglas was renowned for his expertise in growing evergreens that adapted well to the Chicago region. His son Thomas H. started a nursery at Garyanza, Los Angeles County, ca. 1888. Robert spent summers in California, and retired by 1888.

DUNLAP, MATHIAS L. (1814–75) Born in Cherry Valley, New York, Mathias L. Dunlap moved to Illinois in 1836. After working in Chicago as a dry goods clerk, Dunlap bought a farm at Leyden in Cook County, known as Dunlap's Prairie. He started a nursery in 1846, but the new railroad routes were not convenient to his farm. He moved to a farm near Champaign in 1857, and became an important link between Chicago and downstate Illinois. Dunlap was president of the Illinois State Horticultural Society. He helped find a location for Industrial University at Urbana, and was on that school's first board of trustees. Dunlap edited the *Illinois Farmer* for six years, published in Springfield. He was also secretary of the Union Agricultural Society. Dunlap wrote under the name Old Rural for the *Chicago Tribune*.

ELLSWORTH, LEWIS (1805–85) One of the original founders of what would become the American Association of Nurserymen, Lewis Ellsworth started his fifty-acre nursery, the DuPage County Nurseries, in 1849 in Naperville. Ellsworth's nursery, the first in DuPage County, was part of his three-hundred-acre farm, and was started primarily with orchard stock. It grew rapidly into an ornamental tree, shrub, and flower business. Ellsworth was prominent in civic affairs, elected as the first school commissioner (1841) and as a county judge (1839). He was president of the Union Agricultural Society in 1845, and held other leadership positions in various horticultural groups.

FRIENDS OF OUR NATIVE LANDSCAPE This conservation group, formed by Jens Jensen, Sherman Booth, and H. C. Cowles in 1914, included many of Chicago's socially prominent family names as well as kindred spirits in environmentalism. Members included Mr. and Mrs. H. W. Austin, Mr. and Mrs. Avery Coonley, Mr. and Mrs. O. C. Doering, Mr. and Mrs. A. G. Becker, John V. Farwell, W. C. Egan, S. T. Mather, Harriet Monroe, Mr. and Mrs. E. L. Ryerson, and Dwight Perkins.

HILL, DAVID Scottish immigrant David Hill founded his Dundee, Illinois, nursery (simply called D. Hill nurseries) in 1855, one of the earliest and longest-lasting nurseries in the Chicago area. His early specialty was evergreens. Today's Platt-Hill Nurseries descends from this original pioneer in horticulture.

HILL, E. J. Called the "Dean of all Chicago Botanists" by fellow naturalist H. S. Pepoon in his book *Flora of the Chicago Region*, Ellsworth J. Hill was a passionate plant collector. Hill's thistle (*Cirsium hillii*) is named for him.

ILLINOIS STATE AGRICULTURAL SOCIETY Formed ca. 1853, this umbrella organization included virtually all of the county agricultural societies, including those of Kane County and Lake County. The society was one of the first formed when Illinois became a state, evolving from the Illinois Agricultural Association. It exists today as the Illinois Department of Agriculture.

ILLINOIS STATE FLORISTS' ASSOCIATION Organized in March 1905, the association's first board of five directors included Chicagoans J. C. Vaughan and Willis N. Rudd. The group exists today and continues to promote floriculture through workshops and certification programs.

ILLINOIS STATE HORTICULTURAL SOCIETY Recognized as the first horticultural society in Illinois by the *Prairie Farmer*, a fourteen-member group of nurserymen formed the Illinois State Horticultural Society at a meeting on October 15, 1846, in Peoria. The society was officially organized on December 17, 1856. Presidents included Chicago-area residents Lewis Ellsworth (1859), John Kennicott (1862), Arthur

Bryant Sr. (1872), Mathias L. Dunlap (1874), Robert Douglas (1875), and Arthur
Bryant Jr. (1887).

ILLINOIS STATE NURSERYMEN'S ASSOCIATION The Illinois State Nurserymen's
Association was organized in 1925. Many Chicago-area nurserymen have been
involved in leadership positions, including Guy Bryant (son of Arthur Bryant Jr.),
A. M. Augustine, George Klehm, Alvin Nelson, and Henry Klehm. In 2007, the
group changed its name to the Illinois Green Industry Association.

JENNEY, WILLIAM LE BARON (1832–1907) Often recognized for his groundbreaking
work in architecture (including Chicago's 1884–85 Home Insurance Building,
considered the world's first skyscraper), Jenney was also a remarkable landscape
architect. His designs for the West Parks system were influenced by his study of
French design.

JENSEN, JENS (1860–1951) Danish immigrant who arguably exerted one of the most
dramatic changes in Chicago regional landscape architecture. Jensen attained his
credentials through his work in the Chicago park system. His signature use of naturalistic designs, favoring native plants, open "sun clearings," stratified stonework,
and trademark "council rings" earned him a number of commissions for private
residences of Chicago's elite. Jensen also founded or joined a number of local conservation groups, including the Prairie Club and Friends of Our Native Landscape.

JUDD, ORANGE (1822–92) Publisher of many agriculture and horticulture-related
books and periodicals, Orange Judd was editor of the *American Agriculturist* (1853)
and agricultural editor for the *New York Times* (1855–63). Judd eventually moved
from the East Coast to Chicago. He became editor of the *Prairie Farmer* on May 1,
1884, and published his own agricultural paper, the *Orange Judd Farmer*.

KANST, FREDERICK (ca. 1848–1937) First head gardener and landscaper for the South
Parks, having started there ca. 1873 and retiring in 1921. Kanst supervised much of
the plant installation during the Columbian Exposition. He was the first recipient
of the Hutchinson medal, an honor given by the CHS to recognize outstanding
work in horticulture or a related field.

KENNICOTT, JOHN (1800–1863) Born in Montgomery County, New York, Dr. John
Albert Kennicott, one of the Chicago region's earliest and most influential horticulturists, moved to Cook County in 1836. Kennicott founded or was a leader in
virtually all of the important early horticultural societies, including president of the
Illinois State Horticultural Society in 1860, horticultural editor of the *Prairie Farmer*,
founder of the Cook County Agricultural and Horticultural Society in 1856, and first
president of the Northwestern Fruit Growers Association in 1851 and 1852. Today,
Kennicott's homestead, the Grove in Glenview, is a national historic landmark and
includes a nature center as well as Kennicott's restored home.

KENNICOTT, ROBERT (1835–66) Robert Kennicott was the son of horticulturist John
Kennicott. Under his father's mentorship, while exploring the countryside near
his home at the Grove, Robert learned to love natural history. He became a nationally renowned naturalist, and in his short lifetime of thirty years, he cofounded the
Chicago Academy of Sciences, led field expeditions for the Smithsonian Institution,
established a natural history museum at Northwestern University, and surveyed
southern Illinois for the Illinois Central Railroad.

KING, LOUISA (1863–1948) Louisa Yeomans King (aka Mrs. Francis King) was one of America's most influential garden writers at the beginning of the twentieth century. She began her gardening in Elmhurst, Illinois, under the tutelage of her mother-in-law. She was a founder of the Women's National Farm and Garden Association, and in 1913 one of the founders of the Garden Club of America. She was the first woman to receive America's highest horticultural award from the Massachusetts Horticultural Society, the George White Medal.

KLEHM FAMILY John Adam Klehm (1834–1916) moved to Buffalo, New York, in 1851 as a teenager with his brother and widowed mother from Germany. Moving to Arlington Heights, he bought an acre of land on the corner of present-day Miner and Davis streets. He worked as a bricklayer until he took a job grafting cherry trees, which he found to be much more profitable than the brick business. In 1862, he started a nursery business and orchard on 9 acres of land he'd originally bought to grow potatoes.[3] This nursery expanded to 147 acres and was in use for about 50 years; ultimately it was sold to developers of the Scarsdale subdivision. Klehm's three sons, Charles, George, and Henry, all entered the family business upon reaching the age of twelve. With the second generation of nurserymen, the Klehm family started to diversify. Charles Klehm started his own 50-acre apple orchard on land ultimately sold to become the Arlington Park racetrack. Other nurseries and orchards soon followed, including a seven-hundred-acre farm in Barrington and land in Rockford. Charles became a specialist in peonies, a hallmark of the Klehm family today.

KOCH, AUGUST (ca. 1873–1946) Born in Alsace-Lorraine, August Koch began his gardening career in France. In 1898, he began work at the Missouri Botanical Gardens. In 1912, he moved to Chicago to work for the park district, ultimately becoming the park district's chief florist. Koch retired in 1939. A tropical water lily bears his name.

MCCAULEY, LENA A regular contributor to Chicago newspapers, Lena May McCauley wrote the book *The Joy of Gardens*. She was a much-sought lecturer to local garden clubs.

MONINGER, JOHN C. John C. Moninger was the proprietor of a major greenhouse manufacturer in Chicago. Its facilities were at Smith Avenue and Weed Street (1910).

NELSON, SWAIN (1828–1917) Swedish immigrant Swain Nelson first started a nursery near Lincoln Park in 1856 to supply the landscaping for the park and his other clients. He created a second nursery in River Forest and finally, in 1893, established his large nursery in Glenview. Nelson landscaped many private and public residences throughout the Midwest, and served as an adviser for the Columbian Exposition. Swain Nelson and Sons landscaped many of the exhibits at the 1933 World's Fair. The Nelson offices are now part of the Glenview Park District, and a subdivision, Swainwood, in that suburb was designed by Nelson.

NELTNOR, JOHN C. (?–1938) John C. Neltnor, along with C. W. Richmond, opened Grove Place Nurseries in West Chicago in 1870. He became sole proprietor in 1874. Neltnor's five-and-a-half-acre home, Grove Place, was said to be the highest spot in DuPage County, and contained every species of tree native to Illinois. He was a police magistrate and drugstore owner, and publisher of the quarterly *Fruit, Flower, and Vegetable Grower* in the early 1880s.

NORTHWESTERN AGRICULTURAL SOCIETY The Northwestern Agricultural Society was a transitional group organized in 1860, perhaps to persuade the state agricultural society to make Chicago the site of the next state fair. The society purchased the Garden City Race Course six miles south of the city for permanent grounds. Members included John Kennicott, W. B. Egan, and William L. Church.

NORTHWESTERN FRUIT GROWERS ASSOCIATION The Northwestern Fruit Growers Association started in 1851 with thirty-three members, including John Kennicott as first president and Arthur Bryant Sr. as first treasurer. Lewis Ellsworth and Mathias L. Dunlap were also charter members. By 1852, sixty-two people were members. It merged into the Illinois State Horticultural Society ca. 1857.

OGDEN, WILLIAM B. (1805–77) William B. Ogden was the first mayor of Chicago and an enthusiastic patron of horticulture.

OLMSTED, FREDERICK LAW (1822–1903) The Father of American Landscape Architecture, Frederick Law Olmsted designed a blue-ribbon list of parks, natural areas, and estates throughout America. His seminal work with partner Calvert Vaux in New York's Central Park in the 1860s brought international acclaim. Olmsted influenced many of Chicago's major landscapes. The Olmsted and Vaux design for Riverside, Illinois, is considered a masterpiece of community planning. His work in the South Parks helped establish Chicago's world-class park system. As chief landscape architect for the Columbian Exposition, Olmsted's design set the standard for world's fairs to come. Many of Olmsted's protégés subsequently came to Chicago and advanced the city's landscape architecture. Upon the retirement of his father, Frederick Law Olmsted Jr. and his stepbrother, John Charles Olmsted, formed the successor firm, Olmsted Brothers.

PEATTIE, DONALD CULROSS (1898–1964) In 1933, just as Chicago hosted the COP World's Fair, naturalist Donald Culross Peattie moved to the Grove, built by John Kennicott. With his author wife, Louise Redfield Peattie, great-granddaughter of Kennicott, Peattie introduced a new generation to the natural world. A botanist educated at the University of Chicago and Harvard, Peattie authored or coauthored many books on nature. His lyrical prose made botany and environmentalism accessible to the average American and revitalized an environmental spirit.

PEPOON, H. S. (1860–1941) Professor H. S. Pepoon was the head instructor in botany and agriculture at Lake View High School, where he taught for thirty-eight years. A tireless naturalist, he cataloged the plants of the region into his seminal *Flora of the Chicago Region* (1927). Pepoon was an adviser or member of many local native-plant conservation groups.

PERIAM, JONATHAN (1823–1911) Born in Newark, New Jersey, Jonathan Periam moved to Cook County in 1838. Periam was the first superintendent of agriculture for Illinois Industrial University. Periam is best known for his early and long-term success with market gardening. His farm in the Calumet area in 1862 dedicated fifty acres to Chicago market crops. He was the editor of the *Prairie Farmer* (1888) and frequent writer for *Western Rural, Farm Field and Fireside, Farmer's Review,* and *Breeders' Gazette.* He held leadership positions in the Illinois State Horticultural Society, Wisconsin State Horticultural Society, American Pomological Society, and Horticultural Society of Northern Illinois.

PETERSON, P. S. (1830–1903) Per S. Peterson practiced gardening in the 1840s in Sweden and Germany and, after working with some eastern nurserymen, settled in Chicago. He started Peterson Nurseries in 1856, with Rosehill Cemetery as a major customer. His son inherited the business, which continued at least through the 1920s. The North Park Village Nature Center in Chicago is on the site of the old Peterson Nurseries, and Peterson Avenue (6000 north) is named after him.

POEHLMANN BROTHERS Nurserymen in Morton Grove, Illinois. According to Chicago greenhouse manufacturer John C. Moninger, the Poehlmann Brothers glass range was the largest in the world in 1913. August Poehlmann was a vice president of the CHS in 1910.

PRAIRIE CLUB The club was founded in 1908 as an outgrowth of the Outdoor Playground Association. It was officially named the Prairie Club in 1911 at the suggestion of charter member Jens Jensen. Club members sponsored weekly walks in Chicago-area countryside, and were leading forces in establishing the Cook County Forest Preserves, as well as preserving the Indiana Dunes.

PRAIRIE FARMER The *Prairie Farmer* was one of the most influential agricultural and horticultural papers in Chicago and the Midwest for decades, taking on many social and horticultural issues, including Osage orange fencing, school establishment, and park systems. The first issue, called the *Union Agriculturist*, was released in October 1840, with John S. Wright as editor. Because Wright was not himself a farmer, "Farmers, Write for your Paper" was the paper's masthead slogan, and it encouraged a long tradition of reader participation. The newspaper was renamed the *Prairie Farmer* in 1843. The editorship changed over the years and included some of the region's leading business and horticultural personages.

RAYNE, MARTHA (1836–1911) Martha L. Rayne established the nation's first school of journalism for women in Detroit in 1886. She was born in Halifax, Nova Scotia, and began a career in newspapers in the early 1860s. By 1870, she had reached Chicago, freelancing for the *Chicago Tribune* under the pen name Vic. In 1870, she became editor of the *Chicago Magazine of Fashion, Music, and Home Reading*. Rayne's book *What Can a Woman Do?* is one of the earliest comprehensive guides to employment for women outside the home. Gardening, and allied occupations, feature prominently in the author's suggestions.

REINBERG BROTHERS George and Peter Reinberg grew flowers, especially roses, for the wholesale florist business from about the 1880s to the 1920s. Their thirty-two-acre farm, inherited from their father, was located on Chicago's north side on Foster Avenue, between Damen and Hoyne streets.

ROOT, RALPH RODNEY (1884–1964) Ralph Rodney Root was a landscape architect with a loyal following on the North Shore. In 1912, Root created the Division of Landscape Architecture at the University of Illinois, Champaign-Urbana.[4] He authored several books, including *Contourscaping* and *Landscape Gardening* (the latter with Charles Kelley).

RUDD, WILLIS. N. (1860–1925) Born in Worth, Illinois, Willis N. Rudd attended Cornell University. In 1886, he began a long-term association with Mount Greenwood Cemetery Association and was its president from 1906 to 1925. He started a nursery specializing in Rudd's Carnations, and helped establish a Division of Floriculture at

University of Illinois. Rudd's positions included president of the American Carnation Society (1897–98), editor/manager of *American Florist* magazine (1897–98), secretary and treasurer of the CHS from 1896 to 1906, and also served as its president.

SANDERS, EDGAR (1827–1907) Prolific writer and early nurseryman in Chicago. Born in Sussex County, England, Sanders's father was a gardener on an English estate. The eldest of fourteen children, the young Sanders worked as a gardener at English nurseries and estates. In 1853, he came to America and settled in Albany, New York. He worked in James Wilson's nursery and laid out the grounds of Luther Tucker, who had just started the *Country Gentleman*. Sanders wrote for the *Country Gentleman* for six years. In 1857, Sanders moved to Chicago and built his first fifty-foot greenhouse in Lake View. In the mid-1860s, he opened a florist shop at 56 Clark Street at the Sherman House. In early years, he was active in landscaping, and did the plan for Calvary Cemetery and the University of Chicago grounds, although the latter was not carried out. Although interrupted by the Chicago Fire, he stayed in the florist trade until 1882. He wrote for the *Prairie Farmer* from 1857 through 1890, the *Orange Judd Farmer*, the *London Gardeners' Chronicle*, the *American Gardeners' Chronicle*, and the *Florists' Exchange*. In the 1870s, Sanders was a commissioner of highways for Lake View, and its treasurer from 1879 to 1883. Among his many professional associations, he was first temporary president of American Association of Nurserymen in 1876, president in 1884, vice president for Illinois Society of American Florists, and president of Chicago Florists' Club. He also helped organize the Cook County Agricultural and Horticultural Society and the Chicago Gardeners' Society.

SIMONDS, O. C. (1855–1931) Ossian Cole Simonds came to Chicago to study under William Le Baron Jenney in 1878. His work from architecture to the field of landscape gardening. Known for his preference for native plants and naturalistic style, Simonds was long associated with the landscapes at Graceland Cemetery and Lincoln Park, as well as a number of estates throughout the Chicago area and across the country.

SOCIETY OF AMERICAN FLORISTS The SAF formed in Chicago in 1884 with twenty-one members of the American Association of Nurserymen, Florists, and Seedsmen. Its original mission was obtaining hail insurance. Several Chicagoans are in the SAF's Florists' Hall of Fame.

SULZER, CONRAD (1807–73) A Swiss-born horticulturist, Conrad Sulzer arrived in Chicago in 1836. He bought one hundred acres at the southeast corner of Clark and Montrose streets and established one of the earliest nurseries in the Lakeview area. He provided plants for Graceland Cemetery and many homes in the area. His son, Frederick, joined his father in his nursery business. The Conrad Sulzer Regional Library on North Lincoln Avenue in Chicago is named for the senior Sulzer.

UNION AGRICULTURAL SOCIETY The Union Agricultural Society was formed by John S. Wright and others in 1840. William B. Ogden was treasurer. The society included La Salle, Will, Cook, McHenry, and Kane counties. The society sponsored the *Prairie Farmer* magazine.

URE, JOHN Originally, John C. Ure was gardener to Chicago businessman I. N. Arnold. Ure became president of Chicago Gardeners' Society in 1862, and was associated with Grapeton Gardens in 1862.

VASEY, GEORGE (1822–93) English-born Dr. George Vasey moved to Illinois by way of New York in 1840. As a youth, he took an early interest in botany, and began a correspondence and decades-long relationship with Asa Gray of Harvard. Receiving his medical degree from Berkshire Medical Institute in Pittsfield, Massachusetts, in 1846, Vasey returned to Illinois to practice medicine first in Ringwood and then in Elgin.[5] Late in 1858, along with John Kennicott and other like-minded citizens, Vasey founded the Illinois Natural History Society. Vasey wrote a botany column for the *Prairie Farmer* in the late 1850s. In a multipart series titled "The Flora of Illinois," he described major plant families native to the state. In 1871, Vasey was selected as the first botanist for the U.S. Department of Agriculture. He was later appointed the first director of the Smithsonian's National Herbarium.

VAUGHAN, J. C. (1851–1924) Born in Pennsylvania, J. C. Vaughan sold fruit trees in summer to pay for high school education. One of Chicago's most prominent seedsmen, Vaughan entered the seed business in 1876. His wholesale commission florist business in Chicago was near La Salle and Randolph, and he had greenhouses in Western Springs, Illinois, and offices in New York. Vaughan and his sons were founders or leaders in multiple organizations, including the SAF, American Seed Trade Association, CHS, and Chicago Florists' Club. Vaughan helped organize horticultural congresses at the 1893 and 1907 international expositions.

WILDFLOWER PRESERVATION SOCIETY, ILLINOIS CHAPTER Organized in 1913, this early conservation group pledged, "All my influence shall be used to protect wildflowers from destruction by others." The group's motto was "Enjoy, Do not Destroy." Among other activities, it sponsored pageants and plays in Chicago and Ravinia, exhibits at Marshall Field's, and guided trips for underprivileged children through the forest preserves. Members included such prominent figures as H. C. Cowles, Mrs. Charles L. Hutchinson, Mrs. Philip D. Armour, Mrs. William R. Harper, H. S. Pepoon, Mrs. Frank Dudley, Mrs. Charles Walgreen, Lena McCauley, Mrs. Raymond Watts, Floyd Swink, and staff from the Field Museum.

WITTBOLD, GEORGE (1833–1910) Once employed in the King of Hanover's garden, this early florist came to Chicago in 1857. Wittbold started a greenhouse and flower shop at Clark and North streets, then adjacent to the early cemetery in Lincoln Park. When the cemetery relocated to Lake View, Wittbold moved with it. Around 1862, Wittbold was listed as gardener to prominent Chicagoan Ezra McCagg. Wittbold was in partnership with Archibald Williams until 1864 when he bought him out at "old cemetery gate place."

WRIGHT, JOHN (1815–74) Preeminent Chicago booster, Massachusetts-born John S. Wright arrived in Illinois in 1832. He made and lost a fortune in Chicago's real estate. First editor of the *Prairie Farmer*, Wright was the so-called Prophet of the Prairie, and inventor of the reaping rake.

CHAPTER ONE

1. William H. Keating, "The Natural History, Indians, and Disadvantages of Chicago," in *As Others See Chicago*, ed. Bessie Louise Pierce (Chicago: University of Chicago Press, 1933), 33.

2. Olmsted, Vaux and Co., *Preliminary Report upon the Proposed Suburban Village at Riverside, near Chicago* (New York: Sutton, Bowne and Co., 1868), 1.

3. Edward L. Peckham, "Chicago: A Mean Spot," in Pierce, *As Others See Chicago*, 166.

4. John A. Kennicott, "Western Horticultural Review," *Horticulturist* 7 (1852): 541.

5. J. A. Wight, "Chicago, Horticulturally," *Horticulturist* 3 (1849): 459.

6. "Orchard and Garden," *Prairie Farmer*, April 1846, 112.

7. Wight, "Chicago, Horticulturally," 461.

8. Juliette Kinzie, *Wau-Bun: The Early Day in the Northwest* (Chicago: Rand, McNally and Co., 1901), 143.

9. *Prairie Farmer*, January 11, 1868, 1.

10. "Tree Peddling Tramps," *Transactions of the Horticultural Society of Northern Illinois*, 1879, 252–53.

11. Keating, "Natural History," 33.

12. Daniel Baker to S. Sibley, June 1, 1817, and November 19, 1818, Sibley Papers, Chicago History Museum, 948:106, 950:56. Whether these are traditional bog cranberries or another small fruit is unknown. It is possible, on the sandy Lake Michigan

shore or in marshy areas, that traditional cranberries known to New Englanders might have been grown. W. B. Egan apparently had some success with growing upland cranberries (*Prairie Farmer*, August 1849, 233). John Kennicott discussed cranberry culture in the *Prairie Farmer* (October 1855, 313), citing success in Indiana.

13. Edwin O. Gale, *Reminiscences of Early Chicago* (Chicago: Fleming H. Revell Co., 1910), 26–27.

14. Kinzie, *Wau-Bun*, 205.

15. Sometimes spelled "Brookes." Other accounts give the honor of the first greenhouse to one John Goode.

16. Edgar Sanders, "Historical Horticultural Sketch," *Prairie Farmer*, June 1884, 406.

17. There are numerous competing claims as to who brought the first piano to Chicago. As a symbol of culture, many early families credited their pioneer forebears with this mark of refinement.

18. Everett Chamberlin, *Chicago and Its Suburbs* (Chicago: T. A. Hungerford and Co., 1874), 48.

19. Kennicott, "Western Horticultural Review," 541.

20. "The Prospects of Western Farmers," *Prairie Farmer*, 1843, 2–3.

21. Frank J. Heinl, "Pioneer Nurserymen in Illinois," *American Nurseryman*, July 15, 1950, 142–43.

22. Sanders, "Historical Horticultural Sketch," 406.

23. In 1859, for example, Harman Klahre was gardener for William B. Ogden, John Betts for Ezra McCagg, John C. Ure for I. N. Arnold, and L. P. Eastman for J. Y. Scammon. *Prairie Farmer*, May 26, 1859, 328.

24. "Horticultural Society Again," *Prairie Farmer*, October 20, 1866, 258.

25. James T. Gifford, "Ode of Illinois Farmer to Steam Car," *Prairie Farmer*, March 1850, 101.

26. Jonathan Periam, "Interests in Horticulture," *Transactions of the Horticultural Society of Northern Illinois*, 1889, 240.

27. Robert Douglas, "Progress of Horticulture in the Northwest," *Transactions of the Horticultural Society of Northern Illinois*, 1872, 307.

28. John A. Kennicott, "About Strawberries," *Prairie Farmer*, September 1855, 281.

29. Edgar Sanders, "Evergreens for the Prairies," *Prairie Farmer*, August 20, 1864, 118.

30. "Chicago Gardeners' Society," *Prairie Farmer*, April 26, 1860, 262.

31. Chamberlin, *Chicago and Its Suburbs*, 340.

32. Wight, "Chicago, Horticulturally," 462.

33. The men, immigrants all, were Francis Chapronne, who lived on the north branch of the Chicago River; James Bell; Austin Ganer from the Dutch Settlement; William Giles from the Fourth Ward on Lake Street; and Thomas Kelly, also on the north branch.

34. "Private Gardens in Chicago," *Prairie Farmer*, September 1847, 276.

35. Mathias L. Dunlap, "Address before the Illinois State Horticultural Society at Bloomington," December 15, 1858, reprinted in *Prairie Farmer*, 1859, 135.

36. Official Proceedings of the Illinois State Horticultural Society, reprinted in *Prairie Farmer*, 1859, 38.

37. Ibid.

38. Ibid., 39.

39. "Gardeners Wanted," *Prairie Farmer*, March 7, 1861, 152.

40. Chamberlin, *Chicago and Its Suburbs*, 356.

41. Ibid., 394.

42. Ibid., 417, 401, 409.

43. Edgar Sanders, "Thirty Years in Western Horticulture," *Transactions of the Horticultural Society of Northern Illinois*, 1889, 233.

44. Ibid., 234.

45. Jane Addams, *Twenty Years at Hull-House* (New York: Macmillan Co., 1910), 89.

46. James R. Grossman, Ann Durkin Keating, Janice L. Reiff, eds., *The Encyclopedia of Chicago* (Chicago: University of Chicago Press, 2004), 8.

CHAPTER TWO

1. "The First Field Meeting of the Chicago Academy of Sciences," *Prairie Farmer*, July 5, 1860, 1.

2. "Chicago Botanical Society," *Chicago Daily Tribune*, October 4, 1869, 4.

3. Henry H. Babcock to the Board of South Park Commissioners, "Botanical Garden," Report of South Park Commissioners, Chicago, December 1, 1875, 15.

4. William K. Higley and Charles S. Raddin, "The Flora of Cook County, Illinois, and a Part of Lake County, Indiana," *Bulletin of the Chicago Academy of Sciences* 2, no. 1 (1891): vi–viii.

5. *Chicago Medical Journal*, quoted in *Chicago Daily Tribune*, October 15, 1876, 16.

6. The *Pharmacist*, as quoted in the *Chicago Daily Tribune*, August 6, 1876, 10.

7. Report of South Park Commissioners, Chicago, December 1, 1877, 6–7.

8. Higley and Raddin, "Flora of Cook County," viii.

9. "Botany at the University of Chicago," Library of Congress Web site, http://memory.loc.gov/ammem/award97/icuhtml/aepsp3.html.

10. "Orchard and Garden: Chicago Horticultural Society," *Prairie Farmer*, August 1847, 234.

11. C. D. Bragdon, "A Biographical Sketch of the Life of Dr. John A. Kennicott," *Transactions of the Horticultural Society of Northern Illinois*, 1863, 39.

12. "The Nursery Business," *Prairie Farmer*, July 1854, 258.

13. The CHS continued to have quiescent periods. In 1934, Garden Club of Illinois president Mary Dynes wrote, "The passing of Mr. [J. P.] Vaughan was a great blow to all horticultural interests. Failure of the [Chicago Horticultural] Society to find another individual with the same broad, unselfish vision probably was responsible for the cessation in the old Horticultural Society's activities [since Vaughan's death in 1924], and as a Society, it has remained inactive." "Our State Federations," Eighth Annual Garden and Flower Show program, Garden Club of Illinois, 1934, 7. Today, the CHS is very visible and active, with its management of the Chicago Botanic Garden.

14. The Chicago Gardeners' Society held its first monthly exhibition in May 1859. Like the CHS, membership in the Chicago Gardeners' Society waxed and waned until the group apparently dissolved in the early 1860s. Other regional groups had limited life spans, including the Cook County Agricultural and Horticultural Society, Kane County Agricultural Society, and Lake County Agricultural Society.

15. "Northwestern Agricultural Society," *Prairie Farmer*, February 2, 1860, 72.

16. John A. Kennicott, "Illinois State Horticultural Society: Annual Meeting," *Prairie Farmer*, November 7, 1861, 310.

17. Richard P. White, *A Century of Service: A History of the Nursery Industry Associations of the United States* (Washington, DC: American Association of Nurserymen, 1975), 7, 9, 18.

18. "The Nurserymen," *Chicago Daily Tribune*, June 14, 1876, 8.

19. White, *Century of Service*, 27.

20. Ibid., 276, 277.

21. Ibid., 230.

22. Illinois Green Industry Association, http://www.ina-online.org.

23. Frank Ridgway, "Nurserymen Revive 'Plant a Tree' Move," *Chicago Daily Tribune*, January 13, 1923, 13.

24. Accounts disagree in identifying Chicago's first florist. The *Chicago Daily Tribune's* November 19, 1901, obituary for John Goode says he built Chicago's first greenhouse in 1845. Edgar Sanders's writings clearly identify Samuel Brooks as Chicago's first florist, starting in the 1830s. Undoubtedly there were other street sellers of flowers at the time, so formal distinctions were understandably vague.

25. Fritz Bahr, *Commercial Floriculture* (New York: A. T. De La Mare Co., 1924), 19.

26. Benjamin Taylor to Edgar Sanders, June 1, 1888, Edgar Sanders Collection, Chicago History Museum.

27. "Cut Flowers," *Chicago Daily Tribune*, October 2, 1881, 9.

28. "Floral Lore," *Chicago Daily Tribune*, February 19, 1885, 9.

29. "Florists Enjoy Themselves," *Chicago Daily Tribune*, August 20, 1887, 6.

30. "Newest Fad of the Florist," *Chicago Daily Tribune*, June 5, 1897, 16.

31. "Many Want Flower Stands," *Chicago Daily Tribune*, May 6, 1899, 14.

32. "War on Flower Thieves," *Chicago Daily Tribune*, May 24, 1901, 16.

33. "West Parks Face Strike," *Chicago Daily Tribune*, October 3, 1903, 3.

34. "Flowers to Be Supreme," *Chicago Daily Tribune*, August 22, 1903, 4.

35. H. B. Dorner, "Gardening under Glass," *Transactions of the Illinois State Horticultural Society*, 1908, 81.

36. "A City of Flowers," *Chicago Daily Tribune*, September 24, 1907, 8.

37. Paul Potter, "Farm and Garden," *Chicago Daily Tribune*, March 23, 1930, A13.

38. Bahr, *Commercial Floriculture*, 22.

39. "Chicago as a Fruit Center," *Prairie Farmer*, October 10, 1861, 246.

40. Jonathan Periam, "Progress of Horticulture in the Northwest," *Transactions of the Horticultural Society of Northern Illinois*, 1872, 307.

41. Charles Klehm, "Commercial Fruit Growing in Cook County," *Transactions of the Horticultural Society of Northern Illinois*, 1924, 308.

42. W. H. Schuyler, "Report upon Utilizing and Marketing Fruits," *Transactions of the Illinois State Horticultural Society*, 1875, 126.

43. "Chicago Open or Street Markets," *Chicago Herald*, June 30, 1887, as reported in the *Prairie Farmer*, 1887, 523.

44. August Geweke, "Successful Truck Gardening," *Transactions of the Illinois State Horticultural Society*, 1911, 117, 121.

45. "Truck Gardening Profitable Work," *Chicago Daily Tribune*, September 10, 1911, E3.

46. Henry M. Hyde, "Bad Rural Roads to Chicago Boost High Living Cost," *Chicago Daily Tribune*, April 15, 1913, 1.

47. John G. DeLong, "Motor Wagons to Hold Boards," *Chicago Daily Tribune*, February 4, 1912, J3. This reflects the total number registered—not only for the garden trucking industry.

48. Reed L. Parker, "News of the Motor Truck World," *Chicago Daily Tribune*, September 22, 1912, B8.

49. Henry Farrington, "War Shows Value of Auto Truck," *Chicago Daily Tribune*, January 24, 1915, F1.

50. "Boys Need to Help Feed City, Farmers Say," *Chicago Daily Tribune*, April 15, 1917, 3.

51. Frank Ridgway, "Home Gardens Will Soon Come into Action," *Chicago Daily Tribune*, May 26, 1928, 10.

52. "Happy in Jail for 40 Years—as a Gardener," *Chicago Daily Tribune*, December 22, 1940, W1.

53. August Geweke, "The Gardener and His Troubles," *Transactions of the Horticultural Society of Northern Illinois*, 1915, 257–58.

54. "The Prairie Flowers," *Prairie Farmer*, October 1843, 236.

55. "A Sensible Woman," *Prairie Farmer*, November 1846, 308.

56. Frances D. Gage, "Pomological Congress and Cooper Apple," *Western Horticultural Review*, October 1850, 145, as reprinted from the *Ohio Cultivator*.

57. Frances D. Gage, "Letter from Mrs. Gage: The New Home," *Ohio Cultivator* 9 (1853): 173.

58. John Gage, "A Horticultural School," *Prairie Farmer*, August 1850, 244.

59. Frances D. Gage, "Dress for Women," *Prairie Farmer*, November 1860, 330.

60. Frances D. Gage, "Western Women," *Prairie Farmer*, 1862, 330.

61. Ruth Hall, *Prairie Farmer*, August 22, 1863, 122.

62. Ruth Hall, "Spring Styles," *Prairie Farmer*, 1862, 247.

63. Mrs. Observer, *Prairie Farmer*, July 11, 1863, 22.

64. Edgar Sanders, "Calls at Greenhouses," *Prairie Farmer*, March 1884, 167.

65. "Flower Mary Loses Heavily," *Chicago Daily Tribune*, February 4, 1898, 8.

66. Mary Clemmer, *Memorial Sketch of Elizabeth Emerson Atwater, Written for Her Friends* (Buffalo: Courier Co., 1879), 31.

67. Anna Pedersen Kummer, "The Herbarium of the Chicago Academy of Sciences," *Chicago Naturalist* 7, no. 2 (1944): 35.

68. Jane C. Overton, "Women on Committees," *Prairie Farmer*, July 19, 1860, 36–37.

69. "Mrs. Sam Jones Goes to the Meeting of the Illinois State Horticultural Society," *Transactions of the Illinois State Horticultural Society*, 1874, 221–25.

70. *Transactions of the Illinois State Horticultural Society*, 1879, 34–35.

71. "Horticulture for Women," *Prairie Farmer*, 1888, 516.

72. Edgar Sanders, "Women as Florists," *Prairie Farmer*, 1890, 823.

73. *Prairie Farmer*, 1892, 8, 823.

74. Mrs. M. P. Handy, "The Women of the World's Fair City," *Munsey's Magazine*, March 1893, 608–9.

75. Ibid., 613.

76. Ellen Bryant Freeman, "Women as Gardeners," *Transactions of the Horticultural Society of Northern Illinois*, 1894, 289.

77. Edgar Sanders, *Transactions of the Horticultural Society of Northern Illinois*, 1894, 293.

78. Sarah Randolph Frazier, "The Mission of Flowers in School and Home," *Transactions of the Horticultural Society of Northern Illinois*, 1900, 307.

79. "Successful Woman Florist," *Chicago Daily Tribune*, February 27, 1910, E5.

80. Frances Copley Seavey, "The Outside of the Home," *Chicago Daily Tribune*, April 13, 1904, 7.

81. "Plans Suggested for a Perfect Park," *Chicago Daily Tribune*, February 11, 1900, 8.

82. "Mrs. M'Crea, Who Expects to Be Reinstated," *Chicago Daily Tribune*, July 23, 1900, 3.

83. "The Case of Mrs. M'Crea," *Chicago Daily Tribune*, July 24, 1900, 6.

84. "Parlor for Polls Is Ideal," *Chicago Daily Tribune*, November 8, 1911, 4.

85. *Voice of the Fair*, June 21, 1865, 2.

86. "Floral Department," *Voice of the Fair*, June 21, 1865, 2.

87. "Something about South Waukegan," *South Waukegan News Print*, 1892, 15, Hagley Museum and Library Collections, Wilmington, DE.

88. "The Flower Garden," *Prairie Farmer*, April 1845, 97.

89. Mrs. Hillis, "The Cultivation of Flowers," *Transactions of the Horticultural Society of Northern Illinois*, 1874, 297.

90. Frazier, "Mission of Flowers," 300.

91. "Poor of City Get 90 Acres to Till," *Chicago Daily Tribune*, March 24, 1909, 7.

92. "Aliens in Rush for City Farms," *Chicago Daily Tribune*, April 12, 1915, 7.

93. "Outdoor Art League Promotes Home Gardens," *Chicago Daily Tribune*, August 4, 1912, B5.

94. Gertrude Barnum, "The Chicago Woman and Her Clubs," *Graphic*, May 27, 1893, 343.

95. J. H. Prost, "What Chicago Is Doing for Its Trees," *Country Life in America*, August 1910, 18.

96. "Club Women Would Make Gardening Regular Part of Public School Course," *Chicago Daily Tribune*, May 22, 1910, B8.

97. "Fine 'Fifteen Minute Gardens' of Chicago Telephone Girls," *Chicago Daily Tribune*, July 15, 1906, D7.

98. The Woman's World's Fair Chicago Souvenir Program, 1927, 7.

99. Mrs. Jesse F. McDonald, ed., *Fiftieth Year of the Garden Club of Illinois* (Chicago: Garden Club of Illinois, 1977), 85.

100. Ibid.

101. Kate Brewster to D. H. Burnham, May 18, 1932, University of Illinois at Chicago Special Collections, folder 5-136.

102. F. R. Moulton to Mrs. Walter S. Brewster, October 10, 1932, University of Illinois at Chicago Special Collections, folder 5-136.

103. John Stephen Sewell, Director of Exhibits, to General Manager of Century of Progress Fair, interoffice memo, June 24, 1932, University of Illinois at Chicago Special Collections, folder 5-136.

104. Abbie S. Kendall, "Seed Time," Third Annual Chicago Garden and Flower Show program, 1929, 66.

105. Men's Garden Clubs of America Web site, www.tgoa-mcoa.org.

106. Mrs. R. R. Hammond, "Women Gardeners," WGN broadcast reprinted in *Garden Glories*, September 1933, 12.

107. Annual meeting of the Chicago Gardeners Society, as reported in *Prairie Farmer*, December 1859.

108. Frederick Law Olmsted, *The Papers of Frederick Law Olmsted: The Years of Olmsted, Vaux and Company*, vol. 6, ed. Charles E. Beveridge and David Schuyler (Baltimore: Johns Hopkins University Press, 1992), 430–31.

109. Ibid., 430.

110. David Fairchild, "The Barbour Lathrop Bamboo Grove," *Journal of Heredity* 10 (June 1919): 243–49.

111. "A Chrysanthemum Exhibit," *Chicago Daily Tribune*, October 26, 1888, 9.

112. "How Chicago Millionaires Spend Fortunes Cultivating Rare Plants," *Chicago Daily Tribune*, November 1, 1908, G1.

113. "A Love of Rare Flowers," *Chicago Daily Tribune*, May 7, 1888, 1.

114. "Mrs. Selfridge's 2000 Varieties," *Chicago Daily Tribune*, May 10, 1903, A5.

115. "Horticulture in Chicago," *Chicago Daily Tribune*, July 29, 1894, 25.

116. Cathy Jean Maloney, *The Prairie Club of Chicago* (Chicago: Arcadia Publishing, 2001), 28, 50–51.

117. Mildred Jaklon, "Throng Visits Lake Forest Flower Show," *Chicago Daily Tribune*, June 15, 1930, 6.

118. Henry M. Hyde, "Sample Gardens in Public Parks," *Chicago Daily Tribune*, October 3, 1913, 1.

119. Suzanne Carter Meldman, *The City and the Garden: The Chicago Horticultural Society at Ninety* (Chicago: Chicago Horticultural Society, 1981), 10.

120. John A. Kennicott, advertisement in *Prairie Farmer*, 1854, as reprinted in "John A. Kennicott of the Grove" by Erik A. Ernst, *Journal of the Illinois State Historical Society* 74, no. 2 (1981): 114.

121. *Prairie Farmer*, August 22, 1861, 104.

122. *Prairie Farmer*, July 1846, 220.

123. *A Brief History of Our Garden* (Chicago: University of Chicago, n.d.), 7–8.

124. Olmsted Brothers to C. L. Hutchinson, Chair of the Committee of Trustees, March 20, 1902, 3, Correspondence of the Secretary of the Board of Trustees, box 3, folder 3, University of Chicago, Board of Trustees Correspondence. Jean Block's indexing of the university's archives on buildings and grounds has been invaluable.

125. Ibid.

126. "Small War at Botany Hall," *Chicago Daily Tribune*, January 6, 1898, 4.

127. John M. Coulter to W. R. Harper, February 16, 1904, Correspondence of the Secretary of the Board of Trustees, box 3, folder 3.

128. Unrealized was Coulter's plan to establish the largest university botanical garden in the nation. Although the idea was approved by University of Chicago authorities in 1914, there are no remnants today. The plan was to develop ornamental botanical gardens, laid out like the Garfield playground, between Cottage Grove and Maryland avenues and from the Midway Plaisance to 58th Street. Existing apartments and cottages were to be razed to make room for the garden, which would also include artificial ponds, a large fernery, and hidden grottoes. "Botanical Garden Planned by University of Chicago," *Chicago Daily Tribune*, April 2, 1914, 11.

129. Interview with Richard Bumstead, November 19, 2004.

130. L. R. Flook to E. H. Bennett, February 16, 1927, Department of Building and Grounds Records 1892–1932, box 5, folder 4, University of Chicago Library Special Collections.

131. L. R. Flook to Trevor Arnett, December 1, 1924, Department of Building and Grounds Records 1892–1932, box 5, folder 7.

132. L. R. Flook to Mr. Arnett, March 24, 1926, Department of Building and Grounds Records 1892–1932, box 5, folder 7.

133. L. R. Flook to L. R. Steere, November 10, 1930, Department of Building and Grounds Records 1892–1932, box 5, folder 7.

134. E. J. Kraus to L. R. Flook, October 22, 1929, Department of Building and Grounds Records 1892–1932, box 5, folder 4.

135. Jean F. Block, *Hyde Park Houses: An Informal History, 1856–1910* (Chicago: University of Chicago Press, 1978), 206.

136. America's Library of Congress, http://memory.loc.gov/ammem/award97/icuhtml/aep-sp3.html.

137. See the walking tour of the university's botanical garden at http://www.uchicago.edu/docs/gardbroNS/mainfill.htm.

CHAPTER THREE

1. Edward W. Brewster, "Inquiries about Illinois," *Prairie Farmer*, March 1845, 58.

2. Current name: *Juglans cinerea*.

3. Current names/spellings: white pine, *Pinus strobus*; Scotch fir aka Scots pine, *Pinus sylvestris*; silver fir, *Abies alba*; hemlock, *Tsuga canadensis*; Norway spruce, *Picea abies*; Oregon fir, *Pseudotsuga menziesii*; white larch, *Larix decidua*; black larch, *Larix laricina*; red larch, *Larix laricina*; and cedar of Lebanon, *Cedrus libani*.

4. "Culture of Forests," *Prairie Farmer*, February 1849, 67.

5. Current spelling: *Robinia pseudoacacia*.

6. Jonathan Periam, *Transactions of the Horticultural Society of Northern Illinois*, 1880, 261.

7. "Tree Peddlers," *Transactions of the Horticultural Society of Northern Illinois*, 1867, 68.

8. Jonathan Periam, "Report of the Committee on Trees and Planting, " *Transactions of the Horticultural Society of Northern Illinois*, 1864, 72.

9. D. C. Scofield, "Pine Timber," *Transactions of the Horticultural Society of Northern Illinois*, 1875, 214–15.

10. M. L. Knapp, "Arboriculture, No. 6," *Prairie Farmer*, October 1843, 221. Ogden grew up with an appreciation for the monetary value of trees. His father had a significant white pine forest, which produced masts for ships, including those of the famed *North Carolina*.

11. Scofield, "Pine Timber," 216. Nurseryman D. C. Scofield and members of his subcommittee on timber urged Congress to repeal the duties imposed on foreign tree seeds, plants, and lumber, and to pass laws encouraging people to plant trees. Evidently, Scofield and his committee did not feel that the Timber Culture Act of 1873 was sufficient. This law encouraged tree planting in the west by granting land to settlers who planted forty acres of timber. The law was eventually repealed in 1882.

12. Jonathan Periam, "Forest Tree Planting as a Means of Wealth," *Transactions of the Illinois State Horticultural Society*, 1871, 34, 37.

13. J. Sterling Morton, "Arbor Day: Its Origins and Growth," address delivered in Lincoln, NE, April 22, 1887, and reprinted in *Arbor Day: Its History and Observance*, ed. N. H. Egleston (Washington, DC: Government Printing Office, 1896), 22.

14. Egleston, *Arbor Day*, 13. Illinois began celebrating Arbor Day in 1888, by proclamation of Governor Richard J. Oglesby. At that time, twenty-seven states and three U.S. territories officially celebrated Arbor Day. *Prairie Farmer*, April 7, 1888, 218.

15. *Chicago Magazine*, April 15, 1857, 181.

16. Now more popularly known as the "weed tree" box elder.

17. Periam, "Trees and Planting," 68.

18. Ibid., 71.

19. Jonathan Periam, *Home and Farm Manual* (New York: N. D. Thompson and Co., 1884), 576.

20. Periam, "Trees and Planting," 68.

21. "Evergreens from Seed," *Prairie Farmer*, October 27, 1867, 272.

22. D. C. Scofield, "Evergreens," *Transactions of the Horticultural Society of Northern Illinois*, 1879, 236–40.

23. "Stealing Fruit," *Prairie Farmer*, January 1846, 1.

24. John A. Kennicott, "Fruits and Fruit Trees in the Lake Region," *Prairie Farmer*, March 1847, 86.

25. "Fruit Culture in Cook County," *Chicago Daily Tribune*, May 22, 1863, 2.

26. O. S. Willey, "Planting Fruit Trees," *Prairie Farmer*, 1861, 340.

27. H. M. Dunlap, "Illinois Horticulture the World's Columbian Exposition 1893," *Transactions of the Illinois State Horticultural Society*, 1893, 72–82.

28. Ibid., 67.

29. A. W. Brayton, "President's Address," in *Proceedings of the 63rd Annual Convention of the Illinois State Horticultural Society*, November 19, 1918, 47.

30. "Chicago Show," *Transactions of the Illinois State Horticultural Society* 58 (1924): 279–80.

31. "American Trees," *Prairie Farmer*, April 1853, 159.

32. Current name: sycamores.

33. Prost, "What Chicago Is Doing," 18–19.

34. Bahr, *Commercial Floriculture*, 227–33.

35. William C. Egan, "Select List of Vines," in *How to Make a Flower Garden*, ed. L. H. Bailey (Garden City, NY: Doubleday, Page and Co., 1914), 97–105.

36. John A. Kennicott, "Mistakes of Tree Planters," lecture read before the Illinois State Horticultural Society at Bloomington, December 19, 1860.

37. "Trees for Shade," *Prairie Farmer*, June 1853, 219.

38. Ibid.

39. Lewis Ellsworth, "Hints for Transplanting," in 1853 *Catalogue of DuPage Nursery*.

40. Robert Douglas, "Timber Planting," *Transactions of the Horticultural Society of Northern Illinois*, 1873, 285–86.

41. L. R. Bryant, "Report of Committee on Spraying Fruit Trees," *Transactions of the Illinois State Horticultural Society*, 1889, 132–34.

42. "Insects Wither Chicago Lawns and Bring Trees to Sear and Yellow Leaves," *Chicago Daily Tribune*, August 27, 1899, 8.

43. L. A. Day, "The Commercial Manufacture of Insecticides and Spraying Material," *Transactions of the Illinois State Horticultural Society*, 1917, 80–83.

44. "Flowers for Farmers' Wives and Daughters to Cultivate," *Prairie Farmer*, May 1846, 146. Current name: *Taraxacum officinale*.

45. "Flowers for Farmers' Wives and Daughters to Cultivate," *Prairie Farmer*, January 1846, 19.

46. Edgar Sanders, "Definition of Gardeners' Language," *Prairie Farmer*, 1862, 55.

47. Edgar Sanders, "A Chapter on Flower Gardening," in *The Prairie Farmer Annual* (Chicago: Prairie Farmer Co., 1868), 20–23.

48. "Growing Ornamental Grasses," *Prairie Farmer*, May 5, 1888, 302.

49. Current name: *Miscanthus sinensis*.

50. "Gardening Is Easy," *Chicago Daily Tribune*, May 12, 1895, 45.

51. "Artistic Gardens: A Paper for Amateurs," *Chicago Daily Tribune*, June 16, 1895, 46.

52. "Annuals for the Renter to Grow," *Chicago Daily Tribune*, January 26, 1913, J4.

53. "Prairie Flowers Fit for Culture," *Prairie Farmer*, November 1847, 348.

54. Current name: *Chamaecrista fasciculata*.

55. Current name: *Physostegia virginiana*.

56. Current name: *Gentianopsis crinita*.

57. C. Thurston Chase, *Journal of the Illinois State Horticultural Society*, 1862, 30.

58. M. L. Dunlap, "The Status of Horticulture," *Transactions of the Horticultural Society of Northern Illinois*, 1867, 33.

59. Jens Jensen, quoted in *Jens Jensen: Maker of Natural Parks and Gardens* by Robert E. Grese (Baltimore: Johns Hopkins University Press, 1992), 8.

60. O. C. Simonds, "Wild Flowers," *Transactions of the Illinois State Horticultural Society*, 1908, 77.

61. Ibid., 77–78.

62. "Wild Flowers Becoming Extinct," *Chicago Daily Tribune*, May 26, 1912, D1.

63. "The Library Exhibit of Native Wild Flowers," *Chicago Public Library Book Bulletin* 7, no. 8 (1917).

64. *Prairie Farmer*, July 1849, 213.

65. "The Tomato," *Prairie Farmer*, August 1852, 387.

66. "We Need Vegetables," *Prairie Farmer*, February 9, 1860.

67. Jonathan Periam, "The Kitchen Garden," in *The Prairie Farmer Annual* (Chicago: Prairie Farmer Co., 1868), 28.

68. Ibid., 30.

69. "Table Luxuries for Winter," *Prairie Farmer*, August 12, 1865, 1.

70. Edgar Sanders, "Celery: Best Land and Culture," *Prairie Farmer*, 1888, 159.

71. "Now Is the Time to Plant," *Chicago Daily Tribune*, March 30, 1902, 46.

72. Edgar Sanders, "Window Gardening," *Prairie Farmer*, 1861, 356.

73. Edgar Sanders, "Specimen Plants: A Word about Them," *Prairie Farmer*, May 26, 1859, 326.

74. Edgar Sanders, "Training Greenhouse Plants," *Prairie Farmer*, 1859, 278.

75. Edgar Sanders, "Ornamental Hanging Baskets," *Prairie Farmer*, January 10, 1863, 21.

76. *Prairie Farmer*, 1884, 199.

77. Edgar Sanders, "Design for Plant Window," *Prairie Farmer*, 1888, 456.

78. Edgar Sanders, "Window Gardening," *Prairie Farmer*, 1860, 6.

79. "Brooklyn Follows Chicago," *Chicago Daily Tribune*, April 24, 1901, 9.

80. Thomas McAdam, "Landscape Gardening under Glass," *Country Life in America*, December 15, 1911, 11, 13. The author wishes to thank Julia Bachrach for identifying this research article.

81. Jens Jensen, "Natural Parks and Gardens," *Saturday Evening Post*, March 8, 1930, 19.

82. McAdam, "Landscape Gardening under Glass," 13.

83. Ibid., 50.

84. Lorado Taft, "The Sculptor's Interpretation," in *Catalog Guide to Garfield Park Conservatory* (n.p.: West Park Commissioners, 1924).

85. According to a sketch in McAdam, "Landscape Gardening under Glass," 13.

86. The American Association of Botanic Gardens and Arboreta classifies membership by size, based on operating budget. Among those classified as "large," the Morton Arboretum is the second oldest after Harvard's Arnold Arboretum, established in 1872.

87. O. C. Simonds, "Morton Arboretum: Some of the Landscape Features," *Morton Arboretum's Bulletin of Popular Information*, no. 3, May 22, 1925, 13. Special thanks to Scott Mehaffey and Craig Johnson for their invaluable input to this essay.

88. In 1920, just prior to his work on the Morton Arboretum, Simonds wrote a book, *Landscape Gardening*, in which he outlined his principles for designing arboreta, cemeteries, and other public landscapes.

89. Teuscher later went on to become the influential director of the Jardin botanique de Montréal in Canada.

90. Simonds, "Morton Arboretum," 9.

91. Ibid., 11.

92. The landscape along the "ideal mile" of the Lincoln Memorial Highway was designed by Simonds's contemporary, Jens Jensen, and financed through contributions by many Midwest industrialists. It is interesting to note that before the arboretum, Morton commissioned Jensen to design a rock-lined swimming pool and two ponds for the Thornhill grounds, none of which was ever implemented.

93. Simonds, "Morton Arboretum," 11.

94. William Plum, circular, Lombard Historical Society Collections.

CHAPTER FOUR

1. "Flowers for Farmer's Wives and Daughters to Cultivate," *Prairie Farmer*, March 1846, 85.

2. "Private Gardens in Chicago," *Prairie Farmer*, September 1847, 276.

3. John A. Kennicott, "Western Horticultural Review," *Horticulturist* 7 (1852): 540.

4. "What Shall Its Name Be?" *Prairie Farmer*, May 1853, 170.

5. Mount Auburn Cemetery, http://www.mountauburn.org.

6. *Description and Dedication of the Rosehill Cemetery*, brochure, July 28, 1859, 21.

7. Ibid., 13, 11.

8. Ibid., 30.

9. Ibid., 60, 30.

10. Sanders, "Thirty Years in Western Horticulture," 237.

11. *Rosehill Cemetery*, July 28, 1859, 13.

12. Pioneer nurserymen had also contributed to earlier private cemeteries. Edgar Sanders, for example, is said to have designed the plan for Calvary, a Catholic cemetery on the north shore of Lake Michigan.

13. Thomas Bryan to John Withers, May 12, 1860, Thomas Barbour Bryan Letters (1853–89), Chicago History Museum.

14. *Rosehill Cemetery*, 33.

15. Andreas Simon, *Chicago: The Garden City* (Chicago: Franz Gindele Printing Co., 1894), 127.

16. O. C. Simonds, "Graceland at Chicago," *Landscape Architect*, January 1932, 12.

17. John A. Kennicott, "The Flower Garden," *Prairie Farmer*, April 1854, 152–53.

18. John C. Ure, "Ready Hints," *Prairie Farmer*, 1859, 166.

19. "Lawns and Flower Beds," *Prairie Farmer*, 1859, 181.

20. Edgar Sanders, "Bedding Plants, No. 1," *Prairie Farmer*, 1859, 37.

21. "Winter Gardens," *Prairie Farmer*, 1859, 119.

22. John A. Kennicott, *Proceedings of the Illinois State Horticultural Society*, December 15, 1858.

23. "Popular Errors in Ornamental Gardens," *Prairie Farmer*, January 20, 1859, 37.

24. Individuals such as John Kennicott issued a call for troops and opened their home-steads to camping soliders. Chicago horticulturists may have made contact with Frederick Law Olmsted, through the Sanitary Commission.

25. "A Home in the West," *Prairie Farmer*, March 22, 1860, 177.

26. *Chicago Illustrated: Reproductions of Colored Lithographs Published by Jevne* (Chicago: Steam Press of Church, Goodman and Donnelley, [1868?]).

27. Kim Coventry, Daniel Meyer, and Arthur H. Miller, *Classic Country Estates of Lake Forest* (New York: W. W. Norton and Co., 2003), 37–61.

28. John Blair, "Landscape Gardening," in *The Prairie Farmer Annual* (Chicago: Prairie Farmer Co., 1869), 81.

29. John A. Kennicott, "Public Parks and Gardens," *Prairie Farmer*, October 1853, 386–87.

30. Chamberlin, *Chicago and Its Suburbs*, 314–40.

31. L. Schick, *Chicago and Environs* (London: Brentano's, 1892), 144.

32. David Schuyler, "Jacob Weidenmann," in *American Landscape Architecture: Designers and Places*, ed. William Tishler (Washington, DC: Preservation Press, 1989).

33. *Land Owner*, October 1869.

34. H. W. S. Cleveland, *Landscape Architecture as Applied to the Needs of the West* (Chicago: Jansen, McClurg, and Co., 1873; revised edition, Pittsburgh: University of Pittsburgh Press, 1965), 20.

35. H. W. S. Cleveland, *Landscape Gardening in the West* (Chicago: Hazlitt and Reed, 1871), 3.

36. Ibid., 4.

37. "Private Grounds on Drexel Avenue," *Land Owner*, November 1870, 288.

38. Map of Hyde Park, Calumet, Thornton, South Chicago (New York Lithography, Engraving and Printing Co., 1870), Chicago History Museum Collections.

39. Cleveland, *Landscape Gardening in the West*, 3.

40. H. W. S. Cleveland to Edgar Sanders, September 31, 1891, Edgar Sanders Collection, Chicago History Museum.

41. Chamberlin, *Chicago and Its Suburbs*, 313.

42. Olmsted was the primary landscape architect for the world's fair, but many other prominent landscapers were involved, any of whom might have been the designer.

43. William W. Lloyd, "Ravenswood in the Seventies," undated manuscript, Conrad Sulzer Regional Library Collections, Chicago Public Library.

44. Chamberlin, *Chicago and Its Suburbs*, 393.

45. Sanders noted the following florists as among those who lost their shops: Desmond and McCormack (on Dearborn between Washington and Randolph), greenhouse in Hyde Park; Gordon Brothers (on Randolph and Dearborn, later moved to Washington and State); Charles Reissig (Washington and Wabash), home on west side of city; Dr. Farrell (office in Tribune Building), garden and greenhouse in Hyde Park; Samuel Murir (Clark and Lake), greenhouse on South Side; and Edgar Sanders (Dearborn near Tremont House). Sanders observed that the following seed stores were lost: Hovey, Emerson and Stafford, Chicago Seed Store, Fogg and Son, Pettigrew and Reid, Miller and Sons, and Krick Brothers.

46. Ann Durkin Keating, *Building Chicago* (Columbus: Ohio State University Press, 1988), appendix tables 4–10.

47. Mary Green Kircher, *Brookfield, Illinois: A History* (Brookfield: Brookfield History Book Committee, 1994), 50.

48. F. L. Olmsted to S. E. Gross, February 8, 1911, Riverside Public Library Olmsted Collection.

49. In the late 1880s, the Victorian fashion of carpet bedding came under attack by several prominent landscapers who favored a more naturalistic style. This style eschewed geometric cutouts of colorful flowers in the middle of a lawn. Since nurserymen depended on the sale of annuals, this design shift posed a serious threat to their businesses.

50. Periam, *Home and Farm Manual*, 535.

51. Ibid., 536.

52. Jens Jensen, *Siftings* (Baltimore: Johns Hopkins University Press, 1990), 94.

53. See Coventry, Meyer, and Miller, *Classic Country Estates*, for an in-depth description of Lake Forest homes and estates.

54. "Want Fountain in Ghetto," *Chicago Daily Tribune*, July 13, 1907, 3.

55. James [Jens] Jensen, "Parks, Boulevards, and Their Influences," *Chicago Daily Tribune*, November 4, 1900, 37. The article is attributed to "James Jensen," which may have been an Americanized name Jensen used at the time. The picture accompanying the article is definitely of Jens Jensen. His dismissal from the parks resulted from his unwillingness to cooperate in various graft schemes in the park system.

56. "Flowers' Reign Is Ended," *Chicago Daily Tribune*, November 12, 1905, 4.

57. "Successful Candidate for Forester, Who Tells Plans for City Beautiful," *Chicago Daily Tribune*, April 29, 1909, 12.

58. J. H. Prost, "Gardens in Windows, on Porches, and in Door Yards Will Help to Make Chicago the Real 'City Beautiful,'" *Chicago Daily Tribune*, October 24, 1909, 12.

59. Philip R. Kellar, "City Plans Must Bring Homes Worker Has to Be Considered," *Chicago Daily Tribune*, April 27, 1913, E4.

60. Henry M. Hyde, "Poet and Peasant among Approvers of 'We Will' Plan," *Chicago Daily Tribune*, April 7, 1913, 1.

61. J. Willard Bolte, "Making Backyard a Garden Spot," *Chicago Daily Tribune*, April 20, 1913, E6.

62. Henry M. Hyde, "Sample Gardens in Public Parks for City 'Farmer,'" *Chicago Daily Tribune*, October 3, 1913, 1.

63. Helen R. Jeter, *Trends of Population in the Region of Chicago* (Chicago: University of Chicago Press, 1927), 34–35.

64. "Aliens in Rush for City Farms," *Chicago Daily Tribune*, April 12, 1915, 7.

65. Jeter, *Trends of Population*, 35.

66. Frank Ridgway, "Farm and Garden," *Chicago Daily Tribune*, June 14, 1925, A12.

67. "Jews Throng to Harvest Festival," *Chicago Daily Tribune*, October 12, 1908, 11.

68. "Toilers to Farm Empty City Lots," *Chicago Daily Tribune*, April 15, 1912, 22.

69. "Shamrocks Are Grown Here," *Chicago Daily Tribune*, March 14, 1909, 5.

70. "Irish Potatoes in Chicago," *Chicago Daily Tribune*, March 13, 1911, 3.

71. "Curious Drugs Sold in Ghetto," *Chicago Daily Tribune*, August 8, 1909, 14.

72. "Gardener for Half Million People," *Chicago Daily Tribune*, May 20, 1906, D3.

73. Wilhelm Miller, "The Prairie Spirit in Landscape Gardening," *Transactions of the Illinois State Horticultural Society*, 1915, 144.

74. Ibid., 145.

75. The *Highland Park, Illinois: Historic Landscape Survey, Final Report* (Highland Park: Park District of Highland Park and the Illinois Preservation Agency, 1988) documents the early landscapes in that suburb, and particularly highlights Jens Jensen's importance to Midwest landscapes. Prairie landscape architecture is recognized in *Jens Jensen: Maker of Natural Parks and Gardens* by Robert Grese (1992) and *Alfred Caldwell* by Dennis Domer (1997).

76. Warren H. Manning, "The Purpose and Practice of Landscape Architecture," *Transactions of the Indiana Horticultural Society*, 1893, 102. Manning observed that Washington, DC, had one acre of green space for every 150 people, Paris had one for every 13, Vienna one for every 100, and Tokyo one for every 167, whereas Chicago had one acre for every 200 people.

77. Miles Bryant, "Landscape Gardening," *Transactions of the Horticultural Society of Northern Illinois*, 1917, 293.

78. Wilhelm Miller, whose writings are often credited with putting the spotlight on prairie landscaping, was not opposed to modifying his definition of the style. In his 1915 address to the Illinois State Horticultural Society, he revised the recommended percentage of native plants from 90 percent to 10–20 percent of Illinois plants in formal gardens, small city yards, and

downtown parks. He also allowed that foundation plantings could have fewer native plantings than borders, to be consistent with good design principles. "Prairie Spirit in Landscape Gardening," 146.

79. Ralph Rodney Root and Charles Fabens Kelley, *Landscape Gardening* (New York: Century Co., 1914), 168.

80. Lena May McCauley, *The Joy of Gardens* (Chicago: Rand McNally and Co., 1911), 34.

81. Root and Kelley, *Landscape Gardening*, 172–73.

82. F. Cushing Smith, "Landscape Design for Gardens," *American Landscape Architect*, December 1931, 14–16.

83. Bryant, "Landscape Gardening," 285.

84. Kate L. Brewster, *The Little Garden for Little Money* (Boston: Atlantic Monthly Press, 1924), 4.

85. Root and Kelley, *Landscape Gardening*, 190.

86. McCauley, *Joy of Gardens*, 94, 170–71.

87. Frank Ridgeway, *Flower Gardening by Farm and Garden Bureau* (Chicago: *Chicago Tribune*, 1929), 59.

88. John C. Moninger, *The Florist's Scrap Book* (Chicago: John C. Moninger Co., 1908), 121.

89. McCauley, *Joy of Gardens*, 30.

90. Ibid., 31.

91. Frazier, "Mission of Flowers," 305.

92. *Fruit, Flower, and Vegetable Grower*, January–March 1883, 10–11.

93. Martha L. Rayne, *What Can a Woman Do?* (Albany: Eagle Publishing Co., 1893), 177.

94. Brewster, *Little Garden for Little Money*, 32.

95. Bahr, *Commercial Floriculture*, 222.

96. Ridgeway, *Flower Gardening*, 94, 95.

97. Edgar Sanders, "Management of Ferneries," in *The Prairie Farmer Annual* (Chicago: Prairie Farmer Co., 1871), 14.

98. *Fruit, Flower, and Vegetable Grower*, May–June 1883, 42.

99. Louise DeKoven Bowen, *Growing Up with a City* (New York: Macmillan Co., 1926), 41.

100. Edgar Sanders, "Hanging Baskets in Our Home," *Prairie Farmer*, 1884, 663.

101. Ridgeway, *Flower Gardening*, 52.

102. Ibid., 75.

103. "Roof Garden in the Heart of the Ghetto District," *Chicago Daily Tribune*, May 26, 1901, 54.

104. "Chicago Girl's Idea Makes Her Popular with Young Friends," *Chicago Daily Tribune*, August 10, 1901, 13.

105. "Schools Get Roof Yard Idea," *Chicago Daily Tribune*, May 2, 1908, 5.

106. "City Folk Have Bungalows Up Near the Stars," *Chicago Daily Tribune*, January 22, 1928, B1.

107. 1933 Century of Progress World's Fair Official Horticultural program and Souvenir Book.

108. Coventry, Meyer, and Miller, *Classic Country Estates of Lake Forest*, 226–28.

109. Much of the historical material for this essay comes from the research, writings, and personal interview with Darlene Larson, charter member of the Friends of Fabyan organization. The honorific of "Colonel" was reportedly given to Fabyan as a National Guard title for Fabyan's service under Illinois governor Richard Yates.

110. Darlene Larson and Laura Hiebert, "The Fabyan Legacy," in *Geneva, Illinois: A History of Its Times and Places*, ed. Julia Ehresmann (Geneva: Geneva Public Library District, 1977), 155–81.

111. Frank Lloyd Wright, "Concerning Landscape Architecture," an address to Oak Park's Fellowship Club, ca. 1900, in *Frank Lloyd Wright Collected Writings*, vol. 1, 1894–1930, ed. Bruce Brooks Pfeiffer (New York: Rizzoli International, 1992), 54–57.

112. Maginel Wright Barney, *Valley of the God-Almighty Joneses* (New York: Appleton-Century, 1965), 129.

113. Interview with Karen A. Sweeney, AIA, Director of Restoration, Frank Lloyd Wright Preservation Trust, January 6, 2005.

114. Frank Lloyd Wright to *Oak Leaves*, 1940, FLW Home and Studio Landscape Slide Presentation, Frank Lloyd Wright Home and Studio Archives, Oak Park. While Wright mentions Blair's involvement in Humboldt Park, park historians dispute the claim.

115. Alice Hayes and Susan Moon, *Ragdale: A History and Guide* (Lake Forest: Ragdale Foundation 1990), 59.

116. Ragdale Foundation Archives.

117. Hayes and Moon, *Ragdale*, 59.

118. Simon, *Chicago*, 113.

119. William Saunders, of Germantown in Philadelphia, was nationally known as a horticultural author and preeminent landscaper who ultimately earned a pioneering role with the nascent U.S. Department of Agriculture. Saunders was well known among Chicago horticulturists, and occasionally joined their professional meetings. Swain Nelson managed a nursery and landscaping business in the north suburbs of Chicago, and his prodigious design career spanned several decades and included hundreds of residential and commercial clients.

120. Swain Nelson, autobiography as told to his granddaughter, available at http://www.gyllenhaal.org.

121. Ibid. Gold was highly valued as a currency because the worth of paper payments was doubtful in this pre–Civil War era.

122. O. C. Simonds, "Cemetery Gardening: A Modern Type of Landscape Ornamentation," reprinted in Simon, *Chicago*, 115.

123. Ibid., 290–309.

124. Ibid., 116.

125. O. C. Simonds, "Mr. Simonds' List of Shrubs," *Prairie Farmer*, August 1887.

126. Simonds, *Landscape Gardening*, 299.

127. Barbara Geiger, "Nature as the Great Teacher: The Life and Work of Landscape Designer O. C. Simonds," master's thesis, University of Wisconsin, Madison, 1997, 61.

128. Interview with Ted Wolff, November 19, 2004.

129. Simon, *Chicago*, 118.

130. Interview with Ted Wolff, November 19, 2004.

131. Simon, *Chicago*, 119.

132. Ibid., 118.

CHAPTER FIVE

1. Helen Hart Oakes, "Illinois Garden," WGN radio broadcast, reprinted in *Garden Glories*, May 1933, 10–11.

2. Periam, *Home and Farm Manual*.

3. B. O'Neil, "Report upon Home Adornment," *Transactions of the Horticultural Society of Northern Illinois*, 1881, 315–17.

4. "City Life vs. Country Life," *Chicago Daily Tribune*, October 27, 1872, 6.

5. "Our Rich Farmers," *Chicago Daily Tribune*, February 27, 1892, 9.

6. Myrtle Walgreen, *Never a Dull Day* (Chicago: Regnery, 1963).

7. Keating, *Building Chicago*, 189.

8. "Building," *Chicago Daily Tribune*, October 7, 1870, 2.

9. Ibid.

10. Chamberlin, *Chicago and Its Suburbs*, 344.

11. Ibid., 455–58, 422.

12. Grossman, Keating, and Reiff, *Encyclopedia of Chicago*, 527.

13. William A. Radford, "Homes in Color Away Out in the Country," *American Builder*, October 1926, 168.

14. *Historic City*, 43.

15. *North Shore Graphic*, October 12, 1927.

16. Alice Woodward Fordyce, "An American Chantilly," *Country Life*, September 1941, 30–33.

17. For detailed history and descriptions of prominent Lake Forest gardens, see Coventry, Meyer, and Miller, *Classic Country Estates*.

18. H. S. Pepoon, *Flora of the Chicago Region* (Chicago: Lakeside Press, 1927), 63–100.

19. Louella Chapin, *Round About Chicago* (Chicago: Unity Publishing Co., 1907), 90.

20. Ibid., 97.

21. *Chicago Daily Tribune*, December 6, 1924, 17; September 2, 1923, 6; June 28, 1919, 5.

22. "Palos Park Will Hold Flower Show," *Chicago Daily Tribune*, September 12, 1913, 9. Some accounts note that the Improvement Club was founded in 1913, making this ninth annual exhibition perhaps the first time it was sponsored under a reorganized, renamed club.

23. "Women Curtail Cost of Living," *Chicago Daily Tribune*, July 17, 1915, 15.

24. "Palos Park's First Realty Office," *Chicago Daily Tribune*, January 17, 1926, B2.

25. "Grounds and Floor Plan of Suburban Residence," *Chicago Daily Tribune*, September 8, 1929, B3.

26. Editorial, *Oak Leaves*, February 7, 1902, 8.

27. Frank Lloyd Wright, *Oak Park Reporter*, August 23, 1900, as reported by Jean Guarino in *A Historical View of Oak Park, Illinois: Prairie Days to World War I* (Oak Park: Oak Ridge Press, 2000), 1:112.

28. Frank Lipo, nomination of Oak Park Conservatory, National Register of Historic Places application, section 8, pp. 2–3, Historical Society of Oak Park and River Forest Archives, Oak Park.

29. The club subsequently included members of neighboring suburb River Forest and, in 1920, became the Garden Club of Oak Park and River Forest, as it is named today.

30. Mrs. Jack Seligman, "Red Ribbon for Best History Garden Club of Oak Park and River Forest," Historical Society of Oak Park and River Forest Archives, Oak Park.

31. Ibid., 7.

32. Grese, *Jens Jensen*, 200–217.

33. *Oak Leaves*, August 20, 1927, 63.

34. "The Rural Improvement Society," *Wheaton Illinoisian*, February 20, 1885.

35. "Schedule of Premiums," *Wheaton Illinoisian*, July 30, 1885.

36. "Horticultural Notes," *DuPage County Tribune*, February 24, 1911.

37. *Wheaton and Its Homes* (Proust and Burnham, 1892; reprint, Wheaton: Wheaton Historic Preservation Council, 1983), 34–35.

38. Ibid., 46–47.

39. Ibid., 29, 41.

40. "Home Town Club Secures J. H. Frost [*sic*]," *DuPage County Tribune*, April 28, 1911.

41. "City at Land Show," *DuPage County Tribune*, February 10, 1911.

42. "Alcohol Sold in Witch Hazel," *DuPage County Tribune*, September 22, 1911.

43. Sears ledger, February 1890, Kenilworth Historical Society Archives, Kenilworth.

44. Mrs. Grant Ridgway, paper read at the thirty-fifth birthday celebration of the Kenilworth Garden Club, Kenilworth Historical Society Archives.

45. Michael H. Ebner, *Creating Chicago's North Shore* (Chicago: University of Chicago Press, 1988), 227.

46. Ridgway, paper read at the thirty-fifth birthday celebration.

47. Kenilworth Garden Club minutes, September 22, 1922, Kenilworth Historical Society Archives.

48. *Garden Glories*, May 1930.

49. For more detail on Maher and Jensen's work in Kenilworth, see *View from the Path: Jens Jensen in Kenilworth* (Kenilworth: Kenilworth Historical Society, 2004).

50. "Hinsdale the Beautiful," *Campbell's Illustrated Journal*, November 1897, reprinted by Hinsdale Historical Society, 2.

51. Cleveland, *Landscape Gardening in the West*, 6, 10.

52. Chamberlin, *Chicago and Its Suburbs*, 419. Stough, in concert with other real estate investors, developed the adjacent western suburb, Clarendon Hills, once called West Hinsdale. Clarendon Hills is remarkably laid out—with the curvilinear streets visible only in a handful of Chicago suburbs (e.g. Riverside, Highland Park, and Lake Forest). Research and discussions between the author and the Clarendon Hills Historical Society have yet to identify the name of the designer of Clarendon Hills. F. L. Olmsted is sometimes attributed, but no hard evidence exists. Perhaps H. W. S. Cleveland might have been involved, given the timing (late 1860s), his work in Hinsdale's Robbins subdivision, and the fact that he returned to Hinsdale to live with his son, Robert. Cleveland died in Hinsdale in 1900.

53. *Gardening*, November 1893, 70.

54. Hugh Dugan, *Village on the County Line* (Chicago: Lakeside Press, 1949), 146.

55. "Hinsdale the Beautiful," 34–35, 42.

56. "Littleford Landscape Nurseries," *Garden Glories*, March 1933, 9.

57. Oakes, "Illinois Garden," 11.

58. Dynes, "Our State Federations," 7.

59. "Gallery of Local Celebrities," *Chicago Daily Tribune*, May 13, 1900, 39.

60. Margaret Franson Pruter, "Elmhurst," in *DuPage Roots*, ed. Richard A. Thompson (Wheaton: DuPage County Historical Society, 1985), 151.

61. Elmhurst Historical Society exhibition, January 19 to May 12, 2000.

62. "The Garden Clubs of Elmhurst," *Garden Glories*, September 1931.

63. Janice Perkins, "Glen Ellyn," in Thompson, *DuPage Roots*.

64. Ada Douglas Harmon, *The Story of an Old Town: Glen Ellyn* (Glen Ellyn: Glen News Printing Co., 1928), 74.

65. *Wheaton Illinoisian*, December 7, 1894.

66. "Mothers and Babies Aided at Glen Ellyn," *Chicago Daily Tribune*, July 29, 1905, 6.

67. "Many Regain Health at 'The Tribune's' Hospital," *Chicago Daily Tribune*, August 9, 1905, 6.

68. Harmon, *Story of an Old Town*, 86.

69. Ibid., 92.

70. Harmon's watercolors are now in the Sterling Morton Library at the Morton Arboretum in Lisle, Illinois.

71. Perkins, "Glen Ellyn," 163.

72. Marie Seacord, "Have You Injured Tree to Save?" *Chicago Daily Tribune*, September 11, 1910, E3.

73. "Glen Ellyn 'Swatters' Rid Village of Flies," *Chicago Daily Tribune*, July 24, 1912, 6.

74. "Garden Club Projects," *Garden Glories*, March 1933.

75. "Report of Committee on Trees and Tree Planting," *Transactions of the Horticultural Society of Northern Illinois*, 1863, 79.

76. Frederick Law Olmsted to Mary Perkins Olmsted, August 23, 1868, in *Papers of Frederick Law Olmsted*, vol. 6.

77. Virginia Cross West, "We Loved to Play 'Lady,'" in *Tell Me a Story: Memories of Riverside*, ed. Riverside Historical Commission (Riverside: Riverside Historical Commission, 2000), 2:7.

78. Mary Frick, "Coming into Riverside, the Temperature Would Drop," in Riverside Historical Commission, *Tell Me a Story*, 2:45.

79. Ridgeway, *Flower Gardening*, 6.

80. A. T. Andreas, *History of Cook County* (Chicago: A. T. Andreas, 1884), 876.

81. *Riverside in 1871* (Chicago: Riverside Improvement Co., 1871), 43.

82. Olmsted's firm was said to be involved with the Murray house design, and may have been instrumental in the decision against a fence.

83. *Riverside in 1871*, 28.

84. Ibid., 21.

85. Andreas, *History of Cook County*, 876.

86. Ibid., 610.

87. Richard T. Ely, "Pullman: A Social Study," *Harper's Magazine*, February 1885, 452–66. Web version at http://www.library.cornell.edu/reps/docs/pullman.htm, unpaginated.

88. William Adelman, *Touring Pullman* (Chicago: Illinois Labor History Society and Ralph Helstein Fund, 1994), 2.

89. Frederick Francis Cook, *Bygone Days in Chicago* (Chicago: A. C. McClurg and Co., 1910).

90. Ibid., 260–61. James H. Bowen had built his own estate, Wildwood, on the banks of the Little Calumet River, and his own property holdings would most likely have increased with Pullman's purchase.

91. Ibid., 262–63.

92. Richard Schermerhorn Jr., "Nathan Franklin Barrett, Landscape Architect," *Landscape Architecture* 10 (April 1920): 112.

93. Mrs. Duane Doty, *The Town of Pullman: Its Growth with Brief Accounts of Its Industries* (Pullman, IL: T. P. Struhsacker, 1893; reprint, Pullman Civic Organization and Historic Pullman Foundation, 1991), 105.

94. Ely, "Pullman."

95. Doty, *Town of Pullman*, 81.

96. Ibid.

97. Andreas, *History of Cook County*, 624.

98. Ely, "Pullman."

99. *The Chicago Zoological Society Yearbook, 1927* (Chicago: Chicago Zoological Society, 1927), 26.

100. From the statute of the State of Illinois allowing for the establishment of Forest Preserve Districts, 1911.

101. *Forest Preserves of Cook County*, 185.

102. "Chicago's New Zoo Prepares to Get Charter," *Chicago Daily Tribune*, January 6, 1921, clipping from Brookfield Zoo Archives, box 1, General Correspondence (1921–1928), folder 1.

103. Hagenbeck Zoo, http://www.hagenbeck.de. Other zoos, such as the London Zoo, were experimenting with barless enclosures, but were not yet as successful as the Hagenbeck zoo.

104. Originally, there was a grassy circle at the intersection.

105. *Chicago Zoological Society Yearbook 1927*, 25.

106. *Forest Preserves of Cook County*, 101–2.

107. W. Soderstrom, November 16, 1927, report of work performed, Office of Buildings and Grounds (1927–63) Monthly Reports, box 87, Brookfield Zoo Archives.

108. Zoological Park Director's first annual report and the Society's Business Administration 1927, Chicago Zoological Society Minutes (1921–38), box 48, Brookfield Zoo Archives.

109. W. Soderstrom, March 1, 1928, report of work performed, Office of Buildings and Grounds (1927–63) Monthly Reports, box 87.

110. Andrea Friederici Ross, *Let the Lions Roar! The Evolution of Brookfield Zoo* (Chicago: Chicago Zoological Society, 1997), 25.

111. Edward H. Bean to Stanley Field, June 3, 1932, Office of Chairman of Building and Grounds (1932–38), box 90, folder 1, Brookfield Zoo Archives.

112. Edward H. Bean to Graham Aldis, March 1, 1929, General Correspondence (1921–28) C.7.5, box 1, folder 3, Brookfield Zoo Archives.

113. Roy West to Edwin H. Clark, April 25, 1934, Office of Building and Grounds (1928–55), folder 1.

114. Interview with Gail Gorski, Brookfield Zoo Manager of Grounds Operations, January 12, 2005.

115. Jean Guarino, "Cheney Mansion Replaced Early Tear-down," *Pioneer Press*, July 4, 2001, 64.

116. Doug Kaarre, Village of Oak Park, Oak Park Landmark Nomination Report on the C. A. Sharpe House, April 2, 2004, Historical Society of Oak Park and River Forest Archives.

117. Ibid., 4.

CHAPTER SIX

1. John A. Kennicott, "Our Second State Fair," *Prairie Farmer*, August 1854, 306.

2. John A. Kennicott, "Come to the Fair," *Prairie Farmer*, September 1855, 267.

3. John A. Kennicott, "Illinois and U.S. Societies' Fairs," *Prairie Farmer*, 1859.

4. John A. Kennicott, "Our Great Western Fairs," *Prairie Farmer*, August 25, 1859, 1.

5. Ibid.

6. "The State Fair," *Illinois Farmer*, February 1861, 36.

7. "The State Fair," *Prairie Farmer*, September 19, 1861, 198.

8. "The State Fair," *Illinois Farmer*, November 1861, 323.

9. "Brighton," *Illinois Farmer*, November 1861, 335.

10. "State Fair," *Prairie Farmer*, 198. *Anasarea* is an old term for an illness caused by severe swelling.

11. C. Thurston Chase, *Journal of the Illinois State Horticultural Society*, 1862, 30.

12. *History of the North-Western Soldiers' Fair* (Chicago: Dunlop, Sewell and Spalding, 1864), 41.

13. Edgar Sanders, *Prairie Farmer*, September 19, 1863, 181.

14. Ibid., 180.

15. Ibid., 182.

16. *Voice of the Fair*, June 5, 1865, 2.

17. Edgar Sanders, "The Horticultural Department of the Sanitary Fair," *Prairie Farmer*, June 10, 1865, 462. Thanks to Willam Dale for this research.

18. *Prairie Farmer*, September 16, 1865, 201.

19. Ibid., 204, 206.

20. *Prairie Farmer*, May 26, 1866, 360.

21. Wadhams's property would later be the training ground for Louisa King, one of America's preeminent female gardeners and garden writers.

22. Ruth Hall, "Ruth Hall among the Strawberries and Flowers," *Prairie Farmer*, July 7, 1866. 10.

23. "The Interstate Exposition Souvenir" (Chicago: Van Arsdale and Massie, 1873), 25.

24. Ibid., 285.

25. Ibid.

26. Ibid., 289.

27. "Flowers at the Exposition," *Chicago Daily Tribune*, September 2, 1879, 9.

28. Jenney is perhaps best known as the Father of the Modern Skyscraper, with his plan for the Home Insurance Building in Chicago. With this bold stroke, he transformed the skyline of the flat prairie into vertical canyons.

29. "The Chicago Florist Club Objects," *Prairie Farmer*, August 12, 1893, 8. Selection of landscaping companies for individual buildings sometimes caused contention. The Chicago Florists' Club, for example, censured the State of Illinois for giving the business to John Ure, a long-time nurseryman in Chicago.

30. John A. Servas, "History of the Chicago Flower Shows," *Garden Glories*, March 1931, 8.

31. R. J. Pearse, ed., *Proceedings of the First Annual School in Fair Management* (Oklahoma City: Times-Journal Publishing Co., 1924), 63.

32. R. J. Pearse, "Planning a Fair Ground," in *Proceedings of the First Annual School*, 61.

33. J. C. Simpson, "Agricultural and Horticultural Exhibits at Fairs and Expositions," in Pearse, *Proceedings of the First Annual School*, 151.

34. "Beautiful Gardens Growing Vigorously in Sand," in *Horticultural Exhibition Souvenir Book, Century of Progress* (n.p.: Horticultural Souvenir Book Publishers, 1934), 36.

35. Lenox R. Lohr, *Fair Management: The Story of A Century of Progress Exposition* (Chicago: Cuneo Press, 1952), 85.

36. John Thorpe, "Floriculture at the World's Fair," *Gardening*, September 15, 1892, 13.

37. L. R. Bryant, "Observations at the Fair," *Transactions of the Horticultural Society of Northern Illinois*, 1894, 275.

38. L. H. Bailey, "The Columbian Exposition," in *Annals of Horticulture in North America for the Year 1893* (New York: Orange Judd Co., 1894), 104–15.

39. Ibid., 104–9.

40. Ibid., 113.

41. "Revised Plan for Jackson Park," *Garden and Forest*, May 20, 1896, 201–5.

42. J. V. Stauffer, *International Friendship Garden "Favorite Flowers"* (Hammond, IN: North State Publishing Co., 1934).

CHAPTER SEVEN

1. E. A. Popenoe to Edgar Sanders, August 30, 1892, Edgar Sanders Collection, Chicago History Museum.

2. Cincinnati horticulturist John Warder identified the coffee substitute as a chickpea (*Cicer arietinum*). *Prairie Farmer*, March 8, 1862, 145.

3. *Prairie Farmer*, March 29, 1862, 193.

4. *Illinois Farmers' Institute's Department of Household Science Year Book 1913*, ed. Mrs. H. A. McKeene (Springfield: Illinois State Journal Co., 1913), 21.

5. Jessie Marie DeBoth, *Modernistic Recipe-Menu Book* (Chicago: DeBoth Homemakers' Cooking School, 1929), 7–8.

6. Ibid., 89.

7. *The Lakeside Cook Book No. 1* (Chicago: Donnelley, Gassette and Loyd, 1878), 45–46.

8. DeBoth, *Modernistic Recipe-Menu Book*, 290.

9. Edgar Sanders, "The Holly Branch and the Mistletoe," *Prairie Farmer*, 1888, 829.

10. Ellye Howell Glover, *Dame Curtsey's Book of Novel Entertainments for Every Day in the Year* (Chicago: A. C. McClurg and Co., 1920), 121.

11. Nelle S. Mustain, *The New Idea Entertainer for Indoors and Out of Doors* (Los Angeles: L. A. Martin, 1903), 133.

12. Bahr, *Commercial Floriculture*, 157.

13. Glover, *Dame Curtsey's Book*, 54.

14. Ibid., 49.

15. Bahr, *Commercial Floriculture*, 177.

16. Mustain, *New Idea Entertainer*, 98.

17. Edgar Sanders, "Immortal Flowers," in *The Prairie Farmer Annual* (Chicago: Prairie Farmer Co., 1870), 93.

18. Bahr, *Commercial Floriculture*, 154.

19. Alex Laurie, *The Flower Shop* (Chicago: Florists' Publishing Co., 1930), 120.

20. Laurie, *Flower Shop*, 121–22.

21. Glover, *Dame Curtsey's Book*, 220.

22. Thomas Bonner, *Medicine in Chicago, 1850–1950* (Madison, WI: American History Research Center, 1957),15.

23. Ibid., 40.

24. "Fruits and Vegetables as Medicines," *Fruit, Flower, and Vegetable Grower*, January–March 1883, 24.

25. Arthur J. Cramp, ed., *Nostrums and Quackery and Pseudo-Medicine* (Chicago: American Medical Association, 1936), 3:45.

26. Ibid., 96.

27. Ibid., 107.

28. "Picnicing," *Voice of the Fair*, June 16, 1865, 4.

29. "Davis' Rural Garden and Pic Nic Grounds," *Prairie Farmer*, May 30, 1863, 344.

30. "Picnicing," 4.

31. "Davis' Rural Garden and Pic Nic Grounds," 344.

32. Ibid.

33. Suburban Railroad Wildflower Route brochure, author's collection.

34. Scott A. Newman, "Jazz Age Chicago: Summer Gardens and Picnic Groves," in *Jazz Age Chicago*, ed. Scott A. Newman, http://chicago.urban-history.org.

35. J. Ambrose Wight, "Address before the N.W. Fruit Growers' Association," *Prairie Farmer*, April 1854, 140.

36. Chamberlin, *Chicago and Its Suburbs*, 184.

37. John B. Bachelder, *Popular Resorts and How to Reach Them* (Boston: John B. Bachelder, 1875), 317.

38. *Land Owner*, July 1869.

39. Advertisement, *Chicago Daily Tribune*, March 23, 1907, 20.

40. Advertisement, *Chicago Daily Tribune*, May 15, 1914, 12.

41. Frances Kinsley Hutchinson, *Our Country Home* (Chicago: A. C. McClurg and Co., 1907), 2.

42. Bertha and William Jaques are likely a married couple, based on the significant similarities in their separate writings about their country home in Michigan. Neither work, however, references the other person by first name.

43. Bertha Jaques, *Country Quest* (1914; Chicago: Libby Co., 1936), 19–20.

44. Ibid., 124.

45. William K. Jaques, "Various Points of View," *Transactions of the Illinois State Horticultural Society*, n.s., 47 (1913): 157.

46. William C. Egan, "How I Built My Country Home," in Bailey, *How to Make a Flower Garden*, 323–31.

47. Ibid., 327.

CHAPTER EIGHT

1. *Chicago Daily Tribune*, September 26, 1928, 31.

2. Farmers were among the early adopters of the airplane. Handy with machinery, most self-reliant farmers could maintain the early airplanes just as they maintained their own tractors. Airplanes were useful in agriculture to traverse large stretches of open fields and, later, to spray crops with insecticides.

3. Kennicott Brothers Co., http://www.kennicott.com.

4. Grossman, Keating, and Reiff, *Encyclopedia of Chicago*, 931.

5. "National Florist Group Moves Its Headquarters Here," *Chicago Daily Tribune*, June 30, 1935, A12.

6. "Prizes Total $16,000 for the International Horticultural Show," *Chicago Daily Tribune*, August 12, 1936, 15.

7. Suzanne Carter Meldman, *The City and the Garden: The Chicago Horticultural Society at Ninety* (Chicago: Chicago Horticultural Society, 1981), 2.

8. Other versions seen in print are the Chicago Horticultural Society and Garden Center or the Garden Institute and Horticultural Society.

9. "Horticultural Club Proposed for Chicagoans," *Chicago Daily Tribune*, November 8, 1931, F2.

10. "An Intriguing Subject Knits Garden Clubs," *Chicago Daily Tribune*, March 19, 1950, W A5.

11. "Women Break Up Male Monopoly in Park District," *Chicago Daily Tribune*, February 28, 1943, N2.

12. Certainly there were many modern-day "wealthy amateurs," who retained an avid interest and studied horticulture. As with Americans in general, however, increased urbanization distanced most people from basic agricultural knowledge.

13. *Chicago Daily Tribune*, January 22, 1941, 56.

14. "War Gardener for Kaiser Now Helps America," *Chicago Daily Tribune*, April 26, 1942, SW6.

15. "Plans Thicket to Check Glare of Auto's Lamps," *Chicago Daily Tribune*, July 26, 1937, 5.

16. "Work to Save Lawns from Plow," *Chicago Daily Tribune*, March 9, 1941, W1.

17. "Plans 450 Acre Estate Project at Barrington," *Chicago Daily Tribune*, February 2, 1936, 18.

18. "Julia C. Lathrop U.S. Housing to Have 975 Family Units," *Chicago Daily Tribune*, May 3, 1936, 26.

19. "Break Ground for Five Homes in New Project," *Chicago Daily Tribune*, July 23, 1939, C14.

20. Jens Jensen, *Siftings* (Baltimore: John Hopkins University Press, 1990), 110.

APPENDIX THREE

1. Higley and Raddin, "Flora of Cook County," vi–viii.

2. Coventry, Meyer, and Miller, *Classic Country Estates*, 46.

3. Margot Stimely, "Klehm Family," Klehm family personal archives.

4. Landscape Architecture Department History, http://www.landarch.uiuc.edu/history.htm (accessed February 21, 2006).

5. Ed Collins, "Searching for Dr. Vasey," *Chicago Wilderness*, winter 2001, http://www.chicagowildernessmag.org.

RECOMMENDED READING

These books were used in developing this manuscript and are highly recommended for anyone interested in Chicago's history, its horticultural history, and general horticultural history.

Addams, Jane. *Twenty Years at Hull-House*. New York: Macmillan Co., 1910.

Andreas, A. T. *History of Cook County Illinois, from the Earliest Period to the Present Time*. Chicago: A. T. Andreas, 1884.

Bahr, Fritz. *Commercial Floriculture*. New York: A. T. De La Mare Co., 1924.

Bailey, L. H., ed. *How to Make a Flower Garden*. New York: Doubleday, Page and Co., 1914.

Birnbaum, Charles, ed. *Pioneers of American Landscape Design*. New York: McGraw-Hill, 2000.

Bluestone, Daniel M. *Constructing Chicago*. New Haven: Yale University Press, 1991.

Brewster, Kate L. *The Little Garden for Little Money*. Boston: Atlantic Monthly Press, 1924.

Bryant, Arthur Sr. *Forest Trees for Shelter, Ornament, and Profit*. New York: Henry T. Williams, 1871.

Burnham, Daniel H., and Edward H. Bennett. *Plan of Chicago*. Chicago: Commercial Club, 1909.

Chamberlin, Everett. *Chicago and Its Suburbs*. Chicago: T. A. Hungerford and Co., 1874.

Chapin, Louella. *Round About Chicago*. Chicago: Unity Publishing Co., 1907.

Chicago Historical Society. *Prairie in the City: Naturalism in Chicago's Parks, 1870–1940*. Chicago: Chicago Historical Society, 1991.

Cleveland, H. W. S. *Landscape Architecture as Applied to the Needs of the West*. Chicago: Jansen, McClurg and Co., 1873. Revised edition, Pittsburgh: University of Pittsburgh Press, 1965.

Cook, Philip L. *Zion City, Illinois: Twentieth-Century Utopia*. Syracuse: Syracuse University Press, 1996.

Coventry, Kim, Daniel Meyer, and Arthur H. Miller. *Classic Country Estates of Lake Forest*. New York: W. W. Norton and Co., 2003.

Doty, Mrs. Duane. *The Town of Pullman: Its Growth with Brief Accounts of Its Industries*. Pullman, IL: T. P. Struhsacker, 1893. Reprint, Pullman Civic Organization and Historic Pullman Foundation, 1991.

Ebner, Michael H. *Creating Chicago's North Shore*. Chicago: University of Chicago Press, 1988.

Egan, W. C. *Making a Garden of Perennials*. New York: McBride, Nast and Co., 1912.

Farwell, Edith Foster. *My Garden Gate Is on the Latch*. Chicago: La Salle Street Press, 1962.

Gale, Edwin O. *Reminiscences of Early Chicago*. Chicago: Fleming H. Revell Co., 1910.

Grese, Robert E. *Jens Jensen: Maker of Natural Parks and Gardens*. Baltimore: Johns Hopkins University Press, 1992.

Heise, Kenan. *The Chicagoization of America, 1893–1917*. Evanston, IL: Chicago Historical Bookworks, 1990.

Highland Park Historic Preservation Commission. *Highland Park, Illinois: Historic Landscape Survey, Final Report*. Highland Park: Park District of Highland Park and the Illinois Preservation Agency, 1988.

Hutchinson, Frances Kinsley. *Our Country Home*. Chicago: A. C. McClurg and Co., 1907.

Jenkins, Virginia Scott. *The Lawn: A History of an American Obsession*. Washington, DC: Smithsonian Institution Press, 1994.

Jensen, Jens. *Siftings*. Baltimore: Johns Hopkins University Press, 1990. Reprint of 1939 original edition.

Keating, Ann Durkin. *Building Chicago*. Columbus: Ohio State University Press, 1988.

———. *Chicagoland: City and Suburbs in the Railroad Age*. Chicago: University of Chicago Press, 2005.

Kinzie, Juliette. *Wau-Bun: The Early Day in the Northwest*. Chicago: Rand, McNally and Co., 1901.

Lanctot, Barbara. *A Walk through Graceland Cemetery*. Chicago: Chicago Architecture Foundation, 1977.

Laurie, Alex. *The Flower Shop*. Chicago: Florists' Publishing Co., 1930.

Lindberg, Richard. *Ethnic Chicago: A Complete Guide to the Many Faces and Cultures of Chicago*. Lincolnwood, IL: Passport Books, 1993.

Maloney, Cathy Jean. *The Prairie Club of Chicago*. Chicago: Arcadia Publishing, 2001.

Mayer, Harold M., and Richard C. Wade. *Chicago: Growth of a Metropolis*. Chicago: University of Chicago Press, 1969.

McCauley, Lena May. *The Joy of Gardens*. Chicago: Rand McNally and Co., 1911.

McCormick, Harriet Hammond. *Landscape Art, Past and Present*. New York: Charles Scribner's Sons, 1923.

Miller, Donald L. *City of the Century: The Epic of Chicago and the Making of America*. New York: Simon and Schuster, 1997.

Moninger, John C. *The Florist's Scrap Book*. Chicago: John C. Moninger Co., 1908.

Olmsted, Frederick Law. *The Papers of Frederick Law Olmsted*. Edited by Charles E. Beveridge and David Schuyler. Baltimore: Johns Hopkins University Press, 1983.

Pacyga, Dominic A., and Ellen Skerrett. *Chicago: City of Neighborhoods*. Chicago: Loyola University Press, 1986.

Pearse, R. J., ed. *Proceedings of the First Annual School in Fair Management*. Oklahoma City: Times-Journal Publishing Co., 1924.

Pepoon, H. S. *Flora of the Chicago Region*. Chicago: Lakeside Press, 1927.

Periam, Jonathan. *Home and Farm Manual*. New York: N. D. Thompson and Co., 1884.

———. *People's Library Illustrated*. Vols. 1–3. Chicago: Rand, McNally and Co., 1883.

Punch, Walter T., ed. *Keeping Eden: A History of Gardening in America*. Boston: Little, Brown and Co., 1992.

Ranney, Victoria Post. *Olmsted in Chicago*. Chicago: R. R. Donnelley and Sons, 1972.

Root, Ralph Rodney. *Contourscaping*. Chicago: Ralph Fletcher Seymour, 1941.

Root, Ralph Rodney, and Charles Fabens Kelley. *Landscape Gardening*. New York: Century Co., 1914.

Simon, Andreas. *Chicago: The Garden City*. Chicago: Franz Gindele Printing Co., 1894.

Simonds, O. C. *Landscape Gardening*. 1920. Reprint, Amherst: University of Massachusetts Press, 2000.

Sinkevitch, Alice, and Laurie McGovern Peterson. *AIA Guide to Chicago*. Chicago: Harcourt Brace and Co., 1993.

Spicer, Anne Higginson. *Songs of the Skokie and Other Verse*. Chicago: Ralph Fletcher Seymour, 1917.

Tankard, Judith B. *The Gardens of Ellen Biddle Shipman*. New York: Sagapress, 1996.

Tishler, William H., ed. *American Landscape Architecture: Designers and Places*. Washington, DC: Preservation Press, 1989.

———. *Midwestern Landscape Architecture*. Champaign: University of Illinois Press, 2004.

Tucker, David M. *Kitchen Gardening in America*. Ames: Iowa State University Press, 1993.

Wille, Lois. *Forever Open, Clear, and Free*. Chicago: University of Chicago Press, 1991. First published in 1972.

Woodruff, Frank Morley. *The Birds of the Chicago Area*. Natural History Survey Bulletin 6. Chicago: Chicago Academy of Sciences, 1907.

OTHER WORKS CITED

Adelman, William. *Touring Pullman*. Chicago: Illinois Labor History Society and the Ralph Helstein Fund, 1994.

Bachelder, John B. *Popular Resorts and How to Reach Them*. Boston: John B. Bachelder, 1875.

Bailey, L. H. *Annals of Horticulture in North America for the Year 1893*. New York: Rural Publishing Co., 1984.

Barnum, Gertrude. "The Chicago Woman and Her Clubs." *Graphic*, May 27, 1893, 343.

Blair, John. "Landscape Gardening." In *The Prairie Farmer Annual*. Chicago: Prairie Farmer Co., 1869.

Bley, Mrs. J. C. "The Consumer." *Transactions of the Illinois State Horticultural Society*, n.s., 47 (1913): 117–22.

Block, Jean F. *Hyde Park Houses: An Informal History, 1856–1910*. Chicago: University of Chicago Press, 1978.

Bonner, Thomas. *Medicine in Chicago, 1850–1950*. Madison, WI: American History Research Center, 1957.

Bowen, Louise DeKoven. *Growing Up with a City*. New York: Macmillan Co., 1926.

Bragdon, C. D. "A Biographical Sketch of the Life of Dr. John A. Kennicott." *Transactions of the Horticultural Society of Northern Illinois*, 1863, 39.

Brayton, A. W. "President's Address." In *Proceedings of the 63rd Annual Convention of the Illinois State Horticultural Society*, reprinted in *Prairie Farmer*, November 19, 1918, 47.

A Brief History of Our Garden. Chicago: University of Chicago, n.d.

Bryant, Guy A. "The Salome Apple: Its Origin and History." *Transactions of the Horticultural Society of Northern Illinois*, 1912, 261–64.

Bryant. L. R. "Observations at the Fair." *Transactions of the Horticultural Society of Northern Illinois*, 1894, 275.

———. "Report of Committee on Spraying Fruit Trees." *Transactions of the Illinois State Horticultural Society*, 1889, 132–34.

Bryant, Miles. "Landscape Gardening." *Transactions of the Horticultural Society of Northern Illinois*, 1917, 285–93.

Chicago Illustrated: Reproductions of Colored Lithographs published by Jevne. Chicago: Steam Press of Church, Goodman and Donnelley, [1868?].

The Chicago Zoological Society Yearbook, 1927. Chicago: Chicago Zoological Society, 1927.

Clemmer, Mary. *Memorial Sketch of Elizabeth Emerson Atwater, Written for Her Friends*. Buffalo: Courier Co., 1879.

Cleveland, H. W. S. *Landscape Gardening in the West*. Chicago: Hazlitt and Reed, 1871.

Collins, Ed. "Searching for Dr. Vasey." *Chicago Wilderness*, winter 2001. http://www.chicagowildernessmag.org.

Cook, Frederick Francis. *Bygone Days in Chicago*. Chicago: A. C. McClurg and Co., 1910.

Cramp, Arthur J., ed. *Nostrums and Quackery and Pseudo-Medicine*. Vol. 3. Chicago: American Medical Association, 1936.

Croly, J. C. *The History of the Women's Club Movement in America*. New York: H. G. Allen and Co., 1898.

Cushing Smith, F. "Landscape Design for Gardens." *American Landscape Architect*, December 1931, 14–16.

Day, L. A. "The Commercial Manufacture of Insecticides and Spraying Material." *Transactions of the Illinois State Horticultural Society*, 1917, 80–83.

DeBoth, Jessie Marie. *Modernistic Recipe-Menu Book*. Chicago: DeBoth Homemakers' Cooking, 1929.

Denny, John. "The Commission Man." *Transactions of the Illinois State Horticultural Society*, n.s., 47 (1913): 136–48.

Domer, Dennis. *Alfred Caldwell*. Baltimore: Johns Hopkins University Press, 1997.

Dorner, H. B. "Gardening under Glass." *Transactions of the Illinois State Horticultural Society*, 1908, 81.

Douglas, Robert. "Progress of Horticulture in the Northwest." *Transactions of the Horticultural Society of Northern Illinois*, 1872.

———. "Timber Planting." *Transactions of the Horticultural Society of Northern Illinois*, 1873, 285–86.

Dugan, Hugh. *Village on the County Line*. Chicago: Lakeside Press, 1949.

Dunlap, H. M.. "Illinois Horticultural the World's Columbian Exposition 1893." *Transactions of the Illinois State Horticultural Society*, 1893, 72–82.

Dunlap, Mathias L. "The Status of Horticulture." *Transactions of the Horticulture Society of Northern Illinois*, 1867.

Dynes, Mary. "Our State Federations." In Eighth Annual Garden and Flower Show program, Garden Club of Illinois, 1934, 7.

Egan, William C. "How I Built My Country Home: A Concrete Example of Landscape Gardening." In *How to Make a Flower Garden*, edited by L. H. Bailey, 323–31. Garden City, NY: Doubleday, Page and Co., 1914.

———. "Select List of Vines." In *How to Make a Flower Garden*, edited by L. H. Bailey, 97–105. Garden City, NY: Doubleday, Page and Co., 1914.

Egleston, N. H. *Arbor Day: Its History and Observance*. Washington, DC: Government Printing Office, 1896.

Ely, Richard T. "Pullman: A Social Study." *Harper's Magazine*, February 1885, 452–66.

Fairchild, David. "The Barbour Lathrop Bamboo Grove." *Journal of Heredity* 10 (June 1919): 243–49.

Forbes, Stephen A. "What Should the State Require of a Negligent Owner of a Dangerous Orchard?" *Transactions of the Illinois State Horticultural Society*, n.s., 45 (1911): 77.

Fordyce, Alice Woodward. "An American Chantilly." *Country Life*, September 1941, 30–33.

The Forest Preserves of Cook County. Chicago: Clohesey and Co., 1918.

Frazier, Sarah Randolph. "The Mission of Flowers in School and Home." *Transactions of the Horticultural Society of Northern Illinois*, 1900.

Freeman, Ellen Bryant. "Women as Gardeners." *Transactions of the Horticultural Society of Northern Illinois*, 1894, 289.

Frick, Mary. "Coming into Riverside, the Temperature Would Drop." In *Tell Me a Story: Memories of Riverside*, edited by Riverside Historical Commission, 2:45. Riverside, IL: Riverside Historical Commission, 2000.

Gage, Frances D. "Letter from Mrs. Gage: The New Home " *Ohio Cultivator* 9 (1853): 173.

———. "Pomological Congress and Cooper Apple." *Western Horticultural Review*, October 1850, 145. Originally in *Ohio Cultivator*.

Geiger, Barbara. "Nature as the Great Teacher: The Life and Work of Landscape Designer O. C. Simonds." Master's thesis, University of Wisconsin, Madison, 1997.

Geweke, August. "The Gardener and His Troubles." *Transactions of the Horticultural Society of Northern Illinois*, 1915, 257–58.

———. "Successful Truck Gardening." *Transactions of the Illinois State Horticultural Society*, 1911.

Glover, Ellye Howell. *Dame Curtsey's Book of Novel Entertainments for Every Day in the Year*. Chicago: A. C. McClurg and Co., 1920. Originally published 1907.

Grossman, James R., Ann Durkin Keating, and Janice L. Reiff, eds. *The Encyclopedia of Chicago*. Chicago: University of Chicago Press, 2004.

Guarino, Jean. *A Historical View of Oak Park, Illinois: Prairie Days to World War I*. Oak Park, IL: Oak Ridge Press, 2000.

Hambrock, A. G. "The Retailer." *Transactions of the Illinois State Horticultural Society*, n.s., 47 (1913): 148–56.

Handy, Mrs. M. P. "The Women of the World's Fair City." *Munsey's Magazine*, March 1893, 608–9, 613.

Harmon, Ada Douglas. *The Story of an Old Town: Glen Ellyn*. N.p.: Glen News Printing Co., 1928.

Hathaway, Edmund. "Report on Landscape Gardening." *Transactions of the Horticultural Society of Northern Illinois*, 1875, 228.

Hayes, Alice, and Susan Moon. *Ragdale: A History and Guide*. Lake Forest: Ragdale Foundation, 1990.

Heinl, Frank J. "Pioneer Nurserymen in Illinois." *American Nurseryman*, July 15, 1950, 142–43.

Higley, William K., and Charles S. Raddin. "The Flora of Cook County, Illinois, and a Part of Lake County, Indiana." *Bulletin of the Chicago Academy of Sciences* 2, no. 1 (1891).

Hillis, Mrs. "The Cultivation of Flowers." *Transactions of the Horticultural Society of Northern Illinois*, 1874, 297.

Historic City: The Settlement of Chicago. Chicago: City of Chicago Department of Development and Planning, 1976.

History of the North-Western Soldiers' Fair. Chicago: Dunlop, Sewell and Spalding, 1864.

Illinois Farmers' Institute. *Department of Household Science Year Book 1913*. Edited by Mrs. H. A. McKeene. Springfield: Illinois State Journal Co., 1913.

Jaques, Bertha. *Country Quest*. 1914. Reprint, Chicago: Libby Co., 1936.

Jaques, William K. "Various Points of View." *Transactions of the Illinois State Horticultural Society*, n.s., 47 (1913).

Jensen, Jens. "Natural Parks and Gardens." *Saturday Evening Post*, March 8, 1930.

Jeter, Helen R. *Trends of Population in the Region of Chicago*. Chicago: University of Chicago Press, 1927.

Keating, William H. "The Natural History, Indians, and Disadvantages of Chicago." In *As Others See Chicago*, edited by Bessie Louise Pierce. Chicago: University of Chicago Press, 1933.

Kendall, Abbie S. "Seed Time." Third Annual Chicago Garden and Flower Show program, 66. 1929.

Kennicott, John A. "Western Horticultural Review." *Horticulturist* 7 (1852): 540–41.

Kircher, Mary Green. *Brookfield, Illinois: A History*. Brookfield: Brookfield History Book Committee, 1994.

Klehm, Charles. "Commercial Fruit Growing in Cook County." *Transactions of the Horticultural Society of Northern Illinois*, 1924.

The Lakeside Cook Book No. 1. Chicago: Donnelley, Gassette and Loyd, 1878.

Larson, Darlene, and Laura Hiebert. "The Fabyan Legacy." In *Geneva, Illinois: A History of Its Times and Places*, edited by Julia Ehresmann, 155–81. Geneva: Geneva Public Library District, 1977.

"The Library Exhibit of Native Wild Flowers." *Chicago Public Library Book Bulletin* 7, no. 8 (1917).

Lindlahr, Henry, and Anna Lindlahr. *The Nature Cure Cook Book and ABC of Natural Dietetics*. Chicago: Nature Cure Publishing Co., 1915.

Lipo, Frank. Nomination of Oak Park Conservatory. National Register of Historic Places application. Historical Society of Oak Park and River Forest Archives, Oak Park.

Manning, Warren H. "The Purpose and Practice of Landscape Architecture." *Transactions of the Indiana Horticultural Society*, 1893.

McAdam, Thomas. "Landscape Gardening under Glass." *Country Life in America*, December 15, 1911, 10–13.

McDonald, Mrs. Jesse F., ed. *Fiftieth Year of the Garden Club of Illinois*. Chicago: Garden Club of Illinois, 1977.

Meldman, Suzanne Carter. *The City and the Garden: The Chicago Horticultural Society at Ninety*. Chicago: Chicago Historical Society, 1981.

Miller, Wilhelm. "The Prairie Spirit in Landscape Gardening." *Transactions of the Illinois State Horticultural Society*, 1915, 144–45.

Morton, J. Sterling. "Arbor Day: Its Origins and Growth." Address delivered in Lincoln, Nebraska, April 22, 1887. Reprinted in *Arbor Day: Its History and Observance*, edited by N. H. Egleston. Washington, DC: Government Printing Office, 1896.

Mustain, Nelle S. *The New Idea Entertainer for Indoors and Out of Doors*. Los Angeles: L. A. Martin, 1903.

Oakes, Helen Hart. "Illinois Gardens." WGN radio broadcast. Reprinted in *Garden Glories*, May 1933, 10–11.

Olmsted, Vaux and Co. *Preliminary Report upon the Proposed Suburban Village at Riverside, near Chicago*. New York: Sutton, Bowne and Co., 1868.

O'Neil, B. "Report upon Home Adornment." *Transactions of the Horticultural Society of Northern Illinois*, 1881, 315–17.

Peckham, Edward L. "Chicago: A Mean Spot." In *As Others See Chicago*, edited by Bessie Louise Pierce, 166. Chicago: University of Chicago Press, 1933.

Periam, Jonathan. "Forest Tree Planting as a Means of Wealth." *Transactions of the Illinois State Horticultural Society*, 1871, 31–40.

———. "Interests in Horticulture." *Transactions of the Horticultural Society of Northern Illinois*, 1889.

———. "The Kitchen Garden." In *The Prairie Farmer Annual*, 28–33. Chicago: Prairie Farmer Co., 1868.

———. "Progress of Horticulture in the Northwest." *Transactions of the Northern Illinois Horticultural Society*, 1872.

———. "Report of the Committee on Trees and Planting." *Transactions of the Horticulture Society of Northern Illinois*, 1864.

Perkins, Janice. "Glen Ellyn." In *DuPage Roots*, edited by Richard A. Thompson. Wheaton, IL: DuPage County Historical Society, 1985.

Prost, J. H.. "What Chicago Is Doing for Its Trees." *Country Life in America*, August 1910.

Pruter, Margaret Franson. "Elmhurst." In *DuPage Roots*, edited by Richard A. Thompson. Wheaton, IL: DuPage County Historical Society, 1985.

Radford, William A. "Homes in Color Away Out in the Country." *American Builder*, October 1926.

Rayne, Martha L. *What Can a Woman Do?* Albany, NY: Eagle Publishing Co., 1893.

Ridgeway, Frank. *Flower Gardening by Farm and Garden Bureau*. Chicago: Chicago Tribune, 1929.

Ridgway, Mrs. Grant. Paper read at the 35th birthday celebration of the Kenilworth Garden Club. Kenilworth Historical Society.

Riverside in 1871. Chicago: Riverside Improvement Co., 1871.

Ross, Andrea Friederici. *Let the Lions Roar! The Evolution of Brookfield Zoo*. Chicago: Chicago Zoological Society, 1997.

Sanders, Edgar, Collection. Chicago History Museum.

———. "A Chapter on Flower Gardening." In *The Prairie Farmer Annual*. Chicago: Prairie Farmer Co., 1868.

———. "Immortal Flowers." In *The Prairie Farmer Annual*. Chicago: Prairie Farmer Co., 1870.

———. "Management of Ferneries." In *The Prairie Farmer Annual*. Chicago: Prairie Farmer Co., 1871.

———. "Thirty Years in Western Horticulture." *Transactions of the Horticultural Society of Northern Illinois*, 1889.

———. "Our Native Flora." *Transactions of the Horticultural Society of Northern Illinois*, 1894.

Schermerhorn, Richard Jr. "Nathan Franklin Barrett, Landscape Architect." *Landscape Architecture* 10 (April 1920).

Schuyler, David. "Jacob Weidenmann." In *American Landscape Architecture: Designers and Places*, edited by William H. Tishler. Washington, DC: Preservation Press, 1989.

Schuyler, W. H. "Report upon Utilizing and Marketing Fruits." *Transactions of the Illinois State Horticultural Society*, 1875.

Scofield, D. C. "Evergreens." *Transactions of the Horticultural Society of Northern Illinois*, 1879, 236–40.

———. "Pine Timber." *Transactions of the Horticultural Society of Northern Illinois*, 1875, 214–15.

———. "Timber Waste." *Transactions of the Horticultural Society of Northern Illinois*, 1873, 241–44.

Scott, Franklin. *The Semi-Centennial Alumni Record of the University of Illinois*. Champaign: University of Illinois, 1918.

Servas, John A. "History of the Chicago Flower Shows." *Garden Glories*, March 1931.

Sibley Papers. Chicago History Museum.

Simonds, O. C. "Cemetery Gardening: A Modern Type of Landscape Ornamentation." In *Chicago: The Garden City* by Andreas Simon, 114–17. Chicago: Franz Gindele Printing Co., 1894.

———. "Graceland at Chicago." *Landscape Architect*, January 1932.

———. "Morton Arboretum: Some of the Landscape Features." *Morton Arboretum's Bulletin of Popular Information*, May 22, 1925, 13.

———. "Wild Flowers." *Transactions of the Illinois State Horticultural Society*, 1908.

Simpson, J. C. "Agricultural and Horticultural Exhibits at Fairs and Expositions." In *Proceedings of the First Annual School in Fair Management*, edited by R. J. Pearse. Oklahoma City: Times-Journal Publishing Co., 1924.

South Park Commissioners. Report. Chicago, December 1, 1877.

Taft, Lorado. "The Sculptor's Interpretation." In *Catalog Guide to Garfield Park Conservatory*, unpaginated. Chicago: West Park Commissioners, 1924.

Thorpe, John. "Floriculture at the World's Fair." *Gardening*, September 15, 1892.

United States Department of Agriculture. *Report of the Commissioner for the Year 1871*. Washington, DC: U.S. Department of Agriculture, 1872.

University of Chicago. Board of Trustees. Correspondence.

———. Department of Building and Grounds. Records. University of Chicago Library Special Collections.

View from the Path: Jens Jensen in Kenilworth. Kenilworth, IL: Kenilworth Historical Society, 2004.

Walgreen, Myrtle. *Never a Dull Day*. Chicago: Regnery, 1963.

West, Virginia Cross. "We Loved to Play 'Lady.'" In *Tell Me a Story: Memories of Riverside*, edited by Riverside Historical Commission, 2:6–8. Riverside, IL: Riverside Historical Commission, 2000.

Wheaton and Its Homes. Proust and Burnham, 1892. Reprint, Wheaton, IL: Wheaton Historic Preservation Council, 1983.

White, Richard P. *A Century of Service: A History of the Nursery Industry Associations of the United States*. Washington, DC: American Association of Nurserymen, 1975.

Wight, J. A. "Chicago, Horticulturally." *Horticulturist* 3 (1849): 459–62.

Wright Barney, Maginel. *Valley of the God-Almighty Joneses*. New York: Appleton-Century, 1965.

Wright, Frank Lloyd. FLW Home and Studio Archives. Oak Park, IL.